# The Proton in Chemistry

# The Proton in Chemistry

R. P. BELL

Professor of Chemistry
University of Stirling

SECOND EDITION

SPRINGER-SCIENCE+BUSINESS MEDIA, B.V.

© 1973 R. P. Bell

Originally published by Chapman & Hall in 1973

*Filmset by Keyspools Ltd, Golborne, Lancs*

*Fletcher & Son Ltd, Norwich,*

ISBN 978-1-4757-1594-1    ISBN 978-1-4757-1592-7 (eBook)
DOI 10.1007/978-1-4757-1592-7

# Contents

PREFACE                                                         *page* vii

1   Introduction                                                      1
2   Acids, bases, and the nature of the hydrogen ion                  4
3   The investigation of protolytic equilibria in aqueous
    solution                                                         26
4   The effect of the solvent on protolytic equilibria               44
5   The thermodynamics of protolytic equilibria                      72
6   Acid-base strength and molecular structure                       86
7   The direct study of rates of simple proton-transfer
    reactions                                                       111
8   The indirect study of rates of proton transfer                  133
9   Examples of reactions catalysed by acids and bases              159
10  Rates, equilibria, and structures in proton-transfer
    reactions                                                       194
11  Isotope effects in proton-transfer equilibria                   226
12  Kinetic isotope effects in proton-transfer reactions            250

    AUTHOR INDEX                                                    297
    SUBJECT INDEX                                                   307

# Preface

The first edition of this book was based on the lectures which I gave at Cornell University during 1958 as George Fisher Baker Lecturer, and I would like to repeat my warmest thanks to Professor F. A. Long and the other members of the Department of Chemistry for their kindness and helpful advice. The present edition was largely written during the tenure of a Visiting Professorship awarded by the Royal Society and the Israeli Academy of Sciences. I am deeply indebted to both of these bodies and also to the hospitality of the Weizmann Institute of Science, in particular to Professor David Samuel and Professor F. S. Klein of the Isotopes Research Department.

The subject as a whole has expanded greatly since 1959, especially in two fields, namely, the direct study of fast proton-transfer reactions (notably by the relaxation methods pioneered by Eigen), and the experimental and theoretical study of hydrogen isotope effects. In order to keep the size of the book within reasonable bounds it has been necessary to adopt a selective policy, and this is particularly the case in Chapter 9 where I have chosen to treat a few types of reaction in some detail rather than to attempt a more complete coverage. The original chapter on concentrated solutions of acids and bases has been omitted, partly because there now exist several books and review articles on acidity functions and related topics, and partly because the interpretation of reaction velocities in these concentrated solutions has become more rather than less confused with the passage of time. The first edition referred frequently to *Acid-Base Catalysis* (Oxford, 1941), which is now out of print. Although many of the theses argued in some detail in this earlier book have become common currency, I have retrieved some of this material in the present edition.

Chapters 10, 11, and 12 involve a rather more advanced treatment than the remainder of the book, and reflect my particular interests and prejudices. They could well be omitted by readers whose interest in the subject is of a more general character.

In conclusion, my best thanks are due to my wife for typing the whole of the manuscript, and for other assistance.

*University of Stirling*  
*Scotland*  
*October 1972*

R. P. Bell

# 1 Introduction

The proton is most naturally thought of by the chemist as the cation derived from the hydrogen atom. Except in purely descriptive chemistry it is not usually profitable to consider a single element in isolation, still less a particular ion, and the topic chosen for this book therefore requires some justification.

The proton is unique among singly charged ions in having no electrons outside the nucleus, and although this property is shared by some multiply charged cations (e.g., $He^{2+}$, $Li^{3+}$) none of these is important in chemical processes taking place under ordinary conditions. This means that the proton has a radius of the order $10^{-13}$ cm, as compared with $10^{-8}$ cm for all other ions. On account of this small radius it can exert an enormous polarizing power on any molecule or ion in the neighbourhood, and for this reason the free proton is encountered only in a vacuum, or in a very dilute gas. However, we shall see that a wide variety of processes can be regarded as *proton-transfer reactions*, and since these involve only the movement of a nucleus, without any attendant electrons, they represent a particularly simple type of change. In particular, they can take place without any serious disorganization of bonding electrons and without bringing into play forces of repulsion between non-bonding electrons. In the terminology of modern organic chemistry, the proton has low steric requirements. Some reactions do, of course, involve the transfer of a hydrogen atom rather than a proton, but these usually take place under more drastic conditions, for example, in the gas phase at high temperatures or under the influence of radiation or bombardment by particles of high energy. There is rarely any doubt as to whether a reaction involves protons or hydrogen atoms, while for other elements (notably the halogens) it is often necessary to consider the possibility of both heterolytic and homolytic mechanisms.

The simple nature of proton transfers is probably responsible for the fact that such transfers are commonly facile and lead to mobile equilibria. This circumstance underlies the general usefulness of the classical acid-base concept, especially in its quantitative aspects. A large proportion of our knowledge about equilibria in solution relates to protolytic equilibria (dissociation constants and related quantities), and there is no other class of reaction for which such accurate data are available. This has played an important part in the development of theories of ionic solutions, on the one hand, and of the interpretation of substituent effects in organic chemistry, on the other. In solution kinetics, as opposed to equilibria, the proton does not occupy such a special position; in fact, many acid-base reactions take place so rapidly that their velocities cannot be measured by ordinary means, and only recently have they become accessible through modern techniques for studying very fast reactions. It has long been realized, however, that *catalysis by acids and bases* usually involves one or more proton transfers, and the simple nature of the proton appears again in regularities which have been observed in the kinetics of catalysed reactions.

There are two classes of phenomena in which the proton acts as a link between two other atoms. The first of these is in *hydrogen bonding*, which to a first approximation may be attributed to the electrostatic attraction between a proton and an unshared pair of electrons. It is again the absence of orbital electrons which makes the proton unique, and hydrogen bonding can often be regarded as an intermediate stage in a proton-transfer reaction. The second type of binding is in *electron-deficient compounds* such as the boron hydrides, where (in the most recent interpretation) the proton forms part of a three-centre bond involving two electrons. This phenomenon differs fundamentally from hydrogen bonding in the usual sense of the term, and it does not necessarily involve a proton, though it usually does so. The analogy with the other special properties of the proton is a thin one, and this topic will not be treated further.

The masses of nuclei are not usually of primary importance in chemical problems, since the forces concerned depend on electronic and nuclear charges rather than on masses. There are, however, a number of situations in which it is of importance that the proton is the lightest nucleus known, having only one-twelfth of the mass of carbon, the next common element in the Periodic Table. Since hydrogen forms strong bonds with many elements, the combination of low mass and high force constant produces stretching frequencies for bonds of the type X—H which are considerably higher than for any other bonds. This would be of little importance if classical mechanics could be applied to molecular phenomena, but in terms of quantum theory it means that bonds containing hydrogen have large vibrational quanta $h\nu$, amounting to 5–10 kilocalories per mole. Any phenomena involving departures from classical behaviour are therefore particularly prominent in compounds of hydrogen. A notable example of this is in the differences between the isotopes hydrogen, deuterium, and tritium. Isotopic differences depend primarily on differences in zero-point energies ($\frac{1}{2}h\nu$), and these depend in turn on vibrational frequencies and on the ratios of isotopic masses. For most elements the frequencies are low and the mass ratios close to unity, so that isotopes differ only slightly in their chemical behaviour. For hydrogen, on the other hand, the frequencies are high and the masses in the ratio $1:2:3$; hence *hydrogen isotope effects* are very large, both for equilibria and in kinetic problems. Many investigations have been carried out on isotope effects in the equilibria and kinetics of proton-transfer reactions, and the results constitute a severe test of the theory of isotope effects, besides providing information about the mechanism of the reactions concerned.

In the kinetic field there is another way in which the small mass of the proton may be important. It is well known that the behaviour of electrons cannot be accounted for in terms of a particulate model but that it is necessary to take into account the wave nature of the electron; on the other hand, it is usually supposed that the motion of nuclei can be described with sufficient accuracy by the laws of classical mechanics. This is undoubtedly true for most nuclei, but calculation shows that the proton may, on account of its small mass, show considerable deviations from classical behaviour. This phenomenon is often described as the *tunnel effect* and should be detectable experimentally, especially by a detailed analysis of kinetic isotope effects. At present the experimental evidence is meagre, but the problem is an interesting one and will be treated in some detail.

Finally, the nuclear magnetic properties of the proton have recently become of great importance to chemists in the technique known as *proton magnetic resonance*. Because of the magnetic moment of the nucleus, when placed in a magnetic field it has two orientations of different energies, and transitions between these two levels can be brought about by the absorption of radiation in the radio-frequency range. These frequencies can be measured very accurately and will give information about the environment of the protons in the sample being examined, and also about the rate at which different protons can change places with one another. Similar considerations will, of course, apply to other nuclei having a magnetic moment, but a large proportion of the work done so far has dealt with protons, and the method offers great possibilities for investigating equilibria and kinetics in proton-transfer reactions.

# 2 Acids, Bases, and the Nature of the Hydrogen Ion

The exact verbal definition of qualitative concepts is more often the province of philosophy than of physical science. However, the various definitions suggested for acids and bases have been closely linked with the development of physical chemistry and have often served to stimulate experimental work and to further our understanding of chemical processes, and we shall therefore devote some time to this subject. The definitions used in the remainder of this book will be those proposed by Brönsted[1] in 1923, namely, *An acid is a species having a tendency to lose a proton, and a base is a species having a tendency to add on a proton.* This can be represented schematically by $A \rightleftharpoons B + H^+$, where A and B are termed a *conjugate (or corresponding) acid-base pair.*[2] Before examining the consequences of this definition and its relation to more recent concepts we shall consider briefly the previous history of the terms 'acid' and 'base'.

A detailed account of the early history of this subject has been given in an earlier series of Baker Lectures,[3] and only a summary will be attempted here. Like most scientific terms of long standing, the terms 'acid' and 'base' originated in empirical observations of chemical or physical

---

[1] J. N. Brönsted, *Rec. Trav. Chim.*, **42**, 718 (1923).
[2] It is frequently stated that the acid-base definition given here was put forward almost simultaneously by Brönsted and by T. M. Lowry [*Chem. and Ind.*, **42**, 43 (1923)]. However, although Lowry's paper undoubtedly contains many of the ideas underlying this definition, especially for bases, it does not contain an explicit definition, and it is nowhere made clear that Lowry at that time regarded $NH_4^+$ as an acid or $CH_3CO_2^-$ as a base. In fact, in a later paper [*J. Chem. Soc.*, 2562 (1927)], Lowry himself writes, 'More novelty is to be found in the perfectly logical conclusion of Brönsted that the anion of an acid is also a base or proton acceptor, in view of the fact that it can combine with a proton to form a molecule of the undissociated acid'; hence it does not seem justifiable to regard Lowry as one of the originators of the definition. I am indebted to the late Professor E. A. Guggenheim for calling my attention to this point. It is also noteworthy that G. N. Lewis (*Valency and the Structure of Atoms and Molecules*, (Reinhold, New York, 1923), p. 141) gave the same acid-base definition, and wrote, '... we may regard the ammonium ion as an acid'. However, he did not follow up the consequences of this view, and preferred the alternative definition of acids with which his name is usually associated.
[3] P. Walden, *Salts, Acids, and Bases: Electrolytes: Stereochemistry*, Cornell, New York, 1929.

properties, rather than in any theoretical interpretation of the nature of the substances concerned or of their reactions. Etymologically, the English 'acid' and the German 'Säure' both derive from the sour taste of acids in dilute solution (cf. Latin *acetum* 'vinegar', and Latin *acidus*, German *sauer*, Old Norse *suur*, all meaning 'sour'). Soon, however, other properties were added to this; for example, Boyle (end of the seventeenth century) included solvent power and the ability to turn blue vegetable dyes red, and William Lewis (1746) included effervescence with chalk. Acids were also recognized by their power of combining with bases (or alkalis) to form salts, usually with the liberation of water, while bases were chiefly characterized, apart from their part in salt formation, by their negative properties of destroying or reversing the effects caused by acids. The complementary nature of acids and bases was at one time emphasized very strongly, as may be illustrated by a quotation from Gay-Lussac: 'Acidity and alkalinity are properties which are related interchangeably to one another, one of which can be defined only through the other. Thus oil in soaps behaves like an acid, because it satisfies the alkalis. In several ethereal substances alcohol behaves as an alkali, because it satisfies acids.'[4]

Apart from these phenomenological definitions, the first theory of acidic behaviour which is comprehensible in modern terms is that of Lavoisier (end of the eighteenth century), who regarded oxygen as the 'acidifying principle' which converted elements such as carbon, nitrogen, and sulphur into acids like carbonic, nitric, and sulphuric. The assumption that all acids must contain oxygen led to the view that hydrochloric acid, and therefore chlorine, contained oxygen; in fact, hydrochloric acid and chlorine were known as muriatic acid and oxymuriatic acid respectively. It was Davy (1810–1815) who first questioned this interpretation, and the discovery of hydrobromic, hydriodic, and hydrocyanic acids cast further doubt on the oxygen theory. In fact, by 1830 the following oxygen-free acids were known: HF, HCl, HBr, HI, HCN, HSCN, $H_2S$, $H_2Se$, $H_2Te$, $H_2SiF_6$, and $HBF_4$. In spite of this, the oxygen theory was strongly supported by some chemists, notably Berzelius and Gay-Lussac, up to about 1840. A reminder of Lavoisier's views still persists in the word 'oxygen', which is derived through the French *oxygène* from the Greek *οξος*, 'vinegar', *οξυς*, 'sour', and *γενναω*, 'I produce'. There are similar parallels in other languages, for example, in German *Säure*, 'acid', and *Sauerstoff*, 'oxygen', and in Russian *kislota*, 'acid', and *kislorod*, 'oxygen'; the same is true in Hebrew.

[4] J. L. Gay-Lussac, *Gilb., Ann. Phys.*, **48**, 341 (1814).

Davy at first expressed the opinion that 'acidity does not depend upon any particular elementary substance, but upon peculiar arrangement of various substances', a view which we shall meet again in discussing Lewis' definition of acids. It soon became clear, however, that all the substances then commonly accepted as acids did contain hydrogen, and Davy soon recognized hydrogen as the essential element in an acid. Liebig adopted the same idea, especially in relation to organic acids, and in 1838 he defined acids as 'compounds containing hydrogen, in which the hydrogen can be replaced by a metal', a definition which held the field until the advent of the ionic theory, and which would still be regarded as essentially correct. Bases were still regarded as substances that reacted with acids to form salts, and there was no theory as to their constitution corresponding to the hydrogen theory of acids.

The theory of electrolytic dissociation (Ostwald and Arrhenius, 1880–1890) showed that hydrogen compounds with acidic properties were also those which gave rise to hydrogen ions in aqueous solutions, and the application of the law of mass action to the dissociation equilibrium leads to the dissociation constant as a rational quantitative measure of the strength of an acid. Similarly, basic properties were associated with the production of hydroxide ions in solution, and the mutually antagonistic effect of acids and bases was explained in terms of the reaction $H^+ + OH^- \rightarrow H_2O$. This led to the definition of acids and bases as substances giving rise to hydrogen and hydroxide ions respectively in aqueous solution, and this definition was generally accepted for the next thirty or forty years. Many quantitative relations were worked out for dissociation, hydrolysis, buffer solutions, and indicator equilibria, and a satisfactory account was given of a large mass of data.

The success of these quantitative developments helped to obscure some logical weaknesses in the qualitative definitions. For example, it was not clear whether a pure non-conducting substance like anhydrous hydrogen chloride should be called an acid or whether it became one only on contact with water. The definition did not apply directly to non-aqueous solvents, where the ions formed differed from those in water, and this difficulty was particularly acute when it was realized that 'typical' acid-base properties such as neutralization, indicator effects, and catalysis often appeared in solvents such as benzene and chloroform where free ions could barely be detected by conductivity methods. A particular ambiguity appears in the definition of bases, some of which (e.g., metallic hydroxides) contain a hydroxyl group, whereas others (e.g., amines) produce hydroxide ions in solution by abstracting a proton from a water molecule. Some authors[5] distinguished these two classes as 'aquo-bases' and 'anhydro-bases' respectively, but there was no general agreement.

Most of these difficulties and ambiguities are removed by the Brönsted definition in terms of the scheme $A \rightleftharpoons B + H^+$. This is now in such general use that only a few points will be mentioned here. The symbol $H^+$ represents the proton, and not the 'hydrogen ion' of variable nature existing in different solvents, so that the definition is independent of the solvent. Acids need not be neutral molecules such as HCl and $CH_3CO_2H$ but may also be anions ($HSO_4^-$, $CO_2H \cdot CO_2^-$) and cations ($NH_4^+$, $Fe(H_2O)_6^{3+}$). The same is true of bases, where the three classes can be illustrated by $RNH_2$, $H_2O$, $CH_3CO_2^-$, $HPO_4^{2-}$, and $Fe(H_2O)_5OH^{2+}$. Since the free proton cannot exist in solution in measurable concentrations, all actual acid-base reactions are of the type $A_1 + B_2 \rightleftharpoons B_1 + A_2$, where $A_1—B_1$ and $A_2—B_2$ are two conjugate acid-base pairs. This scheme includes reactions formerly described by a variety of names, such as dissociation, neutralization, hydrolysis, and buffer action. One acid-base pair may involve the solvent (in water $H_3O^+—H_2O$ or $H_2O—OH^-$), showing that ions such as $H_3O^+$ and $OH^-$ are in principle only particular examples of an extended class of acids and bases, though of course they do occupy a particularly important place in practice.

The term 'acid' is often used nowadays in a different sense, as first proposed by G. N. Lewis.[6] There has been much controversy, sometimes immoderate in tone, about the relative merits of the Brönsted and Lowry definitions of acids. The question is essentially one of the convenience and consistency of verbal definitions, and not of any fundamental differences in the interpretation of experimental facts; moreover, since the present book is about the proton in chemistry we shall have little occasion to mention non-protonic acids. However, a few comments seem desirable at this point.

A Lewis acid is defined as a species which can accept an electron pair with the formation of a covalent bond, and typical examples are $BF_3$, $SO_3$, and $Ag^+$. Similarly, a base is defined as a species which can donate an electron pair with the formation of a covalent bond; since such species can accept protons, the Brönsted and Lewis definitions of bases will comprise the same species. This is not the case for the definitions of acids, since a Brönsted acid always contains a proton, while a Lewis acid need

---

[5] E.g., A. Werner, *Z. Anorg. Chem.*, **3**, 267 (1893); **15**, 1 (1897); *Ber.*, **40**, 4133 (1907).
[6] G. N. Lewis, *Valency and the Structure of Atoms and Molecules*, Reinhold, New York, 1923.

not and usually does not. The proton itself is, of course, a Lewis acid, and a typical Brönsted acid-base reaction such as

$$NH_3 + CH_3CO_2H \rightleftharpoons NH_4^+ + CH_3CO_2^-$$

is formulated in Lewis nomenclature as $B_1 + B_2A^L \rightleftharpoons B_1A^L + B_2$, where $B_1 = NH_3$, $B_2 = CH_3CO_2^-$, and $A^L$ is a Lewis acid, in this case the proton; it is thus analogous to a non-protonic metathesis such as $Et_2\overset{+}{O}\cdot\overset{-}{B}F_3 + NH_3 \rightleftharpoons \overset{+}{N}H_3\cdot\overset{-}{B}F_3 + Et_2O$. The typical Brönsted acid $CH_3CO_2H$ appears primarily as an adduct between the base $CH_3CO_2^-$ and the Lewis acid $H^+$, and it can only be regarded as a Lewis acid insofar as its reaction with a base involves preliminary formation of a hydrogen bond. However, a hydrogen bond certainly differs markedly from a normal covalent bond, and for this reason protonic acids have sometimes been described as 'secondary acids' by Lewis and his school. Apart from these logical differences the main justification for a separate treatment of proton acids has rested on the quantitative relationships involved. Until recently little information was available on quantitative behaviour of Lewis acids, though several observations indicated that they would not always obey the simple laws which apply to proton-transfer reactions in water and other solvents of high dielectric constant. During the last decade there have been developments in two directions in this field. On the one hand there is now much evidence that equilibria involving Lewis acids often obey simple quantitative laws within a series of closely related systems.[7] On the other hand, the possibility of any unique scale of acidity covering the whole field of interactions between Lewis acids and bases is explicitly denied by the development of the concept of *hard and soft acids and bases*.[8] The aim of this classification is to divide both acids and bases into two categories in such a way that hard acids prefer to react with hard bases, and soft acids with soft bases. The theoretical basis of this classification is obscure, and predictions are at best semi-quantitative, but further developments will be awaited with interest.

This books deals mainly with the quantitative aspects of proton-transfer reactions, especially in aqueous solution. We shall therefore use the terms 'acid' and 'base' only in relation to the definition $A \rightleftharpoons B + H^+$, and shall refer to species such as $BF_3$ as *Lewis acids* or *electron pair acceptors*. In a few instances there may be doubt as to the nature of the

[7] See particularly D. P. N. Satchell and R. S. Satchell, *Chem. Soc. Quart. Rev.*, **25**, 171 (1971).

[8] For summaries see: Symposium on Hard and Soft Acids and Bases, *Chem. and Eng. News.*, **43**, 90 (1965); R. G. Pearson, *Science.*, 151, 172 (1966); *Chem. in Britain*, 103 (1967); *Survey Progr. Chem.*, **5**, 1 (1970); M. J. Frazer, *New Scientist*, 662 (1967).

acidic function, especially when some of the species concerned may be hydrated in solution. For example, the first dissociation of boric acid in aqueous solution is commonly written as $H_3BO_3 \rightleftharpoons H_2BO_3^- + H_3O^+$, representing it as a Brönsted acid. However, the Raman spectra of borate solutions[9] show that the borate ion is almost certainly present as the tetrahedral species $B(OH)_4^-$, and this is confirmed by observations of $^{11}B$ nuclear magnetic resonance spectra of borate solutions;[10] borate ion and $BF_4^-$ give very similar chemical shifts. On this basis the dissociation would be formulated as

$$B(OH)_3 + 2H_2O \rightleftharpoons B(OH)_4^- + H_3O^+, \quad \text{or} \quad B(OH)_3 + OH^- \rightleftharpoons B(OH)_4^-,$$

in which $B(OH)_3$ is acting as an electron pair acceptor or Lewis acid. It was suggested in the first edition of this book[11] that undissociated boric acid in aqueous solution might exist as $\bar{B}(OH)_3(\overset{+}{O}H_2)$, which would restore its status as a Brönsted acid, dissociating according to the scheme $\bar{B}(OH)_3(\overset{+}{O}H_2) + H_2O \rightleftharpoons B(OH)_4^- + H_3O^+$, but there is now good experimental evidence that this suggestion is incorrect. The Raman spectrum of aqueous boric acid resembles that of $BF_3$ and of solid $B(OH)_3$, rather than that of $B(OH)_4^-$, and the $^{11}B$ nuclear magnetic resonance chemical shift is close to that of $B(OCH_3)_3$, either as pure liquid or dissolved in benzene, and very different from that of tetrahedral species such as $B(OH)_4^-$ or $BF_4^-$. Since $B(OCH_3)_3$ is monomeric in benzene solution, this shows clearly that the species in aqueous solutions of boric acid and borates are $B(OH)_3$ and $B(OH)_4^-$, and that a Lewis acid formulation is more correct in this case.[12] Fortunately this kind of ambiguity is not a common occurrence. Since water is present in large excess it does not affect the formal description of equilibria in dilute solution, but it does have a bearing on the interpretation of equilibrium constants in terms of molecular structure (p. 93) and of time-dependent processes (p. 223).

The terms *pseudo-acid* and *pseudo-base* have been widely used in the older literature, before the advent of Lewis' concepts, and it is convenient to consider here the present status of this description. The most commonly

[9] J. O. Edwards, G. C. Morrison, V. F. Ross, and J. W. Schultz, *J. Am. Chem. Soc.*, 77, 266 (1955).
[10] T. P. Onak, H. Landesman, R. E. Williams, and I. Shapiro, *J. Phys. Chem.*, 63, 1533 (1959); W. D. Phillips, H. C. Miller, and E. L. Muetterties, *J. Am. Chem. Soc.*, 81, 4496 (1959); R. J. Thompson and J. C. Davis, Jr., *Inorg. Chem.*, 4, 1464 (1965).
[11] R. P. Bell, *The Proton in Chemistry*, Methuen, London, 1959, pp. 13, 93.
[12] For details of the evidence and further references, see R. P. Bell, J. O. Edwards, and R. B. Jones in *The Chemistry of Boron and its Compounds* (ed. E. L. Muetterties), Wiley, New York, 1966, pp. 209-221.

quoted example of a pseudo-acid is nitromethane, which is a weak acid of dissociation constant about $10^{-10}$. Whereas most acids are neutralized effectively instantaneously by strong bases such as aqueous sodium hydroxide, with nitromethane this process takes place at a measurable rate. This slow neutralization was first observed by Hantzsch,[13] who regarded it as the characteristic property of a pseudo-acid. He also observed later that the absorption spectrum of the nitromethane anion differs considerably from that of the parent substance, which could be interpreted (using modern structural formulae) in terms of the structural change

The anion can be regarded as derived from a 'true acid' or an *aci*-isomer of nitromethane, having the structure

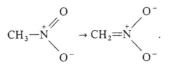

and although this actual substance cannot be isolated or detected, the analogous $C_6H_5CH{=}NO{\cdot}OH$ and several similar compounds can be obtained as pure solids, though they are thermodynamically unstable and revert readily to the normal isomers. Since it was regarded as unthinkable that a neutralization reaction could take place at a measurable speed, Hantzsch supposed that the slow process was the conversion of the ordinary form of nitromethane (a pseudo-acid) into the *aci*-form (a true acid), which was then neutralized rapidly by the alkali, i.e.,

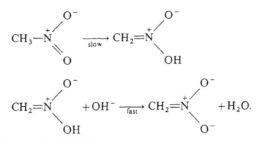

[13] A. Hantzsch, *Ber.*, **32**, 575 (1899).

This reaction scheme as it stands would predict a velocity independent of the alkali concentration, but the observed dependence is restored if we suppose that the isomerization reaction is catalysed by hydroxide ions. This is quite reasonable, since analogous changes, such as keto-enol tautomerism, are known to be very sensitive to catalysis by bases. A similar two-stage process was suggested by Hantzsch for many other neutralizations involving a change in absorption spectra, even when they are effectively instantaneous. Finally, in fact, Hantzsch himself was led to describe almost every acid as a pseudo-acid on the basis of slight optical changes on ionization.[14] Such a wide extension of the term destroys its usefulness, and few other authors have followed him here.

Later developments have largely removed the justification for speaking of a separate class of pseudo-acids, even if restricted to the nitroparaffins and similar substances. In the first place (as was first pointed out clearly by Pedersen[15]), we do not now believe that the neutralization of nitromethane involves the intermediate formation of the *aci*-form, but rather that the removal of a proton from $CH_3NO_2$ to give

is itself a relatively slow process, because of the degree of electronic reorganization involved; the charge on the anion is not associated with the atom from which the proton has been removed. As will be seen later, many of the phenomena of acid-base catalysis can be explained in terms of slow proton transfers of this kind. In fact, Hantzsch's explanation has effectively been turned inside out, since we do not now think that a change such as *aci*-form → *nitro*-form (or keto → enol) commonly takes place by a single-stage migration of a hydrogen atom, and current mechanisms for these changes often involve the anion as an intermediate.

In the second place, it is now realized that the direct observation of a measurable reaction rate with hydroxide ions is an arbitrary criterion to adopt. Thus the neutralization of nitromethane by strong alkalis has a velocity of about 25 $dm^3$ $mol^{-1}$ $s^{-1}$ at 25°C, which is uncomfortably fast for measurements by conventional methods.[16] However, techniques are

---

[14] A. Hantzsch, *Z. Elektrochem.*, **29**, 244 (1923); **30**, 202 (1924); *Ber.*, **58**, 953 (1925).
[15] K. J. Pedersen, *Kgl. Dansk Vid. Selsk. Math-fys. Medd.*, **12** No. 1 (1932); *J. Phys. Chem.*, **38**, 581 (1934).
[16] Hantzsch, and most later workers, made measurements in the neighbourhood of 0°C.

now available for studying reactions in solution with velocity constants up to $10^{11}\,dm^3\,mol^{-1}\,s^{-1}$, a value of $1.5\times 10^{11}$ having been recently given for the reaction $H_3O^+ + OH^- \rightarrow 2H_2O$.[17] There is no doubt that acid-base reactions exist which cover the whole intermediate range of velocities, and the concept of a 'measurable' velocity thus depends entirely on the experimental facilities available.

Equally unsatisfactory is the suggestion of defining a pseudo-acid on the basis of the structural change (or charge shift) which it undergoes on ionization, as revealed by the change in optical properties. In the nitro-paraffins this change is a fairly clear-cut one, consisting of a shift of charge from the carbon to the oxygen atoms, but changes of this kind are now believed to take place in varying degrees over a wide range of acid-base systems. For example, nitramide, $NH_2NO_2$, gives an anion which can be written either as $NH^- \cdot NO_2$ or

and the negative charge in the phenoxide ion is believed to reside partly on the *ortho* and *para* carbon atoms in the ring. Even familiar acids such as carboxylic acids and inorganic oxyacids ionize with a change of struc-ture, since in their anions the negative charge is shared between two or more equivalent oxygen atoms. All of these could be logically termed pseudo-acids, but the term would now include the vast majority of known acids. It seems better on the whole to avoid the use of a term which is difficult to define satisfactorily.

The description 'pseudo-base' could logically have been applied to bases which undergo a structural change on the addition of a proton and are therefore the conjugate bases of pseudo-acids. Examples would be the anions of nitroparaffins, or derivatives of $\gamma$-pyrone, which acts as a base in the following way:

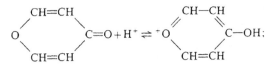

similar changes are responsible for the colour change of many plant pigments (anthocyanines and flavones) with pH. However, the term has

[17] M. Eigen and J. Schoen, *Z. Elektrochem.*, **59**, 483 (1955); M. Eigen and L. De Maeyer, *Z. Elektrochem.*, **59**, 986 (1955).

been more commonly used for a class of organic substances which react with acids (sometimes slowly) to form a cation and water, rather than to add on a proton. The best-known examples are the 'carbinol bases' of various triphenylmethane dyes. Thus the carbinol base of Crystal Violet is $(NMe_2C_6H_4)_3COH$, and it reacts slowly with acids with the loss of a water molecule to give the ion of the dye, which has the structure

$$(NMe_2C_6H_4)_2C = \left\langle \underset{\phantom{x}}{\bigcirc} \right\rangle = \overset{+}{N}Me_2$$

where the quinonoid structure and positive charge can be associated with any one of the three benzene rings. This ion reacts slowly with hydroxide ions to re-form the carbinol base. Similar behaviour is met with in simpler compounds such as the pyrazines and acridines,[18] and in strongly acid solvents such as sulphuric acid analogous changes take place with a number of compounds, e.g.,

$$(C_6H_5)_3COH + 2H_2SO_4 \rightarrow (C_6H_5)_3C^+ + H_3O^+ + 2HSO_4^-$$

$$RCO_2H + 2H_2SO_4 \rightarrow RCO^+ + H_3O^+ + 2HSO_4^-$$
$$\text{(with certain carboxylic acids)}$$

$$HNO_3 + 2H_2SO_4 \rightarrow NO_2^+ + H_3O^+ + 2HSO_4^-.$$

These substances are not acting as bases in the usual sense of the word, since there is no evidence for the existence of their conjugate acids, e.g., $(C_6H_5)_3COH_2^+$, under the specified experimental conditions. It would be convenient to have a term for describing them, but it seems undesirable to retain 'pseudo-base', because of a misleading analogy with 'pseudo-acid', and a more descriptive term would be 'aquo-base', originally used by Werner to distinguish the metallic hydroxides from the 'anhydro-bases' like ammonia and the amines.[5,19]

The hydrogen ion has already been formulated as $H_3O^+$, which we shall term the *hydronium ion*,* and the remainder of this chapter will deal

---

[18] A. Hantzsch and M. Kalb, *Ber.*, **32**, 3116 (1899); J. G. Aston, *J. Am. Chem. Soc.*, **52**, 5254 (1930); **53**, 1448 (1931).

[19] A. Werner, *Neuere Anchauungen auf dem Gebiete der anorganischen Chemie*, 2nd edn., Veweg, Braunschweig, 1909, p. 218.

*Alternative names are *hydroxonium* or *oxonium* ion. The latter is perhaps the most logical, but the term hydronium has been more widely used, and is related to the use of *lyonium* and *lyate* to denote the ions derived from the solvent by the addition or removal of a proton, respectively. It would also be more consistent to write $OH_3^+$ in place of $H_3O^+$, by analogy with $NH_4^+$, $CH_3OH_2^+$, etc. However, this leads to the probably unacceptable formula $OH_2$ for water, and the formulation $H_3O^+$ will be retained in this book.

with the evidence for the existence and structure of $H_3O^+$ and other hydrated forms of the proton. Reviews on this subject have been published by Conway[20] and by Giguère.[21]

The clearest evidence for the existence of $H_3O^+$ comes from the solid hydrates of strong acids such as nitric, perchloric, and sulphuric acid, and the hydrogen halides. Thus it was shown by Volmer[22] that the mono-hydrate of perchloric acid is isomorphous with ammonium perchlorate and gives a very similar set of X-ray reflections. This suggests strongly that it is an ionic crystal $H_3O^+ \cdot ClO_4^-$, and analogous structures are likely for the other hydrates. A definite proof of this, however, had to wait for modern methods of locating hydrogen atoms in crystals, of which the most powerful is proton magnetic resonance. The characteristic frequencies for the transition between different orientations of the proton nuclear magnetic moment are very sensitive to the environment of the proton, and in particular to the proximity of other protons. A group of several protons gives a pattern of frequencies which is characteristic of their number and of their geometrical arrangement, and a quantitative treatment of this pattern yields information about the distances between the protons. Thus Richards and Smith[23] studied the proton magnetic resonance spectra of the solid hydrates $HNO_3 \cdot H_2O$, $HClO_4 \cdot H_2O$, and $H_2SO_4 \cdot H_2O$, and in each case concluded that the crystals contained three protons in an equilateral triangle; alternative structures such as a hydrogen-bonded $H_2O \cdots H-X$ would have given an entirely different type of spectrum. This points to a pyramidal or planar ion $H_3O^+$, and by combining the observed proton–proton distances with reasonable values of the O—H bond length Richards and Smith concluded that it was a flat pyramid with H—O—H about 115°. The detailed interpretation of the proton magnetic resonance results is somewhat complicated by the existence of phase transitions and disorder in the crystals of $H_3O^+ \cdot ClO_4^-$, but later workers[24] were able to confirm that the triangle of protons is equilateral to within a few degrees, and a recent study[25] employing both proton and deuteron resonances was able to define the H—O—H angle as $118.5 \pm 0.7°$ and the O—H distance as $101 \pm 2$ pm. The pyramid is thus

[20] B. E. Conway, in *Modern Aspects of Electrochemistry* (ed. J. O'M. Bockris and B. E. Conway), No. 3, MacDonald, London, 1964, p. 43.
[21] P. A. Giguère, *Rev. Chim. Minérale*, 3, 627 (1966).
[22] A. Volmer, *Annalen*, **440**, 200 (1924).
[23] R. E. Richards and J. A. S. Smith, *Trans. Faraday Soc.*, **47**, 1261 (1951). See also Y. Kakiuchi, H. Shono, H. Matsu, and K. Kigoshi, *J. Chem. Phys.*, **19**, 1069 (1951); *J. Phys. Soc. Japan*, 7, 102 (1952), for $HClO_4 \cdot H_2O$.
[24] E. R. Andrew and N. D. Finch, *Proc. Phys. Soc.*, B, **70**, 980 (1957).
[25] D. E. O'Reilly, E. M. Peterson, and J. M. Williams, *J. Chem. Phys.*, **54**, 96 (1971).

a very flat one, and the O—H distance somewhat longer than that in the water molecule (96 pm); this difference is similar to that found for the N—H distances in $NH_4^+$ and $NH_3$.

These findings are confirmed by X-ray diffraction studies of $HNO_3 \cdot H_2O$, $HNO_3 \cdot 3H_2O$, $HCl \cdot H_2O$, and the two forms of $HClO_4 \cdot H_2O$.[26] Although the positions of the protons could not be determined, H—O—H angles in the range 110–118° could be inferred from the directions of hydrogen bonds formed to other atoms.

Further evidence comes from the infrared spectra of the same solids. Bethell and Sheppard[27] concluded that the spectrum of $HNO_3 \cdot H_2O$ contained bonds characteristic of the ion $H_3O^+$, but there are complications due to overlapping by $NO_3^-$ frequencies. Ferriso and Hornig[28] examined the monohydrates of the four hydrogen halides and detected in each of them four infrared fundamental frequencies characteristic of a pyramidal $H_3O^+$. (A planar ion $H_3O^+$ would have only three frequencies active in the infrared spectrum.) The Raman spectrum of $H_3O^+$ is more difficult to observe, and observations on powdered $H_3O^+ \cdot ClO_4^-$, $H_3O^+ \cdot HSO_4^-$, $H_3O^+ \cdot NO_3^-$, and $(H_3O^+)_2 \cdot SO_4^{2-}$ served only to establish the presence of the anions indicated in these formulae, from which the presence of the $H_3O^+$ cation can be inferred indirectly.[29] Raman frequencies characteristic of $H_3O^+$ itself have been found with single crystals of $H_3O^+ \cdot ClO_4^-$, and show a satisfactory correspondence with the infrared frequencies.[30]

The existence of ionic crystals with the formula $H_3O^+ \cdot X^-$ makes it possible to estimate thermochemically the energy change in the gas reaction $H^+ + H_2O \rightarrow H_3O^+$, i.e., the proton affinity of water, $P_{H_2O}$. This is done by means of the following cycle:

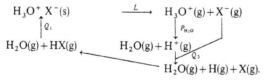

[26] V. Luzzati, *Acta Cryst.*, **4**, 239 (1951); **6**, 157 (1953); Y. K. Yoon and G. B. Carpenter, *Acta Cryst.*, **12**, 17 (1959); F. S. Lee and G. B. Carpenter, *J. Phys. Chem.*, **63**, 279 (1959); C. E. Nordman, *Acta Cryst.*, **15**, 18 (1962). A report [P. Bourre-Maladière, *Compt. Rend.*, **246**, 1063 (1958)] that $H_2SO_4 \cdot H_2O$ contains sulphuric acid molecules has been refuted by I. Taessler and I. Olovsson, [*Acta Cryst.*, **B24**, 299 (1968)], who found good evidence for $H_3O^+ \cdot HSO_4$.
[27] D. E. Bethell and N. Sheppard, *J. Chem. Phys.*, **21**, 1421 (1953).
[28] C. C. Ferriso and D. F. Hornig, *J. Chem. Phys.*, **23**, 1464 (1955).
[29] D. J. Millen and E. G. Vaal, *J. Chem. Soc.*, 2913 (1956).
[30] J. T. Mullhaupt and D. F. Hornig, *J. Chem. Phys.*, **24**, 169 (1956); R. C. Taylor and G. L. Vidale, *J. Am. Chem. Soc.*, **78**, 5999 (1956).

$Q_1$ is the calorimetric heat of formation of the hydrate from HX gas and water vapour, $Q_2$ the heat of dissociation of gaseous HX, and $Q_3$ the sum of the ionization potential of H and the electron affinity of X. The two unknown quantities are thus $L$, the lattice energy of the crystal, and $P_{H_2O}$. In order to obtain the latter it is necessary to make an estimate of the lattice energy, and since the detailed crystal structures of the hydrates are not known this is usually done by analogy with similar crystals. Thus Grimm[31] assumed that $H_3O^+ \cdot Cl^-$ and KCl had the same lattice energies and arrived at the value $P_{H_2O} = 160 \, \text{kcal mol}^{-1}$. A more reliable procedure is that employed by Sherman.[32] He first derives the energy of the process $NH_3 + H^+ \rightarrow NH_4^+$, $P_{NH_3} = 207 \, \text{kcal mol}^{-1}$, by applying a cycle like that given above to the crystals $NH_4Cl$, $NH_4Br$, and $NH_4I$; these have accurately known crystal structures, and their lattice energies can therefore be calculated with fair certainty. He then considers analogous cycles for the crystals $NH_4ClO_4$ and $H_3OClO_4$, making the very reasonable assumption that their lattice energies are the same. Each cycle involves the energy of the process $H^+(g) + ClO_4^-(g) \rightarrow HClO_4(l)$, which is unknown, but this quantity is eliminated by subtraction, and we have finally $P_{NH_3} - P_{H_2O} = Q_1' - Q_1$, where $Q_1'$ and $Q_1$ are the calorimetric heats of reaction for the reactions $NH_3(g) + HClO_4(l) \rightarrow NH_4ClO_4(s)$ and $H_2O(g) + HClO_4(l) \rightarrow H_3OClO_4(s)$. Sherman obtained $P_{H_2O} = 182 \, \text{kcal mol}^{-1}$ by this procedure; a recalculation using modern values of bond energies and electron affinities[33] gave $187 \, \text{kcal mol}^{-1}$, while a more thorough examination of the lattice energies of $H_3O^+ \cdot ClO_4^-$ and $NH_4^+ \cdot ClO_4^-$, with allowance for hydrogen bonding,[34] gave $170 \, \text{kcal mol}^{-1}$.

Although these estimates of $P_{H_2O}$ are subject to some uncertainty, they certainly indicate that $H_3O^+$ is a highly stable entity in the gas phase. For the reaction $H_3O^+ \rightarrow O + H^+ + 2H$, we find $\Delta H = 390 \, \text{kcal mol}^{-1}$, and the same value applies to the process $H_3O^+ \rightarrow O^+ + 3H$,* so that the average bond energy in $H_3O^+$ is about $130 \, \text{kcal mol}^{-1}$, compared with $111 \, \text{kcal mol}^{-1}$ for the water molecule. If we take the value $\Delta H = 170 \, \text{kcal mol}^{-1}$ for the reaction $H_3O^+(g) \rightarrow H_2O(g) + H^+(g)$, this can be converted into a standard free energy change if we can estimate the

[31] H. G. Grimm, *Z. Elektrochem.*, **31**, 474 (1925).
[32] J. Sherman, *Chem. Rev.*, **11**, 164 (1932).
[33] V. Kondratiev and N. D. Sokolov, *Zh. Fiz. Khim.*, **29**, 1265 (1955); F. W. Lampe and J. H. Futtrell, *Trans. Faraday Soc.*, **59**, 1957 (1963).
[34] S. I. Vetchinkin, E. I. Pshenichnov, and N. D. Sokolov, *Zh. Fiz. Khim.*, **33**, 1269 (1959).
*The values for the two processes are fortuitously the same because of the almost identical ionization potentials of the hydrogen and oxygen atoms.

entropy of reaction. $S°(H_2O)$ is of course known experimentally, and $S°(H^+)$ consists only of translational entropy which can be calculated statistically. A reasonable estimate[35] shows that $S°(H_3O^+)$ is almost identical with $S°(H_2O)$, and the final result is $\Delta G°(298, 1\ atm) = 162$ kcal mol$^{-1}$. This corresponds to an equilibrium degree of dissociation of $H_3O^+$ of $10^{-58}$ at 1 atmosphere and 298 K, or of $10^{-13}$ at $10^{-9}$ atm and 1000 K; it thus seems likely that the hydronium ion should exist in ionized gases under a wide range of conditions. This has now been confirmed experimentally in a number of investigations by mass spectrometric and similar techniques, for example in the mass spectrum of glow discharges in water vapour[36] and in field emission studies.[37] The correctness of the assignment of the mass spectra can be checked by using $H_2O$ and $D_2O$. A particularly neat demonstration, independent of any absolute mass scale, is given by Beckey's field emission study[37] of an equimolecular mixture of $H_2O$ and $D_2O$, in which four peaks were observed at consecutive mass numbers with intensities in the ratios 1:3:3:1. These are attributed to the ions $H_3O^+$, $H_2DO^+$, $HD_2O^+$, and $D_3O^+$, where the greater intensity of the second and third arises from the number of ways of distributing two hydrogens and one deuterium (or two deuteriums and one hydrogen) among three positions. It is readily seen that no other ion would produce the same pattern; for example, $H_2O^+$ would give only three peaks with intensities 1:2:1, and $HO^+$ two peaks of equal intensity. It should be mentioned that $H_3O^+$ is by no means the only type of hydrated proton which appears in the mass spectrometer, since ions of composition $H(H_2O)_n^+$ with $n = 1, 2, 3, 4$, and 5 have all been observed, $H(H_2O)_4^+$ being frequently the most abundant. The question of further hydration of the hydronium ion will be considered later in this chapter.

Such observations in the gas phase tell us nothing about the structure of the hydronium ion, but there are various ways in which they can be used to determine stability. There is now a large literature on this subject, and only two examples will be given. In the first type of method the minimum electron energy needed to produce $H_3O^+$ from an organic molecule (i.e., its appearance potential) is determined in the mass spectrometer. If the other products of the reaction can be identified as species with known enthalpies of formation, $\Delta H_f(H_3O^+)$ can be found, and hence $P(H_2O)$ from the relation

$$P(H_2O) = \Delta H_f(H_2O) + \Delta H_f(H_2O) + \Delta H_f(H^+) - \Delta H_f(H_3O^+).$$

[35] Ref. 13, p. 59.
[36] P. F. Knewstubb and A. W. Tickner, *J. Chem. Phys.*, **36**, 674 (1962); **38**, 464 (1963).
[37] H. D. Beckey, *Z. Naturforsch.*, **14a**, 712 (1959); **15a**, 822 (1960).

Thus Van der Raalte and Harrison[38] found the appearance potential of $H_3O^+$ in propan-2-ol to be $310 \pm 3$ kcal mol$^{-1}$. If the reaction is assumed to be $C_3H_7OH + e^- \rightarrow C_2H_2 + CH_3 + H_3O^+ + 2e^-$, this leads to $\Delta H_f(H_3O^+) = 158$ kcal mol$^{-1}$ and hence to $P(H_2O) = 179$ kcal mol$^{-1}$. The results for several other organic molecules are similar. The second type of method, first used by Tal'rose,[39] depends upon the observation (or non-observation) in the mass spectrometer of ion–molecule reactions involving the hydronium ion. For example, Beauchamp and Butterill[40] found that the reaction $C_2H_4 + H_3O^+ \rightleftharpoons C_2H_5^+ + H_2O$ can be observed in both directions. It must therefore be almost thermoneutral, and since $\Delta H_f$ is known for the other species involved, we obtain $\Delta H_f(H_3O^+) = 149$ kcal mol$^{-1}$ and $P(H_2O) = 170$ kcal mol$^{-1}$. Although this type of method involves a search for suitable ion–molecule reactions, it is probably more reliable than the use of appearance potentials, which involves uncertainties about the translational and vibrational energy, of the products and the activation energy of ionization reactions involving rearrangement. However, all the values given for $P(H_2O)$ are in the range 160–190 kcal mol$^{-1}$, with a most probable value of $164 \pm 4$ kcal mol$^{-1}$, in substantial agreement with the results obtained from $H_3O^+ \cdot ClO_4^-$ in the crystalline state.

Since the hydronium ion has a perfectly respectable electronic structure, being isoelectronic with ammonia, there have been numerous attempts to calculate its structure and energy by quantum theory.[41] Although, on a simple view, the presence of an unshared electron pair implies a non-planar structure, earlier treatments by both molecular orbital and valence bond methods predicted a planar ion. However, a molecular orbital treatment by Bishop,[42] involving 33 parameters, gives the most stable configuration as a pyramid with $H—O—H = 115°$, not far from the observed value, and predicts stretching and bending force constants of the right order of magnitude. Quantum theory should be able to predict the proton affinity of water, but this is a very severe test since $P(H_2O)$ is only about 0.3% of the total electronic energy. However, several calcu-

[38] D. Van der Raalte and A. G. Harrison, *Canad. J. Chem.*, **41**, 3118 (1963); see also M. A. Haney and J. L. Franklin, *J. Chem. Phys.*, **50**, 2028 (1969).
[39] V. L. Tal'rose and E. L. Frankevich, *Dokl. Akad. Nauk S.S.S.R.*, **111**, 376 (1956); *J. Am. Chem. Soc.*, **80**, 2344 (1958).
[40] J. L. Beauchamp and S. E. Butterill, *J. Chem. Phys.*, **48**, 1783 (1968); see also J. Long and B. Munson, *J. Chem. Phys.*, **53**, 1356 (1970).
[41] For a summary up to 1963, see J. L. J. Rosenfeld, *J. Chem. Phys.*, **40**, 384 (1964); *Acta Chem. Scand.*, **18**, 1719 (1964). It is interesting to note that theory predicts a positive $\Delta H$ of 40–60 kcal mol$^{-1}$ for the reaction $H_3O^+ + H^+ \rightarrow H_4O^{2+}$; the last species has never been detected experimentally.
[42] D. M. Bishop, *J. Chem. Phys.*, **43**, 4453 (1965).

lations[43] have given approximately the correct value, usually by assuming the observed geometries for $H_2O$ and $H_3O^+$. The most thorough treatment is that of Hopkinson *et al.*,[44] who also calculate proton affinities for 19 other species; their best value for $P(H_2O)$ is 174 kcal mol$^{-1}$, in excellent agreement with experiment.

There is thus good evidence of the existence in the solid state of the species $H_3O^+$, with a stability comparable with that of the ammonium ion, and this ion would be expected to retain its individuality in solution. Some of the earliest evidence on this point came from a study of non-aqueous solutions. For example, Goldschmidt[45] studied the retarding effect of small quantities of water on acid-catalysed esterification reactions in various alcohols and interpreted his results in terms of the equilibrium $H^+(ROH)_n + H_2O \rightleftharpoons nROH + H_3O^+$, it being assumed that the catalytic effect of $H_3O^+$ is small compared with that of $H^+(ROH)_n$. This reaction scheme gave quantitative agreement with the observed results, while any other assumption about the hydration of the proton would destroy this agreement.[46] The same conclusion was reached by other authors, who studied the inhibiting effect of water on the hydrogen-ion catalysed alcoholysis of ethyl diazoacetate,[47] and a similar treatment has been applied to the effect of small quantities of water upon the conductivity of solutions of strong acids in methyl and ethyl alcohol,[48] and to the effect of water additions on the dissociation constant of picric acid in ethyl alcohol.[49] The following observation also gives a clear illustration of the strong tendency of the proton to attach a single molecule of water. Liquid sulphur dioxide alone dissolves very little water, but a non-conducting solution of hydrogen bromide in liquid sulphur dioxide dissolves an equivalent amount of water and becomes a good conductor. When this solution is electrolysed, 1 equivalent of water per

[43] R. Gaspar, I. Tamassy-Lentei, and V. Kruglyak, *J. Chem. Phys.*, **36**, 740 (1962); J. W. Moskowitz and M. C. Harrison, *J. Chem. Phys.*, **43**, 3550 (1965).

[44] A. C. Hopkinson, N. K. Holbrook, K. Yates, and I. G. Cszimadia, *J. Chem. Phys.*, **49**, 3596 (1968).

[45] H. Goldschmidt and O. Udby, *Z. Phys. Chem.*, **60**, 728 (1907); H. Goldschmidt, *Z. Elektrochem.*, **15**, 4 (1909).

[46] It is reasonable to assume by analogy that the 'hydrogen ion' in an alcohol ROH has the formula $ROH_2^+$, hence that the equilibrium can be written $ROH_2^+ + H_2O \rightleftharpoons ROH + H_3O^+$; however, this cannot be deduced from experiments in which the concentration of the alcohol is effectively constant.

[47] G. Bredig, *Z. Elektrochem.*, **18**, 535 (1912); W. S. Miller, *Z. Phys. Chem.*, **85**, 129 (1913).

[48] G. Nonhebel and H. B. Hartley, *Phil. Mag.*, **50**, 734 (1925); L. Thomas and E. Marum, *Z. Phys. Chem.*, **143**, 213 (1929).

[49] P. Gross, A. Jamöck, and F. Patat, *Monatsh.*, **63**, 124 (1933).

Faraday is liberated at the cathode.[50] Moreover, it has been recently shown[51] that these solutions give infrared and Raman spectra closely resembling those of the crystalline hydrates containing the hydronium ion. In particular, they show clearly, especially for $D_3O^+$, the presence of two infrared frequencies in the O—H or O—D stretching range, which is indicative of a pyramidal rather than a planar structure.

It has proved much more difficult to get clear evidence of the existence of recognizable $H_3O^+$ ions in aqueous solutions of acids, and there are three reasons why this is so. Firstly, the characteristic properties of $H_3O^+$ (for example, its spectral frequencies) will not differ very much from those of $H_2O$, which is present in very large excess. Secondly, the ion $H_3O^+$ in aqueous solution will undoubtedly be further hydrated by the more or less firm attachment of additional water molecules, and this will make it difficult to recognize it. Thirdly, there is good evidence that the protons in $H_3O^+$ can exchange very rapidly with those in liquid water, so that the lifetime of any individual $H_3O^+$ ion may be too short to endow it with characteristic properties. This last circumstance makes it impossible to obtain any structural information from studies of proton magnetic resonance in solution, since because of the rapid interchange the observed frequency is averaged over the different kinds of proton present. Many attempts to detect the hydronium ion in aqueous solution by its infrared or Raman spectrum have been unsuccessful. Pure water gives a diffuse vibrational spectrum without any well-defined frequencies, and although this is modified by the addition of acid most workers have failed to detect any new frequencies.[52] However, Falk and Giguère[53] have reported measurements of infrared spectra believed to be characteristic of the $H_3O^+$ ion in solution. They examined very thin layers of concentrated solutions of HCl, HBr, $HNO_3$, $HClO_4$, $H_2SO_4$, and $H_3PO_4$, and of some acid salts of the last two acids. In every case they were able to detect three broad absorption bands at about 1205, 1750, and 2900 cm$^{-1}$ which are in rough agreement with three of the frequencies attributed to $H_3O^+$ in the solid state. Measurements with DCl in $D_2O$ give correspondingly lower frequencies of 960, 1400, and 2170 cm$^{-1}$. Similar measurements have been

[50] L. S. Bagster and B. D. Steele, *Trans. Faraday Soc.*, **8**, 51 (1912); L. S. Bagster and G. Cooling, *J. Chem. Soc.*, 693 (1920).
[51] M. Schneider and P. A. Giguère, *Compt. Rend.*, B, **267**, 551 (1968).
[52] See, e.g., R. Suhrmann and F. Breyer, *Z. Phys. Chem.*, **23B**, 193 (1933).
[53] M. Falk and P. A. Giguère, *Canad. J. Chem.*, **35**, 1195 (1957); **36**, 1680 (1958).

reported subsequently[54] for both the infrared and the Raman spectra of aqueous strong acids.

However, it now seems doubtful whether these spectra of aqueous acids can actually be attributed to the hydronium ion. The detection of a vibrational frequency of 1205 cm$^{-1}$ demands a lifetime of at least $1/(3 \times 10^{10} \times 1205) = 3 \times 10^{-13}$ s. The observed breadth of the infrared bands suggests a lifetime rather less than this, and measurements of proton mobility in ice[55] lead to an estimate of between $0.8 \times 10^{-13}$ and $1.0 \times 10^{-13}$ s for the lifetime of an 'individual' $H_3O^+$ ion.[56]

Although the long-range proton mobility in liquid water is some 30 times smaller, this is because the rate-determining process is now the rotation of water molecules rather than the movement of protons.[13] A theoretical analysis of proton mobility in water[57] gave $2 \times 10^{-13}$ s for the average lifetime of $H_3O^+$, and a similar value follows from ultrasonic measurements for the time constant for reorganization of the water structure.[58] It therefore appears that the effect of acids upon the vibrational spectrum of water is better interpreted as a superposition of a continuum due to the mobile protons upon the normal water spectrum, rather than as a characteristic of the hydronium ion. This is supported by the fact that strong bases such as sodium hydroxide produce almost the same spectrum as strong acids, while salts have very little effect.[59] Recent theoretical calculations of the frequencies expected for a hydronium ion in aqueous solution[60] give values considerably higher than those reported by Falk and Giguère,[53] and the continuous spectrum anticipated for mobile ('tunnelling') protons has been discussed, particularly by Zundel.[61]

The difficulty (or impossibility) of observing its vibrational spectrum does not of course deny the presence of the hydronium ion in aqueous solution, and there are several other pieces of evidence for its existence, two of which will be mentioned here. The first relates to the observed refractivity of the hydrogen ion in aqueous solution,[62] which fits smoothly

[54] C. G. Swain and R. F. W. Bader, *Tetrahedron*, **10**, 182 (1960); C. G. Swain, R. F. W. Bader, and E. R. Thornton, *Tetrahedron*, **10**, 200 (1960); W. R. Busing and D. F. Hornig, *J. Phys. Chem.*, **65**, 284 (1961).
[55] M. Eigen and L. de Maeyer, *Z. Elektrochem.*, **60**, 1037 (1956); *The Structure of Electrolytic Solutions* (ed. W. J. Hamer), Wiley, New York, 1959, p. 64.
[56] M. Eigen, *Angew. Chem.*, **75**, 489 (1963).
[57] B. E. Conway, J. O'M. Bockris, and H. Linton, *J. Chem. Phys.*, **24**, 834 (1956).
[58] L. Hall, *Phys. Rev.*, **73**, 775 (1948).
[59] T. Ackermann, *Z. Phys. Chem* (Frankfurt), **27**, 253 (1961).
[60] R. More O'Ferrall, G. W. Koeppl, and A. J. Kresge, *J. Am. Chem. Soc.*, **93**, 1 (1971).
[61] E. G. Weidemann and G. Zundel, *Z. Phys.*, **198**, 288 (1967); G. Zundel, *Angew. Chem. Internat. Edn.*, **8**, 499 (1969).
[62] K. Fajans and G. Joos, *Z. Phys. Chem.*, **23**, 1, 31 (1924).

into the values for the isoelectronic series $O^{2-}$, $OH^-$, $H_2O$, $H_3O^+$. The second is much more subtle, though it probably constitutes the best evidence for the real existence of $H_3O^+$ in aqueous solution. It concerns the fractionation of the hydrogen isotopes between hydrogen ions and the solvent in $H_2O$, $D_2O$, and their mixtures; the experimental observations can only be accounted for by the assumption that the 'hydrogen ion' contains just three hydrogens whose properties differ appreciably from those in the bulk of the solvent. This subject will be dealt with in Chapter 11.

Like other cations, $H_3O^+$ will be further hydrated in aqueous solution, but for many purposes such hydration can be neglected. However, the presence in the hydronium ion of three hydrogens bearing a positive charge suggests that strong hydrogen bonding of further water molecules will lead to higher hydrates of well-defined geometry and relatively high stability. Particular attention has been given to the ion $H_9O_4^-$ (i.e., $H_3O^+ \cdot 3H_2O$, or $H^+ \cdot 4H_2O$), which can be given the reasonable structure

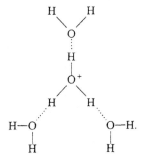

The presence of an ion of this composition in solution was first deduced from the specific heats of aqueous acids,[63] though the argument was based upon a rather artificial picture of water as a mixture of polymers. Supporting evidence comes from the thermodynamic properties of concentrated acid solutions. The rapid rise in the activity coefficients of electrolytes in concentrated solutions can be attributed largely to the removal of water by ionic hydration, with a consequent increase in the true mole fraction of the solute. A quantitative treatment in terms of a reasonable model[64] yields the mean hydration numbers of the ions, and a value close to 4 is found for the hydrogen ion. A similar deduction can be made from the indicator equilibria used in determining the acidity of concentrated

[63] E. Wicke, M. Eigen, and T. Ackermann, *Z. Phys. Chem.* (Frankfurt), **1**, 340 (1954).
[64] E. Glueckauf, *Trans. Faraday Soc.*, **51**, 1235 (1955).

aqueous acids. If the hydration of other species can be neglected, the equilibrium can be written as $H(H_2O)_n^+ + I \rightleftharpoons IH^+ + nH_2O$, where I is an indicator, and it has been shown[65] that the experimental data for solutions of strong acids up to about 8M can be accounted for by assuming a constant value $n = 4$. Finally, it is found[66] that when strong acids are extracted from aqueous solution by organic solvents they frequently take with them four molecules of water per acid molecule, and the same is true for the uptake of acids by ion-exchange resins;[67] this suggests (though it does not prove) the same extent of hydration in the aqueous phase.

As might be expected, the ion $H_9O_4^+$ also turns up in crystals and in the gas phase, though several other hydrates of $H_3O^+$ are observed at least as frequently. The situation for crystals is rather confused. The infrared spectrum of solid $HBr \cdot 4H_2O$ has been interpreted in terms of $H_9O_4^+ \cdot Br^-$,[68] but the structure is an unsymmetrical one, and crystallographic evidence[69] has led to the formulation $H_7O_3^+ \cdot H_9O_4^+ \cdot 2Br^- \cdot H_2O$ for the same compound. Both infrared[70] and crystallographic[71] evidence indicates that $HClO_4 \cdot 2H_2O$ contains the ion $H_5O_2^+$, which has also been reported in the hydrates of a number of metallic complexes and in the dihydrates of HCl and HBr.[72,73] On the other hand, the trihydrates of HCl and HBr appear to contain $H_5O_2^+ \cdot H_2O$ rather than $H_7O_3^+$, and $HAuCl_4 \cdot 4H_2O$ has $H_5O_2^+ \cdot 2H_2O$ rather than $H_9O_4^+$.[71,73,74] The distinction between the different structures is often a fine one; for example, the species $[H_2O \cdots H \cdots OH_2]^+$ would be written as $H_5O_2^+$ if the central hydrogen bond is symmetrical, or nearly so, but as $H_3O^+ \cdot H_2O$ if it is highly unsymmetrical. The hydrogen can rarely be located directly, and the distinction is usually made on the basis of infrared frequencies, or the $O \cdots O$ distance determined crystallographically. However, in a few instances a combination of neutron diffraction, hydrogen and deuterium

[65] R. P. Bell and K. N. Bascombe, *Disc. Faraday Soc.*, **24**, 158 (1957). A similar treatment for concentrated alkaline solution leads to a hydration number of 3 for the hydroxide ion; cf. G. Yagil and M. Anbar, *J. Am. Chem. Soc.*, **85**, 2376 (1963); R. Stewart and J. P. O'Donnell, *Canad. J. Chem.*, **42**, 1681 (1964).

[66] A. H. Laurence, D. E. Campbell, S. E. Wiberley, and H. M. Clark, *J. Phys. Chem.*, **60**, 901 (1956); D. G. Tuck and R. M. Diamond, *J. Phys. Chem.*, **65**, 193 (1961).

[67] E. Glueckauf and G. P. Kitt, *Proc. Roy. Soc.*, A, **228**, (1955).

[68] J. Rudolph and H. Zimmermann, *Z. Phys. Chem.* (Frankfurt), **43**, 311 (1964).

[69] J. O. Lundgren and I. Olovsson, *J. Chem. Phys.*, **49**, 1068 (1968).

[70] A. C. Pavia and P. A. Giguère, *J. Chem. Phys.*, **52**, 3551 (1970).

[71] I. Olovsson, *J. Chem. Phys.*, **49**, 1063 (1968).

[72] R. D. Gillard and G. Wilkinson, *J. Chem. Soc.*, 1640 (1964).

[73] A. S. Gilbert and N. Sheppard, *J. Chem. Soc.*, D, 337 (1971).

[74] J. M. Williams and S. W. Petersen, *J. Am. Chem. Soc.*, **91**, 776 (1969); D. E. O'Reilly, E. M. Peterson, C. E. Scheie, and J. M. Williams, *J. Chem. Phys.*, **55**, 5629 (1971).

nuclear magnetic resonance, and chlorine-35 nuclear quadrupole resonance has made it possible to local the protons directly and to obtain
information about their movements in the ion.[67]

As for the hydronium ion, studies in the gas phase tell us nothing about
the structures of the higher hydrates, but they do provide some information about their energy relations. As already mentioned, a variety of mass-
spectrometric techniques[36,37,75-77] have revealed the presence of $H_5O_2^+$,
$H_7O_3^+$, $H_9O_4^+$, and $H_{11}O_5^+$ (or their deuterium analogues), of which
$H_9O_4^+$ is the most abundant under equilibrium conditions and $H_{11}O_5^+$
is formed to only a small extent. A range of values has been given for the
energy changes involved, but a recent investigation[77] gives $\Delta H = -32$,
$-23$, and $-17\,\text{kcal mol}^{-1}$ for the successive addition of three water
molecules to $H_3O^+$. This corresponds to a mean hydrogen bond energy
of 24 kcal mol$^{-1}$ in $H_9O_4^+$, which is considerably greater than the strength
of a hydrogen bond in water (about 7 kcal mol$^{-1}$), but of course very much
less than the 170 kcal mol$^{-1}$ associated with the formation of $H_3O^+$ from
$H_2O$ and $H^+$.

However we regard the further solvation of the hydronium ion in
solution, interest attaches to the heat evolved in the process

$$H^+(g) + H_2O(\text{liq}) \rightarrow H_3O^+(aq),$$

the heat of solution of the proton, which we shall denote by $Q_s$. This is
experimentally accessible from thermodynamic measurements, for
example from the following cycle, first proposed by Baughan:[78]

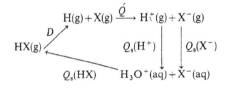

where X is a halogen. $D$ is the dissociation energy of HX gas, and $Q'$
is the sum of the ionization energy of H and the electron affinity of X.
HCl, HBr, and HI are completely dissociated in solution, so that $-Q_s(\text{HX})$
is the measured heat of solution of HX gas. This cycle thus leads to an
unambiguous value for $Q_s(H^+) + Q_s(X^-)$, but $Q_s(H^+)$ can only be obtained
separately by introducing some non-thermodynamic relations based on

[75] P. Kebarle, *Advances in Chemistry*, **72** (*Am. Chem. Soc.*, 1968), p. 24.
[76] M. de Paz, J. J. Leventhal, and L. Friedman, *J. Chem. Phys.*, **49**, 5543 (1968).
[77] M. de Paz, A. G. Giardini, and L. Friedman, *J. Chem. Phys.*, **52**, 687 (1970).
[78] E. C. Baughan, *J. Chem. Soc.*, 1403 (1940).

experiment or on a theoretical model, for example a relation between $Q_s$ and ionic radius. This problem is of course inherent in any attempt to assign values of $\Delta H$ or $\Delta G$ to processes involving individual ions. Many different assumptions have been made, resulting in values of $Q_s(H^+)$ between 253 and 292 kcal mol$^{-1}$, but the most likely estimate is that of Halliwell and Nyburg,[79] who find $Q_s(H^+) = 261 \pm 3$ kcal mol$^{-1}$.

The heat of solution of the hydronium ion is given by

$$Q_s(H_3O^+) = Q_s(H^+) - P(H_2O) = 261 - 170 = 91 \text{ kcal mol}^{-1}.$$

with an uncertainty of about 5 kcal mol$^{-1}$. This lies between the values for $Na^+$ and $K^+$, calculated on the same assumptions (97 and 78 kcal mol$^{-1}$ respectively), and is therefore little greater than would be expected for an ion of this size. Since we do not normally need to specify the hydration of ions like sodium and potassium, the formulation $H_3O^+$ is usually quite sufficient for the hydrogen ion in aqueous solution, and for many purposes the shorthand version $H^+$ will serve.

The *hydroxide ion* behaves in many respects* like a normal anion with a radius similar to that of the fluoride ion. Like fluoride ion, it will be strongly hydrated in aqueous solution, and there is evidence for the existence of a fairly stable species $H_7O_4^-$, or $OH(H_2O)_3^-$. This degree of hydration is consistent with the behaviour of indicators in strongly alkaline solutions,[80] and $H_7O_4^-$ is also found to be the most abundant hydrated species in the gas phase.[81] Experiments on collision-induced dissociation[77] show that the energies needed to remove three successive water molecules are very similar for the two species $H_9O_4^+$ and $H_7O_4^-$.

---

[79] H. F. Halliwell and S. C. Nyburg, *Trans. Faraday Soc.*, **59**, 1126 (1963). These authors give a useful summary of earlier estimates of this quantity. Conway[20] prefers a slightly higher value, but gives an upper limit of 267 kcal mol$^{-1}$. See also N. A. Izmailov, *Zh. Fiz. Khim.*, **34**, 2414 (1960).
* An exception is its abnormally high mobility in aqueous solution, due to the handing on of protons by the process $H—O^- + H—O—H \rightarrow H—O—H + {}^-O—H$.
[80] J. T. Edward and I. C. Wang, *Canad. J. Chem.*, **40**, 399 (1962); G. Yagil and M. Anbar, *J. Am. Chem. Soc.*, **85**, 2376 (1963).
[81] J. L. Moruzzi and A. V. Phelps, *J. Chem. Phys.*, **45**, 4617 (1966).

# 3 The Investigation of Protolytic Equilibria in Aqueous Solution

Since the free proton is never encountered in solution, the equation $A \rightleftharpoons B + H^+$, defining an acid-base pair, does not represent a realizable equilibrium, and all actual acid-base equilibria are of the form, $A_1 + B_2 \rightleftharpoons B_1 + A_2$. This type of reaction is known as a *protolytic* or *proton-transfer* reaction. Any qualitatively sensible concept of acid-base strength would imply that the equilibrium conversion of $A_1 + B_2$ into $B_1 + A_2$ will be more complete the stronger the acid $A_1$ and the base $B_2$ and the weaker the acid $A_2$ and the base $B_1$. Quantitatively, the equilibrium constant $[B_1][A_2]/[A_1][B_2]$ is equal to the ratio of the (hypothetical) constants $[B_1][H^+]/[A_1]$ and $[B_2][H^+]/[A_2]$, and it will therefore measure the ratio of the acid strengths of $A_1$ and $A_2$, or the ratio of the base strengths of $B_2$ and $B_1$. Since these two ratios are equal, there is no point in considering the acid and base strengths separately, and it has become usual to describe the properties of any acid-base pair A–B in terms of the acid strength of A. Thus for the pair $CH_3COOH–CH_3COO^-$ we speak of the acid strength of acetic acid rather than the base strength of acetate ion, and for the pair $NH_4^+–NH_3$ of the acid strength of the ammonium ion rather than the base strength of ammonia.

This approach leads only to the relative strengths of acids (or bases), and it does not seem possible, even in principle, to attach a meaning to the 'absolute strength' of an acid or base. Under these circumstances it is natural to measure strengths relative to some standard acid-base pair $A_0–B_0$, the acid strength for another pair A–B then being given by the equilibrium constant $[B][A_0]/[A][B_0]$. The usual procedure is to use the *solvent* as the source of the standard pair $A_0–B_0$. In aqueous solution there are two possible pairs, $H_3O^+–H_2O$, or $H_2O–OH^-$, and the first of these is commonly used. The rational measure of the strength of any acid A is then $[B][H_3O^+]/[A][H_2O]$, but since the concentration of water is effectively constant in dilute aqueous solutions it is usual to omit it from the equation, thus giving the usual expression for the *acid dissociation constant* $K_c = [B][H_3O^+]/[A]$. This is strictly a constant only in very dilute solution, partly because of the omission of the term $[H_2O]$, but more significantly because at least two of the species A, B, and $H_3O^+$ are ions, and therefore do not obey the simple law of mass action except in very dilute solutions. This can, of course, be formally corrected by replacing concentrations by activities, but for the sake of simplicity we shall in many cases use the simpler concentration constants $K_c$. The symbol $K$ will be reserved for the thermodynamic dissociation constant (equal to $K_c$ in very dilute solution) Similarly, since the observed constants vary through many powers of ten, it is often convenient to use the quantities $pK = -\lg K$ and $pK_c = -\lg K_c$. Obviously, the higher the $pK$, the weaker the acid and the stronger the base in the pair A–B.

We shall now survey briefly the methods available for measuring acid-base strengths, especially in aqueous solution.[1] The most obvious methods involve a direct determination of the concentration of A or B in a solution of known total concentration. The best way to estimate these concentrations without disturbing the equilibrium is to use optical properties such as the absorption spectrum (visible or ultraviolet) or the Raman spectrum. For acids or bases of moderate strength the spectrum of pure A or pure B can be obtained experimentally by adding an excess of hydrogen or hydroxide ions to the solution. If an acid is fairly strong ($K > 10^{-1}$), the reaction $A + H_2O \rightleftharpoons B + H_3O^+$ will go very far to the right, and it may be necessary to use concentrated solutions to get a sufficiently large value of the ratio $[A]/[B]$. The use of concentrated solutions introduces some uncertainty, since the spectrum of any species varies somewhat with environment, especially in its intensity. It appears that optical spectra are considerably influenced in this way, but Raman spectra considerably less so. It has been suggested that nuclear magnetic resonance measurements are particularly suitable for this purpose, since the 'chemical shift' in the observed frequency is determined almost exclusively by the immediate surroundings of the nucleus being observed, i.e., by the bonds and atoms attached to it. Since there is rapid exchange of protons among the acid, water molecules, and hydronium ions, only one peak is observed, its chemical shift representing a weighted mean for these species. Thus in a solution of an acid HX we have the following scheme:

| Species | $H_2O$ | $+ HX$ | $\rightleftharpoons H_3O^+ +$ | $X^-$ |
|---|---|---|---|---|
| Concentration | $(1-x)(1-\alpha)$ | $x(1-\alpha)$ | $\alpha x$ | $\alpha x$ |
| Proton concentration | $2(1-x)(1-\alpha)$ | $x(1-\alpha)$ | $3\alpha x$ | — |

[1] For a general account, see A. Albert and E. P. Serjeant, *Ionization Constants of Acids and Bases*, 2nd edn., Chapman and Hall, London, 1971.

where $x$ is the stoichiometric mole fraction of acid and $\alpha$ its degree of dissociation. The observed chemical shift $\sigma$ (relative to water) is then given by

$$\sigma = \alpha p \sigma_1 + \tfrac{1}{3}(1 - \alpha)p\sigma_2 \tag{1}$$

where $p = 3x/(2 - x)$ and $\sigma_1$ and $\sigma_2$ are the chemical shifts, assumed independent of composition, for $H_3O^+$ and HX respectively. If $\sigma_1$ and $\sigma_2$ are known, by direct measurement or by extrapolation, the degree of dissociation $\alpha$ can be derived from the observed value of $\sigma$.

However, early hopes that the dissociation constants of 'strong' acids ($K > 1$) could be determined reliably from Raman or n.m.r. spectra have not been fulfilled. When $\alpha$ is close to unity the calculated dissociation constant is very sensitive to the value assumed for $\sigma_1$, and this quantity appears to vary with the system being studied, since the n.m.r. frequency for water itself is affected by the nature of the anions present. The estimation of $\sigma_2$ from observations on the pure acid is uncertain because of the possible effect of molecular association, and in general the assumption that $\sigma_1$ and $\sigma_2$ are independent of composition is open to doubt. The use of Raman intensities is probably less open to objection, but problems arise in obtaining reference spectra for the acidic and basic forms, in the overlapping of different bands, and in applying corrections for the varying refractive index of the solutions. Moreover, for both methods the conversion of observed degrees of dissociation into thermodynamic dissociation constants involves the use of measured or calculated activity coefficients and a considerable extrapolation to infinite dilution, thus introducing further uncertainties.

These difficulties will be illustrated by reference to some individual systems. Although the value $K = 24$ derived for nitric acid by both n.m.r. and Raman measurements still appears to be unassailed, the agreement reported in 1957 for a number of other acids[2] is not confirmed by later work. Thus, although perchloric acid was originally assigned[3] $K = 38$, i.e., little stronger than nitric acid, more recent Raman[4,5] and n.m.r.[6] work indicates complete dissociation at least up to concentrations of

[2] O. Redlich and G. C. Hood, *Disc. Faraday Soc.*, **24**, 87 (1957); cf. T. F. Young, L. F. Maranville, and H. M. Smith, in *The Structure of Electrolytic Solutions* (ed. W. J. Hamer), Wiley, New York, 1959, p. 35.
[3] G. C. Hood, O. Redlich, and C. A. Reilly, *J. Chem. Phys.*, **22**, 2067 (1954); G. C. Hood and C. A. Reilly, *J. Chem. Phys.*, **32**, 127 (1960).
[4] K. Heinziger and R. E. Weston, *J. Chem. Phys.*, **42**, 272 (1965).
[5] A. K. Covington, M. J. Tait, and Lord Wynne-Jones, *Proc. Roy. Soc.*, A, **286**, 235 (1965).
[6] R. W. Duerst, *J. Chem. Phys.*, **48**, 2275 (1968).

8–10M. Methanesulphonic acid has been reported to have $K = 73 \pm 6$ and $K = 16 \pm 2$ from Raman[7] and n.m.r.[8] spectra respectively, while two different interpretations[9,10] of indicator equilibria give values of 4 and 13 for the same acid. The position is no better for the somewhat weaker trifluoroacetic acid. A recent careful study[11] gave good agreement between degrees of dissociation derived from Raman and from n.m.r. measurements but because of difficulties in extrapolation the authors were unwilling to specify $K$ more closely than $4 < K < 8$; even this wide range is higher than most other estimates, e.g., 1.8 from earlier n.m.r. measurements,[3] $1.1 \pm 0.3$ from refractometry,[12] and 0.8 or 1.4 from indicator equilibria, according to the details of the treatment. It is thus clear that any reported dissociation constants greater than $1 \text{ mol dm}^{-3}$ must be regarded with suspicion, and 0.2 is probably a better estimate of the upper limit of quantitative measurement in aqueous solution.

If an acid if very weak, it is often more convenient to use a solution of the conjugate base. For example, phenol has $K \sim 10^{-10}$, and it would be necessary to go down to concentrations of about $10^{-6}$M to obtain 1% conversion into phenoxide ion; on the other hand, a $10^{-2}$M solution of sodium phenoxide is hydrolysed to the extent of about 10% according to the equation $C_6H_5O^- + H_2O \rightleftharpoons C_6H_5OH + OH^-$, and the ratio $[C_6H_5O^-]/[C_6H_5OH]$ could be readily determined by optical means. Similarly, the anilinium ion ($K = 2 \times 10^{-5}$) is converted appreciably into aniline according to the equation $C_6H_5NH_3^+ + H_2O \rightleftharpoons C_6H_5NH_2 + H_3O^+$ at $10^{-3}$M, while aniline itself reacts with water to give hydroxide ions to an extent of less than 1% even at a concentration of $10^{-5}$M. From this point of view the most awkward acid-base pairs are those with $K \sim 10^{-7}$, since here either the acid or the base must be diluted to $10^{-5}$M in order to obtain 10% conversion. In equilibria of the type $B + H_2O \rightleftharpoons A + OH^-$ the acid strength of the pair A–B is of course being compared with $H_2O$–$OH^-$ rather than with the usual standard $H_3O^+$–$H_2O$, but the relation between the two is easily expressed in terms of the ionic product of water.

The use of optical properties is of course restricted to a limited selection of acid-base systems, and methods of more general applicability are based on the properties of the hydronium ion, in particular the e.m.f. of cells

[7] J. H. R. Clarke and L. A. Woodward, *Trans. Faraday Soc.*, **62**, 2226 (1966).
[8] A. K. Covington and T. H. Lilley, *Trans. Faraday Soc.*, **63**, 1749 (1967).
[9] K. N. Bascombe and R. P. Bell, *J. Chem. Soc.*, 1096 (1959).
[10] E. Högfeldt, *J. Inorg. Nucl. Chem.*, **17**, 302 (1961).
[11] A. K. Covington, J. G. Freeman, and T. H. Lilley, *J. Phys. Chem.*, **74**, 3773 (1970).
[12] E. Grunwald and J. F. Haley, *J. Phys. Chem.*, **72**, 1944 (1968).

containing a reversible hydrogen electrode or some equivalent, such as a glass electrode. Since we are now measuring a thermodynamic quantity, the free energy, the expressions for the e.m.f. of such cells contain the activity rather than the concentration of the hydronium ion, and this type of measurement therefore leads to the thermodynamic $K$ rather than $K_c$. However, the observed e.m.f. also involves the activities of other ions in the solution in a manner depending on the construction of the cell, and some extrapolation is always necessary to obtain the true value of $K$. Measurements are best made, not on solutions of the acid or the base alone, but on buffer mixtures, i.e., solutions containing comparable amounts of A and B, often prepared by the partial neutralization of A or B by a strong base or a strong acid respectively. Such solutions are insensitive to small quantities of impurities in the materials used, or derived from the vessels or the atmosphere. It is usual to regard the potential of a hydrogen electrode as being determined by the process $\frac{1}{2}H_2 + H_2O \rightleftharpoons H_3O^+ + e^-$, but this is probably unrealistic in a solution of low hydrogen ion concentration. For example, in an alkaline solution the potential-determining process may well be $\frac{1}{2}H_2 + OH^- \rightleftharpoons H_2O + e^-$, and in a buffer solution containing a high concentration of an acid-base pair A–B another possible process is $\frac{1}{2}H_2 + B \rightleftharpoons A + e^-$. However, if all the acid-base pairs are in equilibrium with one another, it is immaterial which is regarded as determining the potential, since it is a thermodynamic necessity that each pair give the same potential.

The oldest method of determining dissociation constants is of course by measurements of electrolytic conductivity. If the solution is prepared from either an uncharged acid or an uncharged base, the equilibria involved are $BH + H_2O \rightleftharpoons B^- + H_3O^+$ and $B + H_2O \rightleftharpoons BH^+ + OH^-$, and since ions appear on only one side of these equations the method can be used even for very weak acids and bases provided that conducting impurities can be kept down to a very low level. If the solution is made up from the salt of a charged acid or base, e.g., $BH^+ \cdot Cl^-$ or $K^+ \cdot B^-$, the corresponding equations are $BH^+ + H_2O + Cl^- \rightleftharpoons B + H_3O^+ + Cl^-$ and $B^- + H_2O + K^+ \rightleftharpoons BH + OH^- + K^+$, and the situation is clearly less favourable for using the conductivity method. However, the ions $H_3O^+$ and $OH^-$ have considerably higher mobilities than any other ions, and the measurement of conductivity can be carried out with very high accuracy, so that the method is still applicable provided that the position of equilibrium is not too far to one side.

Conductivity measurements do not lead directly either to $K$, or to $K_c$, but by making an allowance for the variation of ionic mobilities with environment $K_c$ can be obtained, and hence $K$ by extrapolation to infinite dilution. Since the solutions used have low ionic concentrations, these corrections and extrapolations involve little uncertainty, and the most accurately known dissociation constants are probably those derived from modern conductivity measurements.[13]

Another application of conductivity measurements is less accurate, but is particularly suitable for measuring the dissociation constants of very weak uncharged acids. The change of conductivity produced by adding a considerable quantity of weak acid HX to a dilute solution of sodium hydroxide measures the extent to which the reaction $HX + OH^- \rightleftharpoons X^- + H_2O$ goes from left to right. When HX is very weak it will be impossible to measure the mobility of $X^-$ directly, but since the hydroxide ion has a mobility much greater than that of any other anion, an estimate from anions of similar size involves little error in the final result. Since the above equilibrium involves no change in ionic strength, activity coefficients will cancel out to a good approximation, and $K(HX)$ is obtained by multiplying the observed constant $[X^-]/[HX][OH^-]$ by $K_w$. Very small conductivity changes can be measured accurately, but the extension of the method to extremely weak acids is limited by the medium effect of added HX on the mobility of the sodium and hydroxide ions, which can only be guessed at. This method has been used to determine the dissociation constants of alcohols and diols;[14] an analogous treatment obviously applies to the addition of a very weak uncharged base to a dilute solution of a strong acid.

The p$K$ of a pair $A_1$–$B_1$ is often determined by the addition of a second pair $A_2$–$B_2$ (other than the solvent) of known p$K$, and this can be done in two ways. In the first a buffer solution of $A_1$–$B_1$ is prepared, and a very small amount of $A_2$–$B_2$ is added, the equilibrium ratio $[A_2]/[B_2]$ being measured by an optical method. The ratio $[A_1]/[B_1]$ is little affected by the addition of $A_2$–$B_2$, and the equilibrium constant $[B_1][A_2]/[A_1][B_2]$ is thus known, leading directly to p$K_1$ − p$K_2$. This is of course the familiar indicator method, $A_2$–$B_2$ usually consisting of an organic acid or base absorbing in the visible region. Alternatively, the known system $A_2$–$B_2$ can be used to make a buffer solution of fairly high concentration, to which is added a small amount of $A_1$–$B_1$, the ratio $[A_1]/[B_1]$ being again determined by optical means, usually by ultra-

[13] See, e.g., F. S. Feates and D. J. G. Ives, *J. Chem. Soc.*, 2798 (1956); F. S. Feates, D. J. G. Ives, and J. H. Pryor, *J. Electrochem. Soc.*, **103**, 580 (1956); D. J. G. Ives and P. D. Marsden, *J. Chem. Soc.*, 649 (1965).
[14] P. Ballinger and F. A. Long, *J. Am. Chem. Soc.*, **81**, 2347 (1959); R. P. Bell and D. P. Onwood, *Trans. Faraday Soc.*, **58**, 1557 (1962).

violet absorption measurements. This procedure is naturally limited in its application but has been successful for certain classes of compound, especially aromatic systems where the conjugate acid and base differ considerably in their ultraviolet absorption.[15] The evaluation of the difference $pK_1 - pK_2$ is more reliable if the two systems $A_1$-$B_1$ and $A_2$-$B_2$ are of the same charge type, since activity coefficients will then cancel to a first approximation, and extrapolation to infinite dilution is facilitated.

A special case of the procedure described in the last paragraph consists in measuring the ratio $[A]/[B]$ when a small quantity of A or B is added to a moderately concentrated solution of a strong acid or a strong base; these solutions constitute buffers, since the concentrations of the pair $H_3O^+$-$H_2O$ or $H_2O$-$OH^-$ are large compared with those of the added system. Special attention has been paid to the use of strong acids, to which we shall confine ourselves initially. If the acid concentration is not too high, the unknown $pK$ may be determined by estimating activity coefficients or by an extrapolation procedure. However, this is rarely a satisfactory procedure, and is not applicable to concentrated solutions, so that a more empirical procedure is usually adopted involving the use of *acidity functions*.*

If a small quantity of an uncharged base B (e.g., an indicator) is added to an acid solution, we can define an experimentally measurable function $h_0$ by the equation

$$h_0 = K_{BH^+} \cdot \frac{[BH^+]}{[B]}, \tag{2}$$

where $K_{BH^+}$ is the thermodynamic dissociation constant of $BH^+$, the standard state for activity coefficients being dilute aqueous solution. Equation (2) can be written in the logarithmic form

$$H_0 = -\lg h_0 = pK_{BH^+} - \lg \frac{[BH^+]}{[BH]}, \tag{3}$$

[15] See, e.g., R. J. L. Andon, J. D. Cox, and E. F. G. Herington, *Trans. Faraday Soc.*, **50**, 918 (1954); R. A. Robinson, in *The Structure of Electrolytic Solutions* (ed. W. J. Hamer), Wiley, New York, 1959, p. 253. A general account of this method is given in Ref. 1.
* The first edition of this book (1959) contained a fairly detailed account of acidity functions. However, there now exist several books and reviews on this subject, for example, M. A. Paul and F. A. Long, *Chem. Rev.*, **57**, 1, 935 (1957); R. H. Boyd, in *Solute-Solvent Interactions* (ed. J. F. Coetzee and C. D. Ritchie), New York, 1969; C. H. Rochester, *Acidity Functions*, London, 1970. Only a brief account will therefore be given here, with few references to original papers. Views on reaction rates in concentrated acid or alkaline solutions have become even more confused since 1959, and no attempt will be made to deal with this subject.

and it is the quantity $H_0$, defined by (3), which is usually termed the acidity function. In sufficiently dilute solutions of acids containing no added salt, $K_{BH^+} = [B][H^+]/[BH^+]$, so that $h_0$ and $H_0$ become respectively equal to $[H^+]$ and pH under these conditions. Since B is added in very low concentrations, it will not have any appreciable effect on the properties of the acid solution.

In concentrated solutions $h_0$ differs considerably from $[H^+]$, and similarly $H_0$ from $pH$. For example, in 10-molal solutions of mineral acids $h_0$ is somewhat greater than 1000. This can be represented formally by inserting activity coefficients in (3), giving

$$H_0 = -\lg[H^+] + \lg\frac{f_{BH^+}}{f_B f_{H^+}}. \tag{4}$$

Since we are now dealing with concentrated solutions, the choice of concentration scale becomes of importance; $[H^+]$ can be expressed in either molarities or molalities, with corresponding changes in the values of the activity coefficients.

The usefulness of the acidity function, as defined by Equations (3) and (4), depends upon how far *its value for a given acid solution is independent of the base used*. It is difficult to test this statement directly, since the bases (indicators) which are of value for investigating concentrated acid solutions are so weak that it is difficult to determine $pK_{BH^+}$ accurately. Since $K_{BH^+}$ is a thermodynamic dissociation constant, its determination must involve extrapolation to infinite dilution at some stage. However, less direct tests can easily be made. For example, if 'a' and 'b' are any two acid solutions, which may differ in concentration and also in the nature of the acid, Equation (3) gives

$$(H_0)_a - (H_0)_b = -\left(\lg\frac{[BH^+]}{[B]}\right)_a + \left(\lg\frac{[BH^+]}{[B]}\right)_b. \tag{5}$$

The right-hand side of Equation (5) contains only measurable quantities and should have the same value for different indicators if $H_0$ is independent of B. A convenient method of applying this test is to take a series of acid solutions of different concentrations and to measure $[BH^+]/[B]$ in each solution for a number of different indicators. If $\log[BH^+]/[B]$ is plotted against the acid concentration, (5) implies that different indicators will give a series of parallel curves (not necessarily straight lines).

In order to obtain numerical values for $H_0$ it is necessary to know the value of $pK_{BH^+}$ in Equation (3). For indicators which are moderately strong bases ($pK > 1$) this can be done by making measurements in dilute solutions of strong acids ($< 0.05$M). In these solutions $[H^+]$ can be taken equal to the acid concentration $c$, and the factor $f_{BH^+}/f_B f_{H^+}$ will differ inappreciably from unity; hence $H_0 = -\lg c$ and Equation (3) then gives $pK_{BH^+}$. For somewhat weaker bases ($pK$ between $+1$ and zero) this procedure is no longer applicable, but it is found empirically that the quantity $\lg[BH^+]/[B] - \lg[H^+]$ is a linear function of $c$ for solutions of a strong acid up to about 2M, so that $pK_{BH^+}$ can be obtained as the intercept of such a linear plot at $c = 0$.

For weaker bases such as must be used for concentrated acids ($pK$ negative) neither of these methods is applicable, and $pK$ must be obtained by a stepwise procedure, each indicator being compared in turn with one of higher $pK$. For any two indicators $B_1$ and $B_2$ the observed ratios in a given solution are given by

$$\lg\frac{[B_1H^+]}{[B_1]} - \lg\frac{[B_2H^+]}{[B_2]} = pK_1 - pK_2, \qquad (6)$$

so that if $pK_1$ is known $pK_2$ can be determined. In this way a series of indicators can be established over a wide $pK$ range, though the procedure involves cumulative errors which may become serious as $pK$ becomes more negative.

The assumption that $H_0$ is independent of the base used has been found to be a good approximation for substituted nitroanilines in aqueous sulphuric and perchloric acid, and the values of $H_0$ usually tabulated refer to this series. Very high acidities are reached; for example, in 90% sulphuric acid $H_0$ is about $-9$, corresponding to $h_0 = 10^9$. It was originally hoped that the same $H_0$ scale would apply, at least approximately, to all uncharged bases, in which case the $pK$ value for any pair $SH^+$–$S$ could be determined by applying Equation (3) to the observed values of $[SH^+]/[S]$ in solutions of suitable acidity. However, this hope has proved a vain one, and it now seems clear that the concept of a unique acidity function only applies within a class of closely related bases. Acidity scales, usually for sulphuric acid–water mixtures, have been constructed for tertiary aromatic amines, indoles, pyrroles, azulenes, and ethers (to mention only a few). These scales all differ from one another and from the original $H_0$ scale based on the nitroanilines, the discrepancies reaching 4 logarithmic units in the most acidic solutions. Some work has been done on the acidity function $H_-$, related to indicators of the type BH–B$^-$; this is again different, though the effect of charge type does not appear to be greater than that of chemical nature.

It is thus obvious that acidity functions can be used reliably for p$K$ determinations (at least for p$K$ values more negative than about $-1$) only when the acidity function used is based on measurements with acid-base systems closely resembling the one whose p$K$ is required. This limitation greatly diminishes the usefulness of the acidity function approach. However, a semi-empirical procedure has been devised by Bunnett and Olsen[16] for obtaining p$K$ values of moderate accuracy using only the $H_0$ scale based on nitroanilines, which is now well established[17] for sulphuric acid solutions.

If B represents a nitroaniline base and S a base of different type, then the failure of S to conform to the acidity scale based on B corresponds to the inequality of the ratios $f_{BH^+}/f_B f_{H^+}$ and $f_{SH^+}/f_S f_{H^+}$; cf. Equation (4). Bunnett and Olsen[16] suppose that the variations of these ratios with sulphuric acid concentration are related by the equation

$$\lg \frac{f_{SH^+}}{f_S f_{H^+}} = \beta \lg \frac{f_{BH^-}}{f_B f_{H^+}} \tag{7}$$

where $\beta$ is independent of acid concentration, but depends on the nature of S. ($\beta$ is probably related to the different degrees of hydration of SH$^+$ and BH$^+$, and to differences in the salting out of S and B, but its interpretation is not important in the present context.) We have, by definition,

$$pK_{SH^-} = -\lg \frac{[H^+][S]}{[SH^+]} + \lg \frac{f_{SH^-}}{f_S f_{H^+}}, \tag{8}$$

and when this is combined with (4) and (7) we obtain

$$\lg \frac{[SH^+]}{[S]} - \lg [H^+] = pK_{SH^-} - \beta(H_0 + \lg[H^+]). \tag{9}$$

A plot of the left-hand side of Equation (9), containing only observable quantities, against $H_0 + \lg[H^+]$, where $H_0$ is based on the nitroanilines, should therefore be a straight line of slope $\beta$ and intercept $pK_{SH^+}$. Bunnett and Olsen[16] showed that such plots are in fact almost always linear, confirming the assumption in (7); the values of $\beta$ (corresponding to $1 - \phi$ in Bunnett and Olsen's notation) vary from 0.17 to 3.44, illustrating the

[16] J. F. Bunnett and F. P. Olsen, *Canad. J. Chem.*, **44**, 1899 (1966).
[17] M. J. Jorgenson and D. R. Hartter, *J. Am. Chem. Soc.*, **85**, 878 (1963); C. D. Johnson, A. R. Katritzky, and S. A. Shapiro, *J. Am. Chem. Soc.*, **91**, 6654 (1969); P. Tickle, A. G. Briggs, and J. M. Wilson, *J. Chem. Soc.*, B, 65 (1970). The last two references give values from 15°C to 90°C.

wide variety of behaviour shown by bases of different types. Values of p$K$ obtained by this procedure are subject to some errors of extrapolation, but they are probably reliable to about $\pm 0.3$ unit even in the least favourable cases.

The last few paragraphs refer to the behaviour of weak bases in concentrated aqueous acids. Similar considerations apply to weak acids in concentrated aqueous alkalis, though this field has been much less intensively explored. Most of the acid-base systems investigated are of the type BH–B$^-$, so that the appropriate acidity function is $H_-$; work before 1966 has been summarized in two reviews,[18] since when further results for aqueous solutions of alkali metal hydroxides have been published.[19] Most measurements of $H_-$ in strongly alkaline solutions have employed indicator acids containing NH or NH$_2$ groups, and their use for measuring the p$K$ values for weak acids of other types is likely to involve the same complications as those already noted for concentrated acids; however, there are some indications[20] that CH-acids do follow the same scale. Partly because of solubility problems, much of our knowledge of strongly alkaline solutions relates to non-aqueous or mixed solvents, and will be referred to in Chapter 4.

There are some systems for which ambiguities arise in defining a dissociation constant, independently of the methods used for determining it. One of the long-standing problems of this kind concerns ammonia and the amines. The usual expression for the acid constant is $K_c = [\text{NH}_3^*][\text{H}_3\text{O}^+]/[\text{NH}_4^+]$, where $[\text{NH}_3^*]$ denotes the total concentration of unionized ammonia in the solution. It has commonly been supposed that ammonia exists in aqueous solution partly as the molecule NH$_3$ and partly as 'ammonium hydroxide' NH$_4$OH, so that $K_c$ as usually defined is more properly written

$$[\text{H}_3\text{O}^+][\text{NH}_3 + \text{NH}_4\text{OH}]/[\text{NH}_4^+].$$

If it were possible to determine the extent to which NH$_3$ is converted into NH$_4$OH in solution, then 'true' dissociation constants could be specified for the separate species.* The equilibrium $\text{NH}_3 + \text{H}_2\text{O} \rightleftharpoons$

[18] K. Bowden, *Chem. Rev.*, **66**, 119 (1966); C. H. Rochester, *Chem. Soc. Quart. Rev.*, **20**, 511 (1966).
[19] G. Yagil, *J. Phys. Chem.*, **71**, 1034 (1967).
[20] K. Bowden and R. Stewart, *Tetrahedron*, **21**, 261 (1965).
* By the Brönsted definition NH$_3$ would be the true base, and NH$_4$OH an 'aquobase' (cf. p. 13). Earlier workers concentrated more on the production of hydroxide ions and defined the 'true' basic dissociation constant in terms of the equilibrium NH$_4$OH $\rightleftharpoons$ NH$_4^+$ + OH$^-$.

$NH_4OH$ is presumably highly mobile, and since water is present in large excess, equilibrium measurements in general give no information about the extent of hydration of $NH_3$. Moore and Winmill[21] attempted to solve this problem by measuring the distribution coefficients of ammonia and amines between water and organic solvents over a range of temperature, and their conclusions have often been quoted.[22] These are, however, based on unverifiable assumptions about the distribution coefficients of the individual species, and it is interesting to note that a recent use of the same experimental method has led to conclusions which differ widely from those of Moore and Winmill, though it is doubtful whether they are more trustworthy.[23]

Modern opinion discounts the existence of a definite species such as $NH_4OH$ in aqueous solution. It is not possible to write a covalent structure for $NH_4OH$, and the only alternatives are an ion pair $NH_4^+ \cdot OH^-$ or a hydrogen-bonded structure such as $H—O—H \cdots NH_3$. We do not now believe that ion pairs involving univalent ions exist in appreciable concentrations in aqueous solution, and the energies involved in hydrogen-bond formation are much smaller than ordinary bond energies. Various lines of evidence indicate that the interaction between ammonia and water is a comparatively weak one; thus the solid hydrates $NH_3 \cdot H_2O$ and $2NH_3 \cdot H_2O$ are formed only at low temperatures and are unstable.[24] In a thermodynamic study of these solid hydrates[25] they were formulated as $NH_4^+OH^-$ and $(NH_4)_2O^{2-}$, but their infrared spectra[26] show none of the characteristic frequencies of $NH_4^+$ or $OH^-$ and resemble a superposition of the spectra of $NH_3$ and $H_2O$, with some modifications due to hydrogen bonding. Similarly, the Raman spectrum of ammonia in aqueous solution is very similar to that of anhydrous liquid ammonia,[27] and in general 'ammonium hydroxide' differs markedly from the hydrates of the strong acids. Evidence of this kind has led many authors to doubt the existence of hydrates of ammonia or amines as species whose concentrations can be specified.[28] It is not of course denied that ammonia and the amines interact with

[21] T. S. Moore and T. F. Winmill, *J. Chem. Soc.*, **91**, 1373 (1907); **101**, 1635 (1912).
[22] E.g., N. V. Sidgwick, *Chemical Elements and Their Compounds*, Oxford, 1950, p. 659.
[23] I. B. Khakham, *Zh. Obshch. Khim.*, **18**, 1215 (1948); *Chem. Abs.*, **43**, 6891 (1949).
[24] L. D. Elliott, *J. Phys. Chem.*, **28**, 887 (1924); I. L. Clifford and E. Hunter, *J. Phys. Chem.*, **37**, 101 (1933).
[25] D. L. Hildenbrand and W. F. Giauque, *J. Am. Chem. Soc.*, **75**, 2811 (1953).
[26] R. D. Waldron and D. F. Hornig, *J. Am. Chem. Soc.*, **75**, 6079 (1953).
[27] B. P. Rao, *Proc. Indian Acad. Sci.*, **20A**, 292 (1944).
[28] See, e.g., P. F. van Velden and J. A. Ketelaar, *Chem. Weekblad*, **43**, 401 (1947).

water by hydrogen bond formation, and interaction of this kind is revealed in the thermodynamic properties of their aqueous solutions[29] and in the variation with concentration of proton magnetic resonance frequencies,[30] the latter measurements showing also that these interactions fluctuate very rapidly. Hydrogen bonding in aqueous solution is, however, common to many solutes and is not usually taken into account in specifying the formula or concentration of the solute. In fact, in the equilibrium $NH_3 + H_2O \rightleftharpoons NH_4^+ + OH^-$ (or the analogous one for amines) the ions $NH_4^+$ and $OH^-$ certainly interact more strongly with the solvent than does the molecule $NH_3$, and it would be inconsistent to indicate solvation of the molecule by the formula $NH_4OH$ unless we are also going to express the solvation of the ions. The acid constant of the ion $NH_4^+$ can thus be expressed simply as $[NH_3][H_3O^+]/[NH_4^+]$, where $[NH_3]$ represents the total concentration of unionized ammonia in the solution irrespective of hydration.

A somewhat similar problem arises in defining the acid strength of carbonic acid. When carbon dioxide is dissolved in water, the following equilibria are set up:

$$CO_2 + H_2O \rightleftharpoons H_2CO_3 \rightleftharpoons HCO_3^- + H_3O^+,$$

and we can define two acid constants,

$$K_c(CO_2) = [HCO_3^-][H_3O^+]/[CO_2 + H_2CO_3],$$

and

$$K_c(H_2CO_3) = [HCO_3^-][H_3O^+]/[H_2CO_3].$$

The normal methods of determining acid strengths do not distinguish between $CO_2$ and $H_2CO_3$, and since the equilibrium $CO_2 + H_2O \rightleftharpoons H_2CO_3$ is set up rapidly, the quantity usually given as the first dissociation constant of carbonic acid ($4.45 \times 10^{-7}$ at 25°C) is thus $K_c(CO_2)$. So far the problem appears to be very similar to that of ammonia and the amines, but it is in fact a more amenable one. From the theoretical point of view $H_2CO_3$ (unlike $NH_4OH$) can be given a normal covalent structure, $O = C(OH)_2$, and although it has not been isolated, its organic derivatives are well known; thus it is reasonable to speak of the concentration of $H_2CO_3$ in an aqueous solution. It might be possible in principle to measure the ratio $[H_2CO_3]/[CO_2]$ by some optical method, but this

[29] J. L. Copp and D. H. Everett, *Disc. Faraday Soc.*, **15**, 174 (1953).
[30] H. S. Gutowsky and S. Fujiwara, *J. Chem. Phys.*, **22**, 1782 (1954).

has not been achieved, mainly because this ratio turns out to be a very small one.

All the estimates which have been made depend on the fact that the reversible reaction $H_2O + CO_2 \rightleftharpoons H_2CO_3$ and other reactions involving the $CO_2$ molecule take place much more slowly than the simple acid-base reactions $H_2CO_3 + H_2O \rightleftharpoons HCO_3^- + H_3O^+$ and $H_2CO_3 + OH^- \rightleftharpoons HCO_3^- + H_2O$, although they are still too fast to study quantitatively by conventional methods. It was observed at an early date[31] that when carbon dioxide or carbonates are titrated with phenolphthalein as indicator the colour change is not instantaneous. The earliest estimate of the ratio $[H_2CO_3]/[CO_2]$ was given by Thiel,[32] who assumed that the concentration of carbonic acid in aqueous carbon dioxide was given by the amount neutralized by alkali in less than 0.4 second. Though not accurate, his estimate (0.6%) was of the right order of magnitude, and later workers used essentially similar methods. Faurholt[33] added dimethylamine to aqueous carbon dioxide. This reacts rapidly with carbon dioxide (but not with carbonic acid or its ions) according to the equation $2NHMe_2 + CO_2 \rightarrow NMe_2COO^- + Me_2NH_2^+$, and the remaining carbonic acid is precipitated by adding barium chloride immediately afterward. The success of this method depends upon the fact that the reaction $H_2CO_3 \rightarrow H_2O + CO_2$ does not proceed to a considerable extent before the barium carbonate is precipitated; actually it has a half-time of only a few seconds at 0°C, so that the method is not capable of great accuracy. More reliable values for the equilibrium constant for $CO_2 + H_2O \rightleftharpoons H_2CO_3$ have been obtained by direct determinations of the reaction velocities in the two directions. These can be determined by means of flow methods in conjunction with the measurement of temperature changes, or of changes in pH by indicators[34] or with the glass electrode,[35] or by studying the rate of exchange of the isotope $^{18}O$ between carbon dioxide and water.[36] Finally, the most accurate values for the ratio $[H_2CO_3]/[CO_2]$ have been obtained by studying the conductivity of aqueous carbon dioxide at very high field strengths.[37] This method depends upon the so-called 'dissociation field effect', in

[31] J. W. McBain, *J. Chem. Soc.*, **101**, 814 (1912); D. Vorländer and S. Strube, *Ber.*, **46**, 172 (1913).
[32] A. Thiel, *Ber.*, **46**, 241 (1912).
[33] C. Faurholt, *J. Chim. Phys.*, **21**, 400 (1924); **22**, 1 (1925).
[34] F. J. W. Roughton, *J. Am. Chem. Soc.*, **63**, 2930 (1941).
[35] J. Meier and G. Schwarzenbach *Helv. Chim. Acta*, **40**, 907 (1957).
[36] G. A. Mills and H. C. Urey, *J. Am. Chem. Soc.*, **62**, 1019 (1940).
[37] D. Berg and A. Patterson, *J. Am. Chem. Soc.*, **75**, 5197 (1953); K. F. Wissbrunn, D. M. French, and A. Patterson, *J. Phys. Chem.*, **58**, 693 (1954).

which the degree of dissociation of a weak electrolyte according to the scheme $AB \rightleftharpoons A^+ + B^-$ is increased by the application of a strong field.* The magnitude of this increase can be predicted theoretically in terms of the dissociation constant at low field strengths and the ionic mobilities, and the prediction agrees with experiment for a number of weak electrolytes. In a solution of carbon dioxide the equilibrium $H_2CO_3 + H_2O \rightleftharpoons H_3O^+ + HCO_3^-$ will be affected by the field strength, but not the equilibrium $H_2CO_3 \rightleftharpoons H_2O + CO_2$. Moreover, since the times involved in the measurement are very short (a few microseconds) there is no readjustment of the second equilibrium, and the observed change of conductivity with field strength can be used to obtain the true dissociation constant of $H_2CO_3$. This leads to the values $K(H_2CO_3) = 1.3 \times 10^{-4}$, $K(CO_2) = 4.5 \times 10^{-7}$, $[H_2CO_3]/[CO_2] = 0.0037$ at 25°C, in satisfactory agreement with earlier values.

Carbonic acid is thus a relatively strong acid, comparable with formic acid, and its apparent weakness is due to the small proportion of dissolved carbon dioxide which is in the form $H_2CO_3$. As long as we are dealing with equilibrium properties, the distinction between $CO_2$ and $H_2CO_3$ is unimportant, and it is quite satisfactory to operate with the 'apparent' dissociation constant $K(CO_2)$. However, the distinction does become important as soon as we are dealing with time-dependent phenomena, especially rapid ones, and we shall see later (Chapter 10) that the catalytic effect of solutions containing carbon dioxide or bicarbonate ions can be understood only if we know both the true and apparent dissociation constants.

In the system carbon dioxide–carbonic acid, only the latter would be regarded as an acid in the sense of the Brönsted definition, since carbon dioxide has no proton to lose. There are, however, cases in which two isomeric acids are in equilibrium in solution, each being in equilibrium with the same anion. The commonest example involves keto–enol equilibria. If we write the two isomers as SH (keto) and HS (enol), the equilibrium scheme becomes

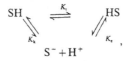

$$S^- + H^+$$

---

*The conductivity of a strong electrolyte is also increased at high field strengths because the ions are moving so fast that the ionic atmospheres are unable to form completely. This is known as the Wien effect; it will also operate in solutions of weak electrolytes, but it is considerably smaller than the dissociation field effect and can be eliminated at sufficiently high field strengths.

omitting the hydration of the hydrogen ion for the sake of brevity. We then have the three equilibrium constants

$$K_t = [HS]/[SH], \quad K_k = [S^-][H^+]/[SH], \quad K_e = [S^-][H^+]/[HS]$$

whence

$$K_t = K_k/K_e.$$

If we measure the acid constant for this system by a static method, equilibrium having been attained, the total concentration of undissociated acid is $[SH] + [HS]$ and the measured equilibrium constant is

$$K' = [S^-]/[H^+]/[SH + HS] = K_k/(K_t + 1) = K_e K_t/(K_t + 1).$$

For example, a dilute aqueous solution of ethyl acetoacetate contains 0.4% of enol (determined by bromine titration) and the measured dissociation constant is $2 \times 10^{-11}$. This is therefore effectively the dissociation constant of the keto form, and the constant for the enol form is $5 \times 10^{-8}$. Since the keto form usually predominates in aqueous solution, the enol form is commonly the stronger acid, but this is not always the case. For example, dimedone (5,5,-dimethylcyclohexane-1,3-dione) exists in aqueous solution as 95% enol and 5% keto, and its apparent dissociation constant is $5.9 \times 10^{-6}$; hence in this case the dissociation constants of the keto and enol forms are respectively $1.2 \times 10^{-4}$ and $5.6 \times 10^{-6}$.[38]

The interconversion of the two isomeric forms of an acid is often a relatively slow process, so that it may be possible to measure the dissociation constants of the two forms separately. Thus when acid is added to an alkaline solution of nitroethane (containing the anion $CH_3CH=NOO^-$), the species produced is the *aci*-form $CH_3CH=NOOH$, which changes only slowly into the *nitro*-isomer $CH_3CH_2NO_2$; hence if pH measurements are made directly after the addition of acid, they can be used to calculate the dissociation constant of the *aci*-form, which is found to be $K_{aci} = 3.9 \times 10^{-5}$. If similar measurements are made after the system has come to equilibrium, the *aci*-form has been transformed almost completely into the *nitro*-form, and we obtain $K_{nitro} = 3.5 \times 10^{-9}$. The same value is obtained if a solution of nitroethane is partly neutralized with sodium hydroxide. By combining the two dissociation constants, we find at equilibrium $[aci]/[nitro] = K_{nitro}/K_{aci} = 9 \times 10^{-5}$, and in this and similar cases the measurement of the two dissociation constants is the best

[38] G. Schwarzenbach and E. Felder, *Helv. Chim. Acta*, **27**, 1701 (1944).

way of estimating the very small proportion of the *aci*-form which is present at equilibrium.[39]

The acid-base properties of the ion $HF_2^-$ pose a special problem. This species is well defined both in the solid state and in solution. Estimates based on the lattice energies of the salts $KHF_2$ and $RbHF_2$ show that the process $HF + F^- \rightarrow HF_2^-$ in the gas phase is exothermic to the extent of $58 \pm 5 \, kcal \, mol^{-1}$,[40] and the equilibrium constant for the same process in aqueous solution has the value 8 at $25°C$.[41] There is no doubt that the species $HF_2^-$ possesses both acidic and basic properties, since it can both lose and gain a proton; for example, in water we have the equilibria $HF_2^- + H_2O \rightleftharpoons H_3O^+ + 2F^-$ and $HF_2^- + H_3O^+ \rightleftharpoons H_2O + 2HF$. We cannot, however adopt the usual procedure of using these equilibria to define the acidic or basic strength of $HF_2^-$, since the equilibrium constants contain the terms $[F^-]^2$ and $[HF]^2$ and thus have physical dimensions differing from those of the usual acid dissociation constants. A logical measure of the acid-base properties of $HF_2^-$ would involve the hypothetical equilibria $HF_2^- + H_2O \rightleftharpoons H_3O^+ + F_2^{2-}$ and $HF_2^- + H_3O^+ \rightleftharpoons H_2O + H_2F_2$. Actually the species $F_2^{2-}$ is unknown, and there is no evidence for the existence of $H_2F_2$ in aqueous solutions, though it is of course present in the vapour and in non-dissociating solvents. We thus have the rather paradoxical situation that $H_2F_2$ may be regarded as a strong acid in aqueous solution (because the equilibrium $H_2F_2 + H_2O \rightleftharpoons HF_2^- + H_3O^+$ is far to the right), while the conjugate species $HF_2^-$ has appreciable basic properties in virtue of the reaction $HF_2^- + H_3O^+ \rightarrow H_2O + 2HF$. Similar difficulties in defining acid-base strengths arise whenever there is not a one-to-one correspondence between acidic and basic species. For example, when a glycol is added to a solution of boric acid, the following equilibrium is set up:

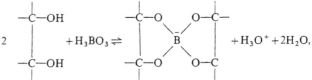

and it is clearly impossible to define the acid-base properties of the system in the usual way. There is no difficulty in representing the equilibrium properties of such systems, but we shall see that special

[39] D. Turnbull and S. H. Maron, *J. Am. Chem. Soc.*, **65**, 212 (1943).
[40] T. C. Waddington, *Trans. Faraday Soc.*, **54**, 25 (1958).
[41] P. R. Patel, E. C. Moreno, and J. M. Patel, *J. Res. Nat. Bur. Stand.*, A, **75**, 205 (1971).

problems arise in connection with time-dependent phenomena such as the kinetics of proton transfer.

The fundamental acid-base equation $A \rightleftharpoons B + H^+$ is formally very similar to the redox equation $R \rightleftharpoons Ox + e^-$, and in both cases the processes realized in practice are obtained by combining two acid-base or redox systems. It is therefore of interest to consider the main differences between the two classes of phenomena. In the first place, equilibrium between two acid-base systems is almost always set up rapidly, whereas two radox systems frequently react very slowly; further, water and similar solvents take part reversibly in acid-base reactions, but are usually indifferent solvents in redox reactions (unless they are irreversibly oxidized or reduced). For this reason acid-base strengths are always expressed in terms of equilibrium with a standard system, usually the solvent, while redox systems are characterized by a potential relative to a standard electrode. Standard redox potentials are of course closely related to equilibrium constants; for example, if a system R–Ox is measured relative to a standard hydrogen electrode, the redox potential is given by $FE = RT \ln K$, where $K$ is the equilibrium constant for the process $R + H_3O^+ \rightleftharpoons Ox + \frac{1}{2}H_2 + H_2O$. In general, however, these equilibria cannot be studied directly, and their equilibrium constants vary through 50–60 powers of ten.[*] In principle acid-base equilibria could equally well be characterized by potentials, since the potential of a hydrogen electrode in solution containing an acid-base system A–B is determined by the process $A + e^- \rightleftharpoons B + \frac{1}{2}H_2$, and if this potential is measured relative to a standard hydrogen electrode the e.m.f. is directly related to the equilibrium constant of the reaction $A + H_2O \rightleftharpoons B + H_3O^+$. These constants can be determined by a number of other methods, and the range of directly accessible values is limited by the acid-base properties of the solvent; thus in aqueous solution the range is limited to about 14 powers of ten. For this reason acid-base strengths are usually specified in terms of equilibrium constants rather than potentials. In aprotic solvents (i.e., solvents lacking appreciable acidic or basic properties) it is impossible to use the solvent as a reference system, and it might then be useful to employ a potential scale, though this has not commonly been done.[42]

---

[*] Because of the slowness of many redox reactions it is possible to study in aqueous solution many redox systems which are thermodynamically capable of oxidising water completely to oxygen or hydrogen peroxide, or of reducing it to hydrogen.
[42] See, however, E. Wiberg, *Z. Phys. Chem.*, **171A**, 1 (1934).

# 4 The Effect of the Solvent on Protolytic Equilibria

Solvent effects on acid-base equilibria are naturally most marked when the solvent itself enters into the equilibrium, as is the case for the conventional definition of acid strength by means of the equilibrium $A + SH \rightleftharpoons B + SH_2^+$ (where SH is the solvent). The existence of such an equilibrium implies that the solvent has some basic properties. Similarly, the occurrence of the reaction $B + SH \rightleftharpoons A + S^-$ (where $S^-$ is the anion derived by abstracting a proton from the solvent) implies that the solvent is acidic. The most important factor determining qualitative behaviour in a wide range of solvents is the acidic or basic nature of the solvent, as determined by its chemical nature. In a preliminary classification we can neglect other factors, notably the effect of dielectric constant on the association of ions or the forces between them.

Only a very few solvents will be considered, chosen so as to illustrate a wide range of behaviour towards acids and bases, and these will include some in which the interaction with the solvent is so small that it is necessary to add a second acid-base pair (for example an indicator) before measurable equilibria are set up. We shall begin with qualitative descriptions, followed by a more quantitative treatment of a few examples. There now exist a number of comprehensive accounts[1] of behaviour in non-aqueous solvents, and few references will therefore be given in this chapter.

The *lower alcohols* are qualitatively very similar to water, in that they are *amphoteric*, forming the ions $ROH_2^+$ and $RO^-$ Substances which dissociate as acids or bases in water will also do so in the alcohols, though to a smaller extent, and the same methods of investigation are applicable, e.g., conductivity, e.m.f., and indicators. A number of acids, e.g., perchloric

---

[1] Notably: (a) *The Chemistry of Non-aqueous Solvents* (ed. J. J. Lagowski), Academic Press, New York and London, 1966–1970; (b) *Non-aqueous Solvent Systems* (ed. T. C. Waddington), Academic Press, New York and London, 1965. Neither of these compilations deals with the alcohols or with the so-called polar aprotic solvents. See also *Chemistry in Non-aqueous Ionizing Solvents* (ed. G. Jander, H. Spandau, and C. C. Addison), Vieweg-Interscience, Braunschweig and London, 1963–1971.

acid and the hydrogen halides, are 'strong' in the alcohols, i.e., their dissociation constants are so high that they appear to be completely dissociated. There appear to be no uncharged bases which are 'strong' in the alcohols, in the sense that they react completely according to the equation $B + ROH \rightleftharpoons BH^+ + RO^-$.* Because of their amphoteric nature the alcohols show self-dissociation according to the scheme $2ROH \rightleftharpoons ROH_2^+ + RO^-$, but the ionic products are much lower than for water, being $10^{-16.7}$ and $10^{-19.1}$ for methyl and ethyl alcohol respectively. We shall see later that these differences can be attributed mainly to the decrease in dielectric constant, which also makes much more difficult a quantitative interpretation of experimental data since it increases the interionic forces. It is possible, however, to obtain information about acids and bases in the alcohols which can be compared directly with data for aqueous solutions, and a large amount of accurate work has been done, especially by conductivity measurements.

Much attention has been paid to solvents which are more strongly acidic than water. Anhydrous *acetic acid* has been widely studied,[2] mostly by indicator measurements or by e.m.f. measurements using a platinum electrode in contact with solid tetrachloroquinone and tetrachlorohydroquinone (chloranil), which serves as a substitute for the hydrogen electrode as does the quinhydrone electrode. Because of its low dielectric constant ($\varepsilon = 6.3$) all salts have dissociation constants less than about $10^{-5}$ and are therefore incompletely dissociated even at very low concentrations. This complicates the quantitative interpretation of the results, but the general picture is clear. Acetic acid is clearly much more acidic than water, and correspondingly it is found that all bases which in water are stronger than aniline give almost identical titration curves with a given acid in acetic acid; this implies that they have reacted almost completely with the solvent according to the equation $B + MeCO_2H \rightarrow BH^+ \cdot MeCO_2^-$. Much weaker bases, for example the nitroanilines, react with anhydrous acetic acid to a considerable extent, and many substances which barely exhibit basic character in water show measurable basic properties in this solvent, for example, urea, the oximes, and triphenylcarbinol.

The ionic product of pure acetic acid ($K_s$) is not known with any certainty, but since the solvent can be prepared with a specific conductivity less than $10^{-8}$ ohm$^{-1}$, $K_s$ must be less than $10^{-13}$, assuming

---

*Very few uncharged bases are strong in aqueous solution, exceptions being guanidine and the amidines, which have not been investigated quantitatively in the alcohols.
[2] See A. I. Popev, Ref. 1(a), Vol. 3, p. 241.

that the limiting mobility for $MeCO_2^- + MeCO_2H_2^+$ is about 40. Since a base as weak as aniline is completely protonated in acetic acid, it follows that this solvent must possess very weak proton-accepting powers, perhaps $10^6–10^8$ times weaker than water. It would therefore be expected to react to a very small extent with most acids, and this is fully borne out in practice. The acids $HClO_4$, HBr, $H_2SO_4$, toluene-*p*-sulphonic, and HCl, all of which are completely dissociated in water, form a series of decreasing strength, judged by conductivity or indicator measurements, and also by the potentials recorded by a chloranil electrode either in solutions of the acids alone, or in titration with a base. Because of the low dielectric constant the concentrations of free ions are very low, but even for the strongest acid, perchloric, only about one-half of the acid is converted into the ion pair $MeCO_2H_2^+ \cdot ClO_4^-$, thus showing that acetic acid is a very weakly basic medium, and not merely one which has a low dissociating power because of its low dielectric constant. Solutions of perchloric or hydrobromic acid in glacial acetic acid have much stronger acidic powers than any aqueous solutions, and they have found practical application in titrating very weak bases such as oximes or amides, which cannot be estimated by titration in aqueous solution.

Anhydrous *formic acid* is more strongly acidic than acetic acid but in general it exhibits a similar behaviour towards uncharged bases.[3] It differs from acetic acid in having a high dielectric constant ($\varepsilon = 62$) and also a high ionic product, $[HCO_2^-][HCO_2H_2^+] = 10^{-6}$, the latter fact showing that it has considerable basic as well as acidic properties. A number of acids (perchloric, sulphuric, and benzenesulphonic) react completely with the solvent, producing completely dissociated products, though hydrochloric acid is a weak acid. The existence of strong, completely dissociated acids makes it possible to measure acidity constants defined by $[B][HCO_2H_2^+]/[A]$, and this has been done especially for systems of the type $RNH_3^+–RNH_2$.

A great deal of work has been done in the strongly acidic solvent *sulphuric acid*,[4] which has a high dielectric constant (110) and also a high ionic product, $[HSO_4^-][H_3SO_4^+] = 1.7 \times 10^{-4}$, the position being further complicated by the existence of another type of dissociation, $2H_2SO_4 \rightleftharpoons H_3O^+ + HS_2O_7^-$, with an equilibrium constant $[H_3O^+][HS_2O_7^-] = 7 \times 10^{-5}$. Because of the large self-ionization, investigation

[3] L. P. Hammett and A. J. Deyrup, *J. Am. Chem. Soc.*, **54**, 4239 (1932); L. P. Hammett and N. Dietz, *J. Am. Chem. Soc.*, **52**, 4795 (1930).
[4] W. H. Lee, Ref. 1(a), Vol. 2, 1967, p. 99; R. J. Gillespie and E. A. Robinson, Ref. 1(b), 1965, p. 117.

by conductivity or e.m.f. measurements is not practicable, and most of the information on the behaviour of solutes comes from cryoscopic measurements. Sulphuric acid is such a strongly acid medium that almost all compounds containing oxygen or nitrogen will accept a proton from it to some degree, thus behaving as bases. For example, not only amines, but also amides, ethers, ketones, and esters give a two-fold freezing point depression, corresponding to complete reaction according to schemes such as

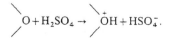

Many substances which behave as acids in hydroxylic solvents exhibit basic properties in sulphuric acid. Thus most carboxylic acids are strong bases, forming the ion $RCOOH_2^+$, though reaction is incomplete for strong acids such as di- and tri-chloroacetic acids, which are thus weak bases in this solvent. Nitro-compounds, sulphones, and sulphonic acids also behave as weak bases, and it is in fact difficult to find substances which are soluble in sulphuric acid without detectable ionization. Since cryoscopic measurements lead only to the total number of solute particles, it is not possible to obtain quantitative measurements of base strength over a wide range, especially since there are complications caused by the self-dissociation of the solvent, and interionic effects, though small, must be taken into account.

Very few substances behave as acids in sulphuric acid, and even such a strong acid as perchloric acid has a dissociation constant of only $10^{-4}$. Both of the ions $H_3SO_4^+$ and $HSO_4^-$ show abnormally high mobilities, presumably because of a proton-jump mechanism analogous to that assumed for hydrogen and hydroxide ions in water. Many substances undergo reactions in sulphuric acid which are more complicated than simple proton transfers, for example,

$$EtOH + 2H_2SO_4 \rightarrow EtOSO_3H + H_3O^+ + HSO_4^-$$

$$(MeCO)_2O + 3H_2SO_4 \rightarrow 2MeCO_2H_2^+ + 3HSO_4^-$$

$$(C_6H_5)_3COH + 2H_2SO_4 \rightarrow (C_6H_5)_3C^+ + 2HSO_4^-$$

$$HNO_3 + 2H_2SO_4 \rightarrow NO_2^+ + H_3O^+ + 2HSO_4^-.$$

Ions like $NO_2^+$ and carbonium ions are of course important in relation to the chemical reactivity of sulphuric acid solutions.

*Hydrogen fluoride*[5] is a solvent of similar acidic strength to sulphuric acid; it also has a high dielectric constant ($\varepsilon = 84$). In contrast to sulphuric acid, its self-dissociation is small, so that conductivity measurements are the usual method of investigation. A wide range of nitrogen and oxygen compounds behave as bases, but acids are usually undissociated. Like sulphuric acid, hydrogen fluoride is sufficiently acidic to protonate some aromatic hydrocarbons to an appreciable extent.

Even more strongly acidic solutions are obtained by adding Lewis acids such as $BF_3$ to anhydrous hydrogen fluoride, when the protonation of any species X is assisted by reaction such as $HF + BF_3 + X \rightarrow BF_4^- + XH^+$. These solutions are able to protonate many hydrocarbons, and in the case of $SbF_5$ even hydrogen fluoride itself is protonated, giving the elusive ion $H_2F^+$ by the reaction $2HF + SbF_5 \rightarrow H_2F^+ + SbF_6^-$. The acidic properties of these mixtures are sometimes described in terms of the very great strength of acids such as $HBF_4$ or $HSbF_6$, but there is no evidence for the existence of these species, and no reasonable valency formula can be written for them; moreover, in the case of $BF_3$ no interaction with hydrogen fluoride can be detected unless a base is added.

There are of course many solvents which are more strongly basic than water, but these have been much less investigated than the acidic ones. Thus, although liquid *ammonia*[6] has been extensively studied, not many of the data relate to acids and bases in the sense in which we are using the terms. Conductivity measurements have been widely used but are difficult to interpret quantitatively because of the fairly low dielectric constant ($\varepsilon = 22$) and the relatively high concentrations which have been used. It is clear, however, that a wide range of acids (e.g., acetic, benzoic, formic, nitric, hydrochloric, and perchloric) react almost completely with the solvent according to the scheme $HX + NH_3 \rightarrow NH_4^+ \cdot X^-$, though the salt produced is incompletely dissociated and the degree of dissociation varies considerably from one case to another. Liquid ammonia also has weak acidic properties, and the dissociation $2NH_3 \rightleftharpoons NH_4^+ + NH_2^-$ takes place to a small extent ($[NH_4^+][NH_2^-] = 10^{-33}$ at $-33°C$), but little is known about its behaviour towards other bases.

There are many other solvents which fall between the extreme cases which we have so far considered, for example, solvents of moderate basicity like the amides or acetonitrile, solvents with basic but no acidic properties such as the ethers, and solvents of moderate acidity such as the

[5] See M. Kilpatrick and J. G. Jones, Ref. 1(a), Vol. 2 (1967), p. 43; H. H. Hyman and J. J. Katz, Ref. 1(b).
[6] J. J. Lagowski and G. A. Moczygemba, Ref. 1(a), Vol. 2 (1967), p. 319; W. J. Jolly and C. J. Hallada, Ref. 1(b), p. 1.

phenols. There are also the so-called *aprotic* solvents, such as the hydrocarbons, which have no appreciable acidic or basic properties and therefore do not take any direct part in acid-base equilibria. Some of these solvents will be dealt with later in connection with relative acid-base strengths in different solvents.

The solvents so far mentioned provide several examples of the *levelling effect*, a term first introduced by Hantzsch. If in any solvent we have three different acids that are dissociated at 0.1M to the extent of 99%, 99.9%, and 99.99%, their dissociation constants will be $10^{+1}$, $10^{+2}$, and $10^{+3}$ respectively, but their solutions will be experimentally indistinguishable from the acid-base point of view, since the solvated proton will be effectively the only acid species present. Thus in water, dilute solutions of the acids $HClO_4$, HI, HBr, and HCl and the sulphonic acids are virtually indistinguishable in their acid properties, though in less basic solvents such as acetic and sulphuric acid or a hydrocarbon they behave very differently. On the other hand, in a basic solvent such as ammonia the list of 'strong' acids is greatly extended, including now many carboxylic acids. Similar levelling effects are observed with bases in acidic solvents. In aqueous solutions, guanidine and amidines are the only common uncharged bases which react completely according to the scheme $B + H_2O \rightarrow BH^+ + OH^-$, but in acetic acid all bases stronger than aniline are completely protonated, while in sulphuric or hydrofluoric acid the list of 'strong' bases includes a wide range of nitrogen and oxygen compounds. In general we can say that any solvent possessing both acidic and basic properties can be characterized by two approximate values $pK'$ and $pK''$. If an acid-base system has $pK$ (in water) $< pK'$, then on dissolving in this solvent it will be converted almost completely into base + solvent cation, whereas if $pK$ (in water) $> pK''$, the base will be converted completely into acid + solvent anion. The range between $pK'$ and $pK''$ represents the range of acid-base pairs whose equilibrium can be usefully studied in the solvent in question, without being 'levelled' in either direction. Figure 1 illustrates the approximate state of affairs for a number of different solvents. Since ether has no acidic properties, no limit of $pK$ is shown on the upper side. Similarly, for the aprotic solvent hexane there is no limit in either direction.

Although information like that contained in Figure 1 makes good sense from the general chemical point of view, it is difficult to obtain any quantitative interpretation of conventional acid or base constants, defined by the equilibria $A + SH \rightleftharpoons B + SH_2^+$ and $B + SH \rightleftharpoons A + S^-$ in terms of the properties of the solvent S. This is partly because these constants are only accessible in solvents in which we can obtain strong, completely dissociated acids and bases in order to prepare solutions with known concentrations of $SH_2^+$ or $S^-$. This limits our consideration to solvents which have appreciable acidic and basic properties, and also to those whose dielectric constant is not too low, and we shall first consider the lower alcohols and formamide.

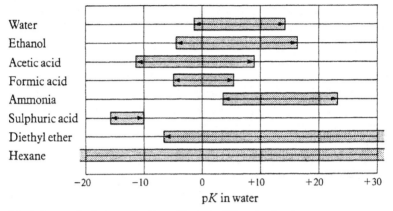

Fig. 1. Range of existence of acids and bases in different solvents.

If we are dealing with an equilibrium of the type $RH + SH \rightleftharpoons R^- + SH_2^+$ or $R + SH \rightleftharpoons RH^+ + S^-$, the right-hand side in each case contains more ions than the left, and the position of equilibrium should depend considerably on the dielectric constant of the medium, quite apart from other factors. In order to minimize this electrostatic effect we must consider equilibria such as $RH^+ + SH \rightleftharpoons R + SH_2^+$ or $R^- + SH \rightleftharpoons RH + S^-$. The first of these is expressed by the conventional acid dissociation constant of a cation acid (e.g., the ammonium ion), while the second is often described as the solvolysis of the anion of a weak acid, and its equilibrium constant is obtained from the conventional dissociation constant of RH and the ionic product of the solvent.

Table 1 contains some values for these constants in water, methanol, and ethanol.[7] Considering first the reactions $RNH_3^+ + SH \rightleftharpoons RNH_2 +$

[7] Most of the values are taken from the compilations of E. Larsson, *Z. Phys. Chem.*, **169A**, 207 (1934), and of L. D. Goodhue and R. M. Hixon, *J. Am. Chem. Soc.*, **56**, 1329 (1934). More accurate data exist for monosubstituted benzoic acids [J. H. Elliott and M. Kilpatrick, *J. Phys. Chem.*, **45**, 454, 566 (1941)] and anilines [M. Kilpatrick and C. A. Arenberg, *J. Am. Chem. Soc.*, **75**, 3812 (1953)], and these have been used to fill in gaps in the table. High accuracy is not important for present purposes. The values used for the ionic products of the alcohols are $pK_s = 16.7$ and $19.1$ for methanol and ethanol respectively [G. Brière, B. Crochon, and N. Felici, *Compt. Rend.*, **254**, 4458 (1962)].

$SH_2^+$, we see that the differences $\Delta_I$ and $\Delta_{II}$ are roughly constant at about 1.1 over a wide range of acid strengths. This could be expressed by saying that methanol and ethanol are less basic than water by a factor of 10 in their behaviour towards cation acids, and since the effect of dielectric

Table 1. ACID-BASE CONSTANTS IN WATER AND THE ALCOHOLS

$K_1 = [SH_2^+][RNH_2]/[RNH_3^+]$,    $K_2 = [S^-][RCO_2H]/[RCO_2^-]$,
$\Delta_I = pK_1(MeOH) - pK_1(H_2O)$,    $\Delta_{II} = pK_1(EtOH) - pK_1(H_2O)$,
$\Delta_{III} = pK_2(MeOH) - pK_2(H_2O)$,    $\Delta_{IV} = pK_2(EtOH) - pK_2(H_2O)$.

| Primary amines | $pK_1$ | | | $\Delta_I$ | $\Delta_{II}$ |
|---|---|---|---|---|---|
| | $H_2O$ | MeOH | EtOH | | |
| n-Butylamine | 10.6 | 11.8 | 12.1 | 1.2 | 1.5 |
| Ammonia | 9.3 | 10.8 | 10.4 | 1.5 | 1.1 |
| o-Chlorobenzylamine | 8.8 | 10.1 | — | 1.3 | — |
| p-Toluidine | 5.2 | 6.7 | 6.2 | 1.5 | 1.2 |
| m-Toluidine | 4.8 | 5.9 | 5.9 | 1.1 | 1.1 |
| Aniline | 4.7 | 6.2 | 5.7 | 1.5 | 1.5 |
| o-Toluidine | 4.5 | 5.9 | 5.6 | 1.4 | 1.4 |
| α-Naphthylamine | 4.3 | 5.5 | 5.3 | 1.2 | 1.0 |
| p-Chloroaniline | 4.0 | 4.9 | 4.7 | 0.9 | 0.7 |
| p-Bromoaniline | 3.9 | 4.8 | 4.5 | 0.9 | 0.6 |
| m-Chloroaniline | 3.6 | 4.6 | 4.2 | 1.0 | 0.6 |
| m-Bromoaniline | 3.6 | 4.4 | 4.2 | 0.8 | 0.6 |
| o-Chloroaniline | 1.9 | 3.5 | 3.3 | 1.6 | 1.4 |
| | | | Mean | 1.2 | 1.0 |

| Carboxylic acids | $pK_2$ | | | $\Delta_{III}$ | $\Delta_{IV}$ |
|---|---|---|---|---|---|
| | $H_2O$ | MeOH | EtOH | | |
| Acetic | 9.3 | 7.4 | 8.8 | −1.9 | −0.5 |
| Phenylacetic | 9.7 | — | 8.8 | — | −0.9 |
| p-Toluic | 9.7 | 7.4 | 8.8 | −2.3 | −0.9 |
| Benzoic | 9.2 | 7.6 | 8.7 | −1.6 | −0.5 |
| p-Bromobenzoic | 10.1 | 8.0 | 9.5 | −2.1 | −0.6 |
| m-Chlorobenzoic | 10.2 | 8.1 | 9.6 | −2.1 | −0.7 |
| m-Nitrobenzoic | 10.5 | 8.3 | 9.9 | −2.2 | −0.6 |
| Salicylic | 11.0 | 8.8 | 10.7 | −2.2 | −0.3 |
| 3,5-Dinitrobenzoic | 11.2 | 9.3 | 11.0 | −1.9 | −0.2 |
| o-Nitrobenzoic | 11.8 | 9.1 | 10.5 | −2.7 | −1.3 |
| 2,4-Dinitrobenzoic | 12.6 | 10.2 | 11.8 | −2.4 | −0.8 |
| Dichloroacetic | 12.7 | 10.3 | 11.8 | −2.4 | −0.9 |
| | | | Mean | −2.1 | −0.7 |

constant should be small for this reaction this effect can be thought of in chemical terms. To obtain comparable values we must allow for the fact that in defining the equilibrium constants for the reaction $RNH_3^+ + SH \rightleftharpoons RNH_2 + SH_2^+$ the concentration of the solvent is omitted from the constant, and if we divide by these concentrations (56, 25, and 17 mol dm$^{-3}$ for $H_2O$, MeOH, and EtOH respectively) we find that the basic strengths of the three molecules are in the ratio 1:0.13:0.33. This result is in the opposite direction to what would be expected in terms of the electron-releasing power of the alkyl groups, as shown by their effect on the basic strength of alkylamines relative to ammonia. These conclusions must not be taken too seriously, since we are comparing pure liquids, in which the position is complicated by association. Qualitatively, however, the same conclusions are reached in experiments where the species $H_2O$, MeOH, and EtOH are present in dilute solution in acetic acid. Kolthoff and Bruckenstein[8] measured the basic strength of these molecules with respect to perchloric acid, using an indicator method, and obtained the ratios 1:0.13:0.22. This shows surprisingly good agreement with the above figures, especially since the products $ROH_2^+ \cdot ClO_4^-$ are certainly incompletely dissociated in acetic acid. Similar information follows from a study of the equilibrium $ROH_2^+ + H_2O \rightleftharpoons ROH + H_3O^+$ which is set up when a little water is added to an alcoholic solution of a strong acid, but this is much more difficult to interpret quantitatively, since of the two bases which we are comparing, one is present in dilute solution and the other as an almost pure liquid. It seems certain, however, that the basic properties of the alcohols are not in accord with the simple picture of electron release by alkyl groups.[9]

A similar treatment can be applied to the data for the equilibrium $RCO_2^- + SH \rightleftharpoons RCO_2H + S^-$. The values of $\Delta_{III}$ and $\Delta_{IV}$ in Table 1 are each reasonably constant, and would suggest that methanol and ethanol are respectively about 100 and 5 times as acidic as water, again showing an effect in the opposite direction to that expected. It seems doubtful, however, whether these figures can be accepted. Hine and Hine[10] have used indicators to study the equilibrium i-PrO$^-$ + ROH $\rightleftharpoons$ RO$^-$ + i-PrOH for dilute solutions of water, methanol, and ethanol in propan-2-ol and thus arrive at the values 1, 3.3, and 0.8 for the relative acidities of the

[8] I. M. Kolthoff and S. Brukenstein, *J. Am. Chem. Soc.*, **78**, 1 (1956).
[9] Other factors which may be operative have been discussed by P. Bartlett and J. D. McCollum [*J. Am. Chem. Soc.*, **78**, 1441 (1956)], who conclude from a combination of kinetic and indicator measurements that propan-2-ol is a very much weaker base than water.
[10] J. Hine and M. Hine, *J. Am. Chem. Soc.*, **74**, 5267 (1952).

molecules $H_2O$, MeOH, and EtOH. Although dissociation may not be complete in this solvent ($\varepsilon = 20$), these ratios seem more probable than $1:100:5$ derived from Table 1, and it is likely that the effect of dielectric constant is not really eliminated in the reaction $RCO_2^- + SH \rightleftharpoons RCO_2H + S^-$. Measurements of the equilibrium $RO^- + H_2O \rightleftharpoons ROH + OH^-$ also suggest that the acidities of $H_2O$, MeOH, and EtOH do not differ by more than a power of ten at the most, though they are difficult to interpret quantitatively.

In equilibria of the type $RCO_2H + SH \rightleftharpoons RCO_2^- + SH_2^+$ and $RNH_2 + SH \rightleftharpoons RNH_3^+ + S^-$, in which charges appear only on one side of the equation, there is of course a considerable effect of dielectric constant, such reactions occurring to a much smaller extent in the alcohols than in water. In terms of the constants in Table 1 these equilibrium constants are given by

$$K_4 = [RCO_2^-][SH_2^+]/[RCO_2H] = K_s/K_2$$

$$K_5 = [RNH_3^+][S^-]/[RNH_2] = K_s/K_1$$

where $K_s$ is the ionic product of the solvent, and the concentration of the solvent is omitted as usual in the equilibrium expressions. The effect of solvent is then represented by the differences

$$\Delta_V = pK_4(MeOH) - pK_4(H_2O) = pK_s(MeOH) - pK_s(H_2O) - \Delta_{III}$$

$$\Delta_{VI} = pK_4(EtOH) - pK_4(H_2O) = pK_s(EtOH) - pK_s(H_2O) - \Delta_{IV}$$

$$\Delta_{VII} = pK_5(MeOH) - pK_5(H_2O) = pK_s(MeOH) - pK_s/H_2O) - \Delta_I$$

$$\Delta_{VIII} = pK_5(EtOH) - pK_5(H_2O) = pK_s(EtOH) - pK_s(H_2O) - \Delta_{II}.$$

If we take $pK_s(H_2O) = 14.0$, $pK_s(MeOH) = 16.7$, $pK_s(EtOH) = 19.1$, the data in Table 1 give the following average values:

$$\Delta_V = 4.8, \qquad \Delta_{VI} = 5.8, \qquad \Delta_{VII} = 1.5, \qquad \Delta_{VIII} = 4.1.$$

These values involve both the changes in dielectric constant and in the acid-base properties of the solvent, and cannot be interpreted in any simple way, though their sign is in each case what would be expected from the electrostatic effect. If we wish to examine the electrostatic effect independent of the acid-base properties of the solvent, we must consider the equilibrium $RCO_2H + R'NH_2 \rightleftharpoons RCO_2^- + R'NH_3^+$, not involving the solvent. The equilibrium constant $K$ for this reaction is given in terms of the constants in Table 1 as $K = K_s/K_1K_2$, and if we again take the average values we find

$$\Delta_{IX} = pK(\text{MeOH}) - pK(\text{H}_2\text{O}) = 3.6$$

$$\Delta_{X} = pK(\text{EtOH}) - pK(\text{H}_2\text{O}) = 4.8.$$

It is of interest to see how far these figures can be accounted for on a simple electrostatic basis. The electrical free energy of a pair of separated ions of charges $+e$ and $-e$ and radius $r$ in a medium of dielectric constant $\varepsilon$ is given by $e^2/\varepsilon r$, and if we apply this to two media of dielectric constants $\varepsilon'$ and $\varepsilon''$ we have

$$-\Delta G^\circ = kT \ln \frac{K'}{K''} = -\frac{e^2}{r}\left(\frac{1}{\varepsilon'} - \frac{1}{\varepsilon''}\right).$$

Applying this to the data for water and ethanol, we find for the mean ionic radius $r = 150$ pm. Similarly, the data for water and methanol give $r = 130$ pm. Although these are of the right order of magnitude, they are smaller than would be anticipated, though it is clearly difficult to define the effective radius of unsymmetrical ions like those concerned here. On the same basis we can attribute most of the differences in the ionic products of $\text{H}_2\text{O}$, MeOH, and EtOH to the differences in their dielectric constants, since we have seen that the differences in their acidic and basic strengths are not large.

The generalized picture just given for the behaviour of acids and bases in water and the alcohols involves some simplification. In the first place, there are clearly variations among the individual acids and bases in Table 1 which exceed the experimental error. These individual variations would have been much greater if we had not restricted ourselves to the two series of closely similar compounds, carboxylic acids and primary amines. For example, the phenols differ from the carboxylic acids in their values of $\Delta pK$. Similarly, there are considerable differences in the behaviour of primary, secondary, and tertiary amines, the last class often differing from the first two in the sign of $\Delta pK$. These individual deviations are considered in the next section. In the second place, the three solvents considered are of the same chemical type, and the macroscopic dielectric constant becomes an even less adequate measure of the solvent effect if we compare solvents of differing chemical character. This may be illustrated by the behaviour of acids and bases in *formamide*,[11] which is a solvent having a dielectric constant ($\varepsilon = 110$) somewhat greater than water and possessing both weakly acidic and weakly basic properties. Several acids are strong in formamide, and it was possible to determine $pK_s$ (16.8) from

[11] F. H. Verhoek, *J. Am. Chem. Soc.*, **58**, 2577 (1936).

e.m.f. measurements, so that the acid-base strengths could be referred to the solvent species $SH_2^+$–SH–S$^-$. Constants were obtained for twelve carboxylic acids and eight amines; when these are compared with the corresponding constants in water, the differences show individual variations somewhat greater than those in Table 1, but no trend with acid strength. The mean differences are as follows, $\Delta$ in each case representing $pK(\text{formamide}) - pK(H_2O)$:

$$RNH_3^+ + SH \rightleftharpoons SH_2^+ + RNH_2, \qquad \Delta_{XI} = -0.6$$

$$RCO_2^- + SH \rightleftharpoons S^- + RCO_2H, \qquad \Delta_{XII} = 0.9$$

$$RCO_2H + SH \rightleftharpoons SH_2^+ + RCO_2^-, \qquad \Delta_{XIII} = 1.9$$

$$RNH_2 + SH \rightleftharpoons S^- + RNH_3^+, \qquad \Delta_{XIV} = 3.4$$

$$RCO_2H + R'NH_2 \rightleftharpoons RCO_2^- + R'NH_3^+, \qquad \Delta_{XV} = 2.5.$$

The values of $\Delta_{XI}$ and $\Delta_{XII}$ indicate that formamide is a somewhat stronger base and a weaker acid than water, which is reasonable on chemical grounds. On the other hand, $\Delta_{XIII}$ and $\Delta_{XIV}$ are each 2.5 units greater than $\Delta_{XI}$ and $\Delta_{XII}$ respectively, showing that when the reaction involves the production of charge formamide is considerably less effective in reacting with both acids and bases. This is not at all what would be expected electrostatically, since the dielectric constant of formamide is somewhat greater than that of water, which should favour the production of charge. The same anomaly is shown in the value of $\Delta_{XV}$, which does not involve participation of the solvent. On the other hand, freezing point measurements for salts in formamide[12] show that their activity coefficients are close to those found in water, so that the above anomalies cannot be due to incomplete dissociation. There is of course ample evidence that the dielectric constant does not give a satisfactory account of the effect of solvent upon the free energy of ions, and this may be particularly the case in dealing with charged acids and bases, which depart very far from the picture of spherical ions. Thus in considering the ion $SH_2^+$, we are making a sharp distinction between the interaction of the proton with one solvent molecule (regarded as a chemical effect) and its interaction with the remainder of the solvent (represented by the dielectric constant term). In fact, as we have already seen for the hydronium and hydroxide ions (Chapter 2), the interaction of ions with the solvent is often much more specific than this, frequently involving the formation of hydrogen bonds.

[12] E. N. Vasenko, *Zh. Fiz. Khim.*, **21**, 361 (1947); **22**, 999 (1948); **23**, 959 (1949).

Thus, in addition to its acid-base properties and its dielectric constant, it is often necessary to consider the ability of the solvent to form hydrogen bonds as an additional characteristic which determines its behaviour towards dissolved acids and bases. As long as we are comparing solvents of very similar chemical type, such as water and the alcohols, it may not be necessary to consider hydrogen bonding in addition to bulk dielectric constants, but it becomes an important factor as soon as solvents of different chemical nature are considered.

Most of the equilibria discussed so far involve solvent ions, and for many purposes it is convenient to consider the equilibrium between two added acid-base systems. In this way the acidity or basicity of the solvent is eliminated, and the problem now involves only the less intimate interactions with the solvent. We have already mentioned reactions of the type $RCO_2H + R'NH_2 \rightleftharpoons RCO_2^- + R'NH_3^+$, but it is more useful to evaluate constants for the equilibria $RCO_2H + R'CO_2^- \rightleftharpoons RCO_2^- + R'CO_2H$ or $RNH_3^+ + R'NH_2 \rightleftharpoons RNH_2 + R'NH_3^+$: these measure the *relative strength* $K_r$ of two acids in the same class and should give a rough cancellation of electrostatic effects. These relative strengths can be measured more accurately than the individual values of acid-base constants involving the solvent, and it is often possible to measure relative strengths when nothing quantitative is known about the acid-base properties of the solvent, which may even be completely inert in this respect. Thus $K_r$ can sometimes be measured directly by spectrophotometric methods or, more frequently, by using an indicator in conjunction with the two separate systems. Provided that there is no association of the ionic species produced, $K_r$ is given directly as the ratio of the two equilibrium constants with the indicator. Alternatively $K_r$ can be derived from the e.m.f. of cells containing a hydrogen electrode (or some substitute such as the glass or quinhydrone electrode) in conjunction with a standard electrode. The difference between the e.m.f. of two similar cells (which may, for example, contain buffer solutions of two different carboxylic acids) will give reliable values for $K_r$, even when it is difficult to interpret the e.m.f. of a single cell because of uncertainties about standard electrode potentials and liquid junction potentials. Most of the reliable values of $K_r$ in non-aqueous solvents have been obtained by these two methods.

The approximate constancy of the last two columns in Table 1 shows that to a first approximation relative acid strengths are the same in water, methanol, and ethanol, provided that we remain within a group of similar substances, e.g., carboxylic acids or amine cations. However, there are certainly variations considerably greater than experimental error, and we must now consider these. Kilpatrick and his co-workers have made accurate measurements of relative strengths of benzoic acids[13] and of anilinium ions[14] the results of which are given in Tables 2 and 3. Although the values of $K_r$ run roughly parallel in each solvent, $K_r$ is certainly not independent of the solvent in spite of the fact that we are considering series of closely related compounds. Some of the variations are large, but they do not depend on the value of $K_r$, or in any obvious way on the nature of the substituent. There is some regularity, however, in that $K_r$ usually varies monotonically with the dielectric constant, though the magnitude and direction of this variation are different for different compounds.

As was first pointed out by Wynne-Jones,[15] this behaviour can be accounted for, at least formally, on an electrostatic basis. For an equilibrium $A + B_0 \rightleftharpoons A_0 + B$, if $A$ and $A_0$ have a positive charge $ze$, an

Table 2. RELATIVE STRENGTHS OF SUBSTITUTED BENZOIC ACIDS IN HYDROXYLIC SOLVENTS AT 25°C

Values of lg $K_r$ relative to benzoic acid

| Substituents | $H_2O$ | $(CH_2OH)_2$ | $CH_3OH$ | $C_2H_5OH$ | n-$C_4H_9OH$ |
|---|---|---|---|---|---|
| | $(\varepsilon = 78.5)$ | $(\varepsilon = 37.7)$ | $(\varepsilon = 31.5)$ | $(\varepsilon = 24.2)$ | $(\varepsilon = 17.4)$ |
| $p$-OH | $-0.36$ | $-0.45$ | $-0.53$ | $-0.55$ | $-0.57$ |
| $p$-OMe | $-0.27$ | $-0.32$ | $-0.36$ | $-0.32$ | $-0.36$ |
| $p$-Me | $-0.17$ | $-0.17$ | $-0.18$ | $-0.18$ | $-0.19$ |
| $m$-Me | $-0.06$ | $-0.09$ | $-0.09$ | $-0.06$ | $-0.10$ |
| $p$-F | $-0.06$ | $-0.17$ | $-0.18$ | $-0.23$ | $-0.22$ |
| $m$-OH | $0.10$ | $-0.03$ | $-0.11$ | $-0.16$ | $-0.16$ |
| $o$-OMe | $0.11$ | $0.18$ | $0.17$ | $0.24$ | $0.28$ |
| $p$-Cl | $0.22$ | $0.30$ | $0.34$ | $0.42$ | $0.39$ |
| $p$-Br | $0.23$ | $0.37$ | $0.42$ | $0.47$ | $0.42$ |
| $o$-CH$_3$ | $0.30$ | $0.05$ | $0.09$ | $0.02$ | $0.00$ |
| $m$-F | $0.34$ | $0.45$ | $0.51$ | $0.53$ | $0.42$ |
| $m$-I | $0.35$ | $0.49$ | $0.55$ | $0.62$ | $0.57$ |
| $m$-Cl | $0.38$ | $0.52$ | $0.59$ | $0.63$ | $0.59$ |
| $m$-Br | $0.39$ | $0.54$ | $0.60$ | $0.65$ | $0.58$ |
| $m$-NO$_2$ | $0.72$ | $0.93$ | $1.05$ | $1.17$ | $1.10$ |
| $p$-NO$_2$ | $0.78$ | $0.97$ | $1.02$ | $1.17$ | $1.14$ |
| $o$-F | $0.94$ | $0.69$ | $1.00$ | $1.03$ | $0.93$ |
| $o$-Cl | $1.28$ | $1.14$ | $1.21$ | $1.12$ | $1.08$ |
| $o$-I | $1.34$ | $1.11$ | $1.19$ | $1.08$ | $1.04$ |
| $o$-Br | $1.34$ | $1.20$ | $1.27$ | $1.16$ | $1.09$ |
| $o$-NO$_2$ | $2.03$ | $1.74$ | $1.83$ | $1.77$ | $1.78$ |

[13] J. H. Elliott and M. Kilpatrick, *J. Phys. Chem.*, **45**, 454, 466, 473 (1941).
[14] Kilpatrick and Arenberg, in Ref. 7.
[15] W. F. K. Wynne-Jones, *Proc. Roy. Soc.*, A, **140**, 440 (1933).

Table 3. RELATIVE STRENGTHS OF SUBSTITUTED ANILINIUM
IONS IN HYDROXYLIC SOLVENTS AT 25°C

Values of lg $K_r$ relative to anilinium ion

| Substituent | $H_2O$ | MeOH | EtOH |
|-------------|--------|-------|-------|
| $p$-Me      | −0.49  | −0.54 | −0.57 |
| $m$-Me      | −0.09  | −0.19 | −0.29 |
| $p$-F       | 0.06   | 0.32  | 0.35  |
| $o$-Me      | 0.20   | 0.10  | 0.10  |
| $p$-Cl      | 0.77   | 1.00  | 1.07  |
| $m$-F       | 1.20   | 1.46  | 1.47  |
| $m$-Cl      | 1.26   | 1.53  | 1.55  |
| $o$-Cl      | 1.96   | 2.40  | 2.43  |
| $m$-NO$_2$  | 2.13   | 2.91  | 3.24  |
| $p$-NO$_2$  | 3.60   | 4.52  | 5.21  |
| $o$-NO$_2$  | 4.88   | 5.64  | 6.44  |

electrostatic calculation gives for the values of $K_r$ in media of dielectric constants $\varepsilon$ and $\varepsilon'$

$$\ln\frac{K_r}{K_r'} = \frac{e^2}{2kT}\left(\frac{1}{\varepsilon} - \frac{1}{\varepsilon'}\right)\left\{z^2\left(\frac{1}{r_{A_0}} - \frac{1}{r_A}\right) - (z-1)^2\left(\frac{1}{r_{B_0}} - \frac{1}{r_B}\right)\right\} \tag{7}$$

For uncharged acids $z = 0$, and for cation acids $z = 1$, so that (7) becomes

$$\ln K_r = \ln K_r^{\infty} \pm \frac{e^2}{2kT\varepsilon}\left(\frac{1}{r_0} - \frac{1}{r}\right) \tag{8}$$

where $K_r^{\infty}$ is the value of $K_r$ in a hypothetical medium of infinite dielectric constant. The positive and negative signs refer to cationic and uncharged acids respectively, and the radii are those of the ions concerned. Equation (8) predicts that for a given acid in different solvents lg $K_r$ should be a linear function of $1/\varepsilon$, the slope depending on the values of $r_0$ and $r$.

Figure 2 shows a plot of lg $K_r$ against $1/\varepsilon$ for selected acids from Table 3.[16] If we omit n-butanol, in which ions are probably appreciably associated, the linear relation is fairly well obeyed, and the observed slopes can be accounted for by Equation (8) by assuming that $r$ and $r_0$ differ by a few per cent. Positive and negative slopes are about equally common, which at first sight seems unreasonable, since a substituent

[16] Further plots, covering all the acids studied, are given by J. H. Elliott and M. Kilpatrick, *J. Phys. Chem.*, **45**, 466 (1941).

should always increase the overall size of the benzoate ion. However, the effective radius will be determined more by the charge distribution in the neighbourhood of the carboxyl group than by the overall size of the ion, and variations in both directions could be explained on this basis, though it is difficult to account for the individual variations. It seems better to regard the slopes as empirical parameters for each substance which are useful in correlating the effect of dielectric constant, rather than as related to any easily definable radius.

The data for anilinium ions (Table 3) do not include enough solvents to test Equation (8). Some of the changes in $K_r$ are much greater than for carboxylic acids and would demand rather large changes in effective radius if they are to be explained electrostatically.

Much larger discrepancies appear if we compare solvents of differing chemical types and acids which differ more among themselves than the series of benzoic acids or anilinium ions. Both points are illustrated by the

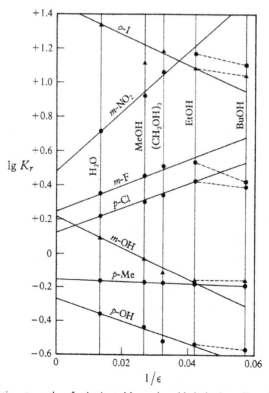

Fig. 2. Relative strengths of substituted benzoic acids in hydroxylic solvents.

results of Verhoek[11] for solutions in formamide, some of which are given in Table 4. We have already seen that for equilibria involving the solvent the effect of changing from water to formamide cannot be explained in any simple way, and the same is true for relative strengths in the two solvents.

Since water and formamide have high and similar dielectric constants, the electrostatic effect represented by (8) will be small; in fact, even if $r$ and $r_0$ are respectively 400 and 200 pm, Equation (8) predicts a difference of only 0.1 in $\lg K_r$ in the two solvents. The differences in Table 4 are much larger than this and could not be accounted for by any plausible values for the radii. In comparing solvents as different as water and formamide, it is clear that more specific solute–solvent interactions must play a part, for example, hydrogen bonding of both ions and uncharged species. It is noticeable that some of the largest discrepancies in Table 4 refer to acid-base pairs whose possibilities for hydrogen bonding differ considerably from those of the standard acid. Thus succinic and salicylic acids possess $-CO_2H$ and $-OH$ groups additional to those in benzoic acid. Similarly,

Table 4. RELATIVE ACID STRENGTHS IN WATER AND FORMAMIDE AT 20°C

| Carboxylic acids | Values of $\lg K_r$ relative to benzoic acid | |
| --- | --- | --- |
| | Water | Formamide |
| | ($\varepsilon = 80.4$) | ($\varepsilon = 111.5$) |
| Trimethylacetic | −0.82 | −1.15 |
| Propionic | −0.68 | −0.84 |
| Succinic | −0.01 | 0.33 |
| *m*-Chlorobenzoic | 0.38 | 0.51 |
| *m*-Nitrobenzoic | 0.75 | 0.87 |
| Salicylic | 1.21 | 1.83 |
| Monochloroacetic | 1.33 | 1.63 |
| *o*-Nitrobenzoic | 2.00 | 1.95 |
| $\alpha,\beta$-Dibromopropionic | 2.07 | 2.15 |
| Dichloroacetic | 2.89 | 3.48 |
| Amine cations | Values of $\lg K_r$ relative to anilinium | |
| Piperidinium | −6.44 | −6.98 |
| Triethylammonium | −6.28 | −5.89 |
| Benzylammonium | −4.68 | −5.67 |
| Pyridinium | −1.23 | −0.38 |
| *p*-Chloroanilinium | 0.66 | 0.84 |
| *o*-Chloroanilinium | 2.01 | 2.53 |
| 2,4-Dichloroanilinium | 2.55 | 3.02 |

in the amine series there are large deviations when we go from the standard aniline, a primary aromatic amine, to amines which are secondary or tertiary or aliphatic.

The quantity $K_r^\infty$ in Equation (8) was termed by Wynne-Jones[15] the 'intrinsic strength' (relative to the standard chosen) and corresponds to the intercepts in Figure 2. In principle it represents the relative strength divorced from external electrostatic effects, and it has been suggested that it should be used in discussing the effect of structure on acid-base strength. However, since Equation (8) applies only to water and the alcohols, it seems probably that $K_r^\infty$ still contains a considerable measure of solute–solvent interaction, and it is doubtful whether it is better suited to structural interpretations than $K_r$ measured in a given solvent.

So far it has been assumed that there is no ionic association in the solutions considered, and it is in fact only on this assumption that relative acidities have a unique meaning, even in a single solvent. The problem has been clearly stated by Kolthoff and Bruckenstein[17] in connection with investigations of acetic acid solutions. If an uncharged base B is in equilibrium with an acid HX, the possible equilibria are

$$HX + B \rightleftharpoons BH^+X^- \rightleftharpoons BH^+ + X^-$$

where $BH^+X^-$ is an ion pair, and the following equilibrium constants can be written down:

$$K_i^{BHX} = \frac{[BH^+X^-]}{[B][HX]}, \qquad K_d^{BHX} = \frac{[BH^+][X^-]}{[BH^+X^-]},$$

$$K_D^{BHX} = K_i^{BHX} K_d^{BHX} = \frac{[BH^+][X^-]}{[B][HX]}. \tag{9}$$

The assumption which we have made so far is that the concentration of the ion pair is very small, so that $K_i$ is very small, $K_d$ very large, and $K_D$ is measurable directly either by optical or electrical means. Under these conditions the relative strengths of two acids HX and HY, using $BH^+$ as the standard acid, are given by

$$\frac{K_D^{BHX}}{K_D^{BHY}} = \frac{[X^-][HY]}{[Y^-][HX]},$$

which is independent of B; i.e., in determining the relative strengths of two acids we may either compare them directly or compare them

[17] I. M. Kolthoff and S. Bruckenstein, *J. Am. Chem. Soc.*, **78**, 1, 10 (1956).

separately with any third system $BH^+$–$B$ and take the ratio of the strengths.

The position is different if the ion pair is almost undissociated, so that $K_d$ and $K_D$ are both very small and $K_i$ is the quantity measured by an optical method. The relative strengths of the two acids HX and HY are now given by

$$\frac{K_i^{BHX}}{K_i^{BHY}} = \frac{[BH^+X^-][HY]}{[BH^+Y^-][HX]} = \frac{[X^-][HY]}{[Y^-][HX]} \cdot \frac{K_d^{BHY}}{K_d^{BHX}}.$$

This is no longer independent of B, since there is no reason to expect either that $K_d^{BHY} = K_d^{BHX}$ or that their ratio is independent of the nature of B. Similarly, we cannot directly investigate the equilibrium $HX + Y^- \rightleftharpoons HY + X^-$ without introducing some cation, say $Z^+$; the equilibrium is then really $HX + Z^+Y^- \rightleftharpoons HY + Z^+X^-$ and its position will depend upon the nature of Z.

Similar considerations apply to acids of other charge types, and in general if investigations are made in a solvent where ions are associated, *the apparent relative strength of two acids will depend upon the base used for comparison*, and also on the nature of other ions present in solution even when these have no acidic or basic properties. The above treatment applies to the investigation of equilibria by optical means (e.g., by means of indicators), but the same conclusion holds if electrical methods are used for measuring the concentrations or activities of the free ions. For example, if we measure the concentrations of $BH^+$ and $X^-$ by conductivity measurements, the 'dissociation constant' obtained will be

$$[BH^+][X^-]/([B]+[BH^+X^-])([HX]+[BH^+X^-]),$$

which is related to $K_D$ through the constant $K_d^{BHX}$, and thus the relative strengths obtained will again depend upon the nature of B. This is essentially the same difficulty that arises in comparing the strengths of Lewis acids (electron acceptors), although, as we shall see, the deviations observed are much smaller.

Under conditions of ion pair formation a reaction such as $HX + B \rightarrow BH^+X^-$ bears a much closer resemblance to the formation of an adduct between a Lewis acid and a base (cf. Chapter 2) than do acid-base reactions in dissociating solvents; however, it does not satisfy the criterion of the formation of an electron-pair bond, and the same is true if the interaction involves a hydrogen bond rather than the complete transfer of a proton.

These predictions are borne out by the indicator measurements of Kolthoff and Bruckenstein[17] in anhydrous acetic acid. This has a dielectric constant $\varepsilon = 6.3$, and an electrostatic calculation[18] predicts that for ionic radii in the range 300–500 pm ion pairs will have dissociation constants in the range $10^{-5}$–$10^{-7}$, as is confirmed by conductivity measurements, so that ionic association will be considerable over any convenient range of concentrations. Correspondingly, it is found that the relative strengths of acids depend upon the bases with which they are reacting. For example, for reaction with urea $K_i(HCl):K_i$(toluene-*p*-sulphonic acid)$:K_i(HClO_4) = 1:2.3:330$, while for reaction with the indicator *p*-naphtholbenzein the corresponding ratios are $1:2.8:1500$. Kolthoff gives other examples of the same kind of behaviour. The effect of concentration upon the equilibrium position will of course also depend upon whether the ions are associated. For example, if we add a small concentration $c$ of a basic indicator to an excess of a weak acid HX, concentration $a$, then if $BH^+X^-$ is dissociated the equilibrium is governed by $cx^2/a(1-x) = K$ ($x$ dependent on $c$), while if $BH^+X^-$ is undissociated, $x/a(1-x) = K$ ($x$ independent of $c$).

The complications due to ion pairing in non-aqueous solvents can be avoided to some extent by the use of *dipolar aprotic solvents*,* which have been much investigated recently. These solvents are unable to act as hydrogen bond donors (i.e., they have at most very weak acid properties), but they have large dipole moments and polarizabilities, and hence moderately high dielectric constants. Examples are acetonitrile ($\varepsilon = 38$), dimethyl sulphoxide ($\varepsilon = 49$), sulpholane ($\varepsilon = 38$), nitrobenzene ($\varepsilon = 35$), nitromethane ($\varepsilon = 36$), dimethylformamide ($\varepsilon = 38$), and propylene carbonate ($\varepsilon = 65$). Some of them can act as bases and hydrogen bond acceptors, but this is not true of nitrobenzene and nitromethane. Unlike the solvents which we have considered so far, they show little specific tendency to solvate anions, which has important consequences for the reactivity of many anions towards organic compounds.[19]

Although in dilute solutions the electrostatic association of ions is small in these solvents, cations containing hydrogen often associate strongly with anions by hydrogen bonding, there being little competition from hydrogen bonding by the solvent. For example, in nitrobenzene

[18] N. Bjerrum, *Kgl. Danske Vid. Selsk. Math.-fys. Medd.*, **7**, No. 9 (1926); R. M. Fuoss and C. A. Kraus, *J. Am. Chem. Soc.*, **55**, 1919 (1933).
* The term *aprotic* is often used in a more restricted sense to denote solvents such as the hydrocarbons which are essentially devoid of any acid-base properties, and which act neither as donors nor as acceptors in hydrogen bond formation.
[19] For a review, see A. J. Parker, *Chem. Soc. Quart. Rev.*, **16**, 163 (1962).

($\varepsilon = 35$) the dissociation constants of tetraalkylammonium picrates are in the range $10^{-1}$–$10^{-2}$, but those of $NH_4^+ Pic^-$, $BuNH_3^+ Pic^-$, $Bu_2NH_2^+ Pic^-$, and $Bu_3NH^+ Pic^-$ are all close to $10^{-4}$; in methanol ($\varepsilon = 32$) the dissociation constants of all these picrates are similar.[20] A similar phenomenon is that known as *homoconjugation*, in which an anion or a cation associates with its conjugate or base by means of hydrogen bonding, i.e., $A^- + HA \rightleftharpoons (AHA)^-$ or $BH^+ + B \rightleftharpoons (BHB)^+$. This kind of association is often important in aprotic solvents, and sometimes more complex aggregates such as $A(HA)_2^-$ are formed, for example in nitromethane.[21] It may therefore require a detailed analysis of indicator, e.m.f., and conductivity data to obtain a complete understanding of acid-base equilibria in solvents of this type.

The most thoroughly investigated solvent of this class is *acetonitrile*, largely owing to the work of Kolthoff[22] and Coetzee.[23] Acetonitrile is a much weaker base and a very much weaker acid than water, its ionic product being only $3 \times 10^{-29}$. Perchloric acid dissociates completely and simply, but the 'strong' acids HBr, $H_2SO_4$, and HCl react incompletely according to the scheme $2HA + MeCN \rightleftharpoons MeCNH^+ + HA_2^-$. Since the equilibrium constant $[HA_2^-]/[HA][A^-]$ can be deduced from conductivity measurements, it is possible to obtain the conventional p$K$ values p$K$(HBr) = 5.5, p$K$($H_2SO_4$) = 7.3, and p$K$(HCl) = 8.9; thus acetonitrile acts as a differentiating solvent for these 'strong' acids. Carboxylic acids and phenols are also very much less dissociated than in water, the average value of p$K$(MeCN) – p$K$($H_2O$) being 14.0. Uncharged bases (amines) are also very much less protolysed in acetonitrile than in water; in terms of the acidic dissociation of the amine cations, the average value of p$K$(MeCN) – p$K$($H_2O$) is 7.3. We may note that $14.0 - 7.3 = 6.7$ represents the effect of passing from water to acetonitrile on the p$K$ of an equilibrium of the type $RCO_2H + R'NH_2 \rightleftharpoons RCO_2^- + R'NH_3^+$. Although this equilibrium does not formally involve the solvent, the effect is very much greater than $\Delta$p$K \sim 2$ which would be predicted electrostatically from the difference in dielectric constants, showing once more the importance of specific solvation effects.

[20] C. R. Witschonke and C. A. Kraus, *J. Am. Chem. Soc.*, **69**, 2471 (1947).
[21] H. van Looy and L. P. Hammett, *J. Am. Chem. Soc.*, **81**, 3872 (1959).
[22] See particularly I. M. Kolthoff, S. Bruckenstein, and M. K. Chantooni, *J. Am. Chem. Soc.*, **83**, 3927 (1961); I. M. Kolthoff and M. K. Chantooni, *J. Am. Chem. Soc.*, **87**, 4428 (1965); *J. Phys. Chem.*, **70**, 856 (1966); I. M. Kolthoff, M. K. Chantooni, and S. Bhowmik, *J. Am. Chem. Soc.*, **88**, 5430 (1966).
[23] For a review see J. F. Coetzee, *Progr. Phys. Org. Chem.*, **4**, 45 (1967).

Much attention has also been given recently to acid-base behaviour in *dimethyl sulphoxide*.[24] This solvent ($\varepsilon = 49$) is a much stronger base and hydrogen bond acceptor than acetonitrile, but also a much weaker acid, so that its ionic product, $10^{-31}$, is not very different. It is possible to prepare solutions of known concentrations of lyonium and lyate ions, using respectively toluenesulphonic acid and the caesium salt of the solvent, and since the glass electrode functions satisfactorily over some 25 units of pH, a wide range of weak acids has been investigated; indicator and conductivity measurements give reasonably concordant results. In dilute solutions there are only minor complications due to homo-conjugation, except with alkoxides, and conventional p$K$ values can be obtained. These are somewhat higher than those in water, the average value of p$K(\mathrm{Me_2SO}) - $p$K(\mathrm{H_2O})$ for 28 acids being 2.5, though there are large individual variations. However, dimethyl sulphoxide differs greatly from water in the very wide range of acidity which can be obtained within the limits of dilute solutions in this single solvent, especially on the alkaline side. Thus a 0.01M solution of the caesium salt of dimethyl sulphoxide has an effective pH some 27 units greater than a 0.01M solution of a strong acid, compared with a difference of 10 units in water. Because of the small tendency of dimethyl sulphoxide to solvate anions, solutions in it of hydroxides and alkoxides have very strongly basic properties, which persist even when considerable quantities of water or alcohol have been added, and solutions of this kind have been widely used to study very weak acids.[25]

It might be expected that acid-base equilibria would show particularly simple behaviour in *completely aprotic solvents*, such as hydrocarbons and their halogen derivatives, in which interactions with the solvent are reduced to a minimum.* In such systems it is of course impossible to observe the reaction of acids and bases with the solvent, and it is necessary to investigate a solution containing two acid-base pairs, sometimes with the addition of a third to act as an indicator. However, there are

[24] C. D. Ritchie and R. E. Uschold, *J. Am. Chem. Soc.*, **89**, 1721, 2752, 2960 (1967); E. C. Steiner and J. M. Gilbert, *J. Am. Chem. Soc.*, **87**, 382 (1965); E. C. Steiner and J. D. Starkey, *J. Am. Chem. Soc.*, **89**, 2751 (1967); I. M. Kolthoff, M. K. Chantooni, and S. Bhowmik, *J. Am. Chem. Soc.*, **90**, 23 (1968).
[25] For a review, see K. Bowden, *Chem. Rev.*, **66**, 119 (1966).
*Some solvents usually included in this class are not completely devoid of specific interactions with acids and bases; for example, chloroform can form weak hydrogen bonds with proton acceptors, and aromatic hydrocarbons become protonated at very high acidities. However, these interactions can usually be neglected provided that extremes of acidity or basicity are avoided.

several reasons why simple behaviour is the exception rather than the rule.[26] For the low dielectric constants involved, usually in the range $\varepsilon = 2\text{--}6$, an electrostatic treatment[18] predicts that the dissociation constants of ion pairs will be $10^{-8}$ or less. There is also much theoretical and experimental evidence[27] that these ion pairs associate into larger aggregates, except in extremely dilute solutions. The frequently used carboxylic acids are largely dimerized in these solvents, and their degree of dimerization is often unknown. Finally, it is now known that amines, the most commonly used bases, can form several types of complexes both with carboxylic acids and with phenols. In addition to ion pairs of the types $BH^+A^-$ and $BH^+(AHA)^-$, which owe some of their stability to hydrogen bonding between cation and anion, a combination of thermodynamic and spectroscopic measurements has shown that stable complexes involving hydrogen bonding without proton transfer, i.e., $B\cdots HA$, are formed in many systems.[28] The predominant type of bonding depends on the nature of the acid and of the bases and also on the solvent. In some systems broad general infrared absorption is observed,[29] suggesting that the proton is oscillating rapidly between two positions; this is reminiscent of the broad absorption bands observed for aqueous solutions of strong acids and alkalis (see p. 21).

It is thus clear that there are many complications in interpreting acid-base equilibria in aprotic solvents. Nevertheless, with a suitable choice of system there are many cases in which a simple equilibrium of the type $B + HA \rightleftharpoons BH^+A^-$ can be established and studied accurately. For example, Figure 3 shows the results of spectrophotometric measurements of the equilibrium between triethylamine and dinitrophenols in benzene solution. The good agreement with straight lines of unit slope establishes the nature of the equilibrium, and values of the constant $K_{BHA} = [BH^+A^-]/[B][HA]$ are given by the intercepts of these lines. The same

[26] Comprehensive reviews have been published by M. M. Davis, to whom much recent work is due. See M. M. Davis, *Nat. Bur. Stand. Monographs*, **105** (1968); Ref. 1(a), Vol. 3 (1970), p. 1.

[27] See, e.g., R. M. Fuoss and C. A. Kraus, *J. Am. Chem. Soc.*, **55**, 2387 (1933); **57**, 1 (1935); F. M. Batson and C. A. Kraus, *J. Am. Chem. Soc.*, **56**, 2017 (1934); D. A. Rothrock and C. A. Kraus, *J. Am. Chem. Soc.*, **59**, 1699 (1937); D. T. Copenhafer and C. A. Kraus, *J. Am. Chem. Soc.*, **73**, 4457 (1951); H. S. Young and C. A. Kraus, *J. Am. Chem. Soc.*, **73**, 4732 (1951); A. A. Maryott, *J. Res. Nat. Bur. Stand.*, **41**, 1 (1948).

[28] See, e.g., G. M. Barrow and E. A. Yerger, *J. Am. Chem. Soc.*, **71**, 5211 (1954); **77**, 4474, 6206 (1955); R. P. Bell and J. E. Crooks, *J. Chem. Soc.*, 3513 (1962); S. Bruckenstein and A. Saito, *J. Am. Chem. Soc.*, **87**, 698 (1965); S. Bruckenstein and D. F. Unterecker, *J. Am. Chem. Soc.*, **91**, 5741 (1969).

[29] D. F. de Tar and R. W. Novak, *J. Am. Chem. Soc.*, **92**, 1361 (1970).

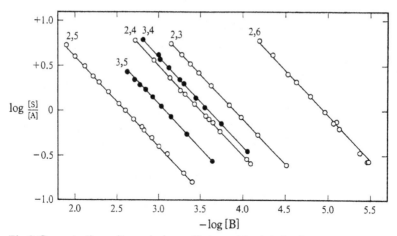

Fig. 3. Concentrations of ion pairs in equilibrium with triethylamine and six isomeric dinitrophenols in benzene solution. A, B, and S represent acid, base, and salt respectively. [Reproduced by permission from M. M. Davis, *J. Am. Chem. Soc.*, **84**, 3623 (1962).]

author has reported constants for the reaction of 52 carboxylic acids with diphenylguanidine in benzene, the indicator Bromophthalein Magenta E being present as a reference acid.[30]

We have already seen that in ionizing solvents differing considerably from water there is at best only a rough parallelism between relative acid strengths in water and in the non-aqueous solvent; it has also been shown that when ion association is present the apparent relative strength of two acids depends in principle upon the choice of the acid-base pair used for comparison. When measurements are made in aprotic solvents of low dielectric constant, the first of these sources of irregularity is diminished, since the pairing of the ions reduces their specific interactions with the solvent. There is therefore often a fairly close parallelism between the values of $K_{BHA}$ as defined in the last paragraph and the p$K$ values of the same acids in water, but this is only true for a series of acids of the same charge and very similar chemical type; for example, phenols cannot be compared with carboxylic acids, or primary amines with secondary or tertiary amines. Thus comparison of a series of substituted benzoic acids in water and in benzene give the following relations:[30]

---

[30] M. M. Davis and H. B. Hetzer, *J. Res. Nat. Bur. Stand.*, **60**, 569 (1958); M. M. Davis and M. Paabo, *J. Org. Chem.*, **31**, 1804 (1966).

meta-Substitution,     $\log K_{BHA} = 14.37 - 2.17 \, pK(H_2O)$

para-Substitution,     $\log K_{BHA} = 12.96 - 1.80 \, pK(H_2O)$

ortho-Substitution,    $\log K_{BHA} = 10.06 - 1.30 \, pK(H_2O)$

These relations do not correspond to unit slopes, and differ considerably from one another.

Returning to solvents in which ionic association is absent, or can be allowed for, it is of interest to express the effect of change of solvent in terms of activity coefficients. Within a given solvent, activity coefficients are usually defined so that $f_i \to 1$ at infinite dilution in the solvent in question, and deviations of $f_i$ from unity arise from the deviations from the laws of dilute solution, for example, because of interionic attraction. We shall suppose that all the equilibrium constants have been extrapolated to infinite dilution, so that this type of activity coefficient can be omitted. When we deal with changes of solvent, all activity coefficients must be referred to infinite dilution *in a given solvent*, which for our purposes is most conveniently taken as water. We can then define an activity coefficient for a species 'i' in any solvent by the expression $\Delta G^\circ = RT \ln f_i^\circ$, where $\Delta G^\circ$ is the free energy change involved in transferring one mole of 'i' from a dilute solution in the solvent in question to a solution of equal concentration in the standard solvent (water).* This kind of activity coefficient has been termed a *degenerate activity coefficient*[31] and is equal to the distribution coefficient of the species 'i' between water and the solvent concerned, assuming dilute solutions in each. If an acid-base equilibrium $A_1 + B_2 \rightleftharpoons A_2 + B_1$ has an equilibrium constant $K^\circ$ in water, its equilibrium constant $K$ in any other solvent is given by

$$K/K^\circ = f_{A_1}^\circ f_{B_2}^\circ / f_{A_2}^\circ f_{B_1}^\circ. \tag{10}$$

Equation (10) expresses the relative strengths of $A_1$ and $A_2$ in water and in the solvent concerned, and the condition for these relative strengths to be the same is that the function $f_{A_1}^\circ f_{B_2}^\circ / f_{A_2}^\circ f_{B_1}^\circ$ shall be unity.

At least two of the species $A_1$, $B_1$, $A_2$, and $B_2$ must be ions, and the individual activity coefficients of these species will have no thermodynamic significance. However, the right-hand side of Equation (10) is always physically significant, and the same will be true for certain

---

* The definition of $f_i^0$ depends upon the concentration scale used, which must of course be the same as is used in defining the equilibrium constants. When dealing with more than one solvent, it is most convenient to use molarities (moles per litre).
[31] E. Grunwald and B. J. Berkowitz, *J. Am. Chem. Soc.*, **73**, 4939 (1951). The term *medium activity coefficient* has also been suggested, and is perhaps more descriptive.

combinations involving ionic activity coefficients. Thus if $A_1$ and $A_2$ have positive charges $z_1$ and $z_2$ respectively, then if $z_1 = z_2$ the ratios $f^\circ_{B_1}/f^\circ_{B_2}$ and $f^\circ_{A_1}/f^\circ_{A_2}$ are physically significant, and if $z_1 + z_2 = 1$ the same is true of the products $f^\circ_{A_1}f^\circ_{B_2}$ and $f^\circ_{A_2}f^\circ_{B_1}$. On the other hand, it is never possible to attach a meaning to the ratios $f^\circ_{A_1}/f^\circ_{B_1}$ or $f^\circ_{A_2}f^\circ_{B_2}$.

These points can be illustrated by taking particular cases. For the equilibrium $R_1CO_2H + R_2CO_2^+ \rightleftharpoons R_2CO_2H + R_1CO_2^-$ the quantities $f^\circ(R_1CO_2H)$ and $f^\circ(R_2CO_2H)$ are separately determinable, and so is the ratio $f^\circ(R_1CO_2^-)/f^\circ(R_2CO_2^-)$. For example, the latter could be determined by measuring the solubilities of suitable sparingly soluble salts $R_1CO_2^-X^+$ and $R_2CO_2^-X^+$ in water and in the solvent considered. Similarly, for $R_1NH_3^+ + R_2NH_2 \rightleftharpoons R_2NH_3^+ + R_1NH_2$ the measurable quantities are $f^\circ(R_1NH_2)$, $f^\circ(R_2NH_2)$, and $f^\circ(R_1NH_3^+)/f^\circ(R_2NH_3^+)$, while for $R_1CO_2H + R_2NH_2 \rightleftharpoons R_2NH_3^+ + R_1CO_2^-$ we can determine $f^\circ(R_1CO_2H)$, $f^\circ(R_2NH_2)$, and the product $f^\circ(R_2NH_3^+)f^\circ(R_1CO_2^-)$.

Equation (10) does not at first sight apply to the effect of solvent upon conventional dissociation constants of the type $A + S \rightleftharpoons B + SH^+$, since it implies that both $A_1$–$B_1$ and $A_2$–$B_2$ exist in the same form in different solvents. We can, however, define a quantity $f^\circ_{H^+}$ (relative to dilute solutions in water) so that $RT \ln f^\circ_{H^+}$ represents the standard free energy change for the process $H_3O^+$ (in water) $+ SH$ (in SH) $\rightleftharpoons H_2O$ (in water) $+ SH_2^+$ (in SH), i.e., for the transfer of a proton from one solvent to another. (A similar scheme would apply for the transfer of any ion if solvation were taken explicitly into account.) Further, since the solvent activities are conventionally put equal to unity in defining dissociation constants, we can write for the ratio of dissociation constants of an acid A in any solvent and in water,

$$K_d/K_d^\circ = f^\circ_A/f^\circ_{H^+}.f^\circ_B. \tag{11}$$

For an uncharged acid $f^\circ_A$ is measurable, and so is the product $f^\circ_{H^+}f^\circ_{B^-}$.* Similarly, for the dissociation of a cation acid the measurable quantities are $f^\circ_B$ and $f^\circ_{A^+}/f^\circ_{H^+}$.

Various attempts have been made to split up Equations (10) and (11) still further, so as to obtain individual ionic activity coefficients.[32] However, such a procedure necessarily involves some extrathermodynamic

---

*For example, $f^0_{H^+} \cdot f^0_{Cl^-}$ could be derived from the e.m.f. of the cell $H_2 | HCl | AgCl \cdot Ag$ in the two solvents, and $f^0_{B^-}/f^0_{Cl^-}$ from the solubilities of suitable salts $Y^+B^-$ and $Y^+Cl^-$, whence $f^0_{H^+}f^0_{B^-} = f^0_{H^+}f^0_{Cl^-} \cdot f^0_{B^-}/f^0_{Cl^-}$.

[32] A recent comparison of the results of various assumptions shows a conspicuous lack of agreement: see A. J. Parker and R. Alexander, *J. Am. Chem. Soc.*, **90**, 3313 (1968).

assumptions, and since all the experimental results used or predicted involve the activity coefficients in thermodynamically acceptable combinations it seems that the values obtained for individual ions must be essentially arbitrary, though sometimes convenient.

If Equation (10) is applied to the relative strengths of two acids of similar structure and charge type, it seems likely that the whole of the right-hand side will deviate less from unity than will any of the combinations into which it can be dissected. For example, in the equilibrium $R_1CO_2H + R_2CO_2^- \rightleftharpoons R_2CO_2H + R_1CO_2^-$ we may suppose that one of the radicals, $R_1$, contains a group which is absent in $R_2$ and which interacts very differently with water and with the solvent concerned. This means that each of the ratios $f°(R_1CO_2H)/f°(R_2CO_2H)$ and $f°(R_1CO_2^-)/f°(R_2CO_2^-)$ will deviate considerably from unity. On the other hand, their deviations are likely to be in the same direction and will tend to cancel out when we consider the ratio

$$f°(R_1CO_2H)f°(R_2CO_2^-)/f°(R_2CO_2H)f°(R_1CO_2^-).$$

This kind of cancellation probably accounts for the approximate constancy of relative strengths which is sometimes observed even between solvents of very different character.

We have already seen how individual differences in relative strengths can be attributed to differences in ionic size, on a continuous dielectric picture, or to specific hydrogen bonding properties when a more detailed view is taken. The latter is certainly the more important when comparing solvents of different chemical types, and will apply to the interaction of the solvent with both ions and molecules. Any pair of species will of course also interact through dispersion forces, and although these will not normally produce marked individual variations, it has been suggested[33] that large effects may occur in special cases. For example, in the dissociation of picric acid the anion possesses a type of polarizability not present in the undissociated molecule; this involves the displacement of negative charge between the phenolic oxygen and the nitro groups (often represented by resonance structures), and is shown by intense absorption in the visible or near-ultraviolet part of the spectrum. In consequence, in passing from water to a solvent in which the immediate environment of the ion contains polarizable material,* the picrate anion should be stabilized relative to the undissociated acid by dispersion forces, with a

[33] E. Grunwald and E. Price, *J. Am. Chem. Soc.*, **86**, 4517 (1964).
* Because of the very short range of dispersion forces, it is the immediate environment of the ion which is important rather than the bulk polarizability of the solvent.

consequent abnormally high degree of dissociation. This is borne out by the following values for the ratio of the dissociation constants of picric and trichloroacetic acids in various solvents: water 0.5, methanol 13, ethanol 29, n-butanol 190. Grunwald and Price[33] give some other examples in which this effect may be significant,[34] but it is not likely to be of common occurrence.

A great deal of experimental work has been carried out on acid-base equilibria in *mixed solvents*, especially mixtures of water with organic solvents. The presence of two solvent species introduces a number of complications. In the first place, there are now a number of different acidic and basic species derived from the solvent. Thus in aqueous alcohol we have as acids $H_2O$, EtOH, $H_3O^+$, and $EtOH_2^+$, and as bases $H_2O$, EtOH, $OH^-$, and $EtO^-$. In the second place, the composition of the solvent can now vary in the neighbourhood of an ion (and to a smaller extent near an uncharged molecule) by a preferential solvation effect, so that the macroscopic properties of the solvent will be even less relevant than they are with pure solvents.* For these reasons the problem of mixed solvents will not be discussed here.

---

[34] See also D. W. Fong and E. Grunwald, *J. Phys. Chem.*, **73**, 3909 (1969).
* It might be thought that the same problem is present in pure solvents, many of which are associated and can be thought of as mixtures of different polymerized molecules. However, these molecules are all in mobile equilibrium with one another so that the solvent behaves thermodynamically as a single species, thus differing from a mixture of two solvents.

# 5 The Thermodynamics of Protolytic Equilibria

So far we have dealt with equilibrium constants at a single temperature, and these are of course related to the standard free energy change by the equation

$$-RT \ln K = \Delta G°. \tag{12}$$

If values of $K$ are available over a range of temperatures, we can deduce the enthalpy change $\Delta H$ from

$$\Delta H = -T^2 \frac{d}{dT}\left(\frac{\Delta G°}{T}\right) = RT^2 \frac{d \ln K}{dT}, \tag{13}$$

the standard entropy change from

$$\Delta S° = -d\Delta G°/dT = (\Delta H - \Delta G°)/T, \tag{14}$$

and the change in molar heat capacity from

$$\Delta C_p = d\Delta H/dT = Td\Delta S°/dT = -Td^2\Delta G°/dT^2. \tag{15}$$

If it is assumed that $\Delta C_p$ remains constant over the temperature range investigated, the above equations can be integrated, giving

$$\Delta H = \Delta H_0 + \Delta C_p T, \qquad \Delta S° = \Delta S_0° + \Delta C_p \ln T,$$

$$\ln K = -\frac{\Delta H_0}{RT} + \frac{\Delta C_p}{R} \ln T + \frac{(\Delta S_0° - \Delta C_p)}{R} \tag{16}$$

Equation (16) is a three-constant equation of the form $\ln K = A/T + B \ln T + C$, and the three constants can be evaluated if $K$ is known with sufficient accuracy over a range of temperature, thus giving values of $\Delta H_0$, $\Delta S_0°$, and $\Delta C_p$, and hence also the values of $\Delta H$ and $\Delta S°$ at some standard temperature such as 25°C. The best method of treating the experimental data has been discussed by many authors, and several equations differing in form from (16) have been suggested to represent the variation of $K$ with temperature. Most of these are empirical, and the available data are not of sufficient accuracy to distinguish between them. The thermodynamic functions calculated from the various equations usually agree well in the experimental range of temperature, and many authors have used the equation $\ln K = A/T + B + CT$, which is more convenient than (16) for numerical computation. This last equation corresponds to the assumption that $\Delta C_p$ is directly proportional to the absolute temperature, and over the range of accessible temperatures this is usually experimentally indistinguishable from a constant $\Delta C_p$, as assumed in Equation (16).

Ives and his collaborators[1] have inferred temperature variations in $\Delta C_p$ from extremely accurate conductometric measurements of the dissociation constants of carboxylic acids, but this demands high experimental accuracy, very pure materials, and a considerable degree of sophistication in the statistical treatment of the experimental results. Most investigations of dissociation constants yield at best an average value of $\Delta C_p$ for the temperature range studied, and even this is often uncertain. The precision of thermodynamic parameters derived from dissociation constants will always decrease rapidly in the series $\Delta G°$, $\Delta H, \Delta S°, \Delta C_p$.

In principle it should be possible to obtain more accurate values of $\Delta H$ by calorimetric measurements; $\Delta S°$ can then be obtained by combining $\Delta H$ with an accurate measurement of $K$ at a single temperature, and $\Delta C_p$ follows if $\Delta H$ has been measured at more than one temperature. Extensive work of this kind has been recently carried out by Christensen and his collaborators[2] for carboxylic acids and amines in water. However, the calorimetric method also has its pitfalls, since a series of apparently accurate measurements[3] for phenols, anilines, and benzoic acids were subsequently deemed unreliable[4] on the basis of a comparison with dissociation constants measured spectrophotometrically over a range of temperatures.

---

[1] See, for example, F. S. Feates and D. J. G. Ives, *J. Chem. Soc.*, 2798 (1956); D. J. G. Ives and P. D. Marsden, *J. Chem. Soc.*, 649 (1965); D. J. G. Ives and D. Prasad, *J. Chem. Soc.*, B, 1652 (1970); D. J. G. Ives and P. G. N. Moseley, *J. Chem. Soc.*, B, 1655 (1970).
[2] J. J. Christensen, J. L. Oscarson, and R. M. Izatt, *J. Am. Chem. Soc.*, **90**, 5949 (1968); J. J. Christensen, R. M. Izatt, and L. D. Hansen, *J. Chem. Soc.*, A, 1212 (1969); J. J. Christensen, M. D. Slade, D. E. Smith, R. M. Izatt, and J. Tsang, *J. Am. Chem. Soc.*, **92**, 4164 (1970); also earlier papers.
[3] W. J. Canady, H. M. Papee, and K. J. Laidler, *Trans. Faraday Soc.*, **54**, 502 (1958); H. M. Papee, W. J. Canady, T. W. Zawidzki, and K. J. Laidler, *Trans. Faraday Soc.*, **55**, 1734, 1738, 1743 (1959).
[4] D. T. Y. Chen and K. J. Laidler, *Trans. Faraday Soc.*, **58**, 480 (1962).

It is not easy to interpret directly the absolute values of thermodynamic functions for conventional acidity constants involving the solvent. As usually written, they have the dimensions of a concentration (e.g., $[X^-][H_3O^+]/[HX]$) so that the values of $\Delta G°$ and $\Delta S°$ are dependent upon the concentration units employed. If we include the solvent concentration, we then have the problem of interpreting the properties of the bulk solvent, which in the case of water is still a largely unsolved problem. It is more satisfactory to consider relative acidity constants, where the species in the equilibrium $A_1 + B_2 \rightleftharpoons A_2 + B_1$ are all in dilute solution. The values of all the thermodynamic functions are then independent of the concentration scale and should be more readily interpreted. Such values can be obtained by subtraction from the published thermodynamic data for conventional acidity constants, and Table 5 gives the results of such a treatment, comparison with one standard system being made for each class of acid concerned.*

In a symmetrical reaction such as $R_1CO_2H + R_2CO_2^- \rightleftharpoons R_2CO_2H + R_1CO_2^-$ it is tempting to guess that the entropy change is close to zero, and hence $\Delta G°$ close to $\Delta H$. Inspection of Table 5 shows that this is by no means the case, since the term $T\Delta S°$ is just as important as $\Delta H$ in determining the value of $\Delta G°$, and hence the equilibrium constant. Moreover, the separate contributions of $\Delta H$ and $T\Delta S°$ to $\Delta G°$ vary in an apparently erratic manner from one acid to another. Although such behaviour is not unknown for non-ionic equilibria and non-polar solvents, the large effects observed suggest strongly that the entropy changes involve an interaction of the ions (and perhaps also the uncharged molecules) with the highly polar and associated aqueous medium. The same conclusion is reached by examining the values of $\Delta C_p$, which is less sensitive than is $\Delta S°$ to changes in the types of motion executed by the system. Thus the complete freezing of an internal rotation corresponds to only $\frac{1}{2}R = 1$ cal mol$^{-1}$, and the changes of several units found in the last column of Table 5 must represent fairly drastic alterations in the system.

---

* Table 5 was compiled in 1958, and includes most of the data of sufficient accuracy and reliability then available, mostly obtained from studies of the e.m.f. of cells. It could now be considerably extended, especially by the addition of values derived from modern conductivity measurements and from calorimetry, and in a few instances the thermodynamic parameters could be replaced by more accurate ones. However, the picture would not be changed by such a revision, and the original table has therefore been retained for the purpose of our semi-quantitative discussion. Extensive tables of thermodynamic parameters for protolytic reactions in water have been recently compiled by Christensen[2] and by Larsen and Hepler [J. W. Larsen and L. G. Hepler, in *Solute-Solvent Interactions* (ed. J. F. Coetzee and C. D. Ritchie), Dekker, New York, 1969].

Table 5. THERMODYNAMIC DATA FOR PROTOLYTIC REACTIONS IN WATER AT 25°C

*Carboxylic acids.* Reaction $RCO_2H + MeCO_2^- \rightleftharpoons RCO_2^- + MeCO_2H$
(p$K$ refers to $[RCO_2^-][H_3O^+]/[RCO_2H]$)

| Acid | Reference | p$K$ | kcal mol$^{-1}$ | | | cal mol$^{-1}$ deg$^{-1}$ | |
| | | | $\Delta G°$ | $\Delta H$ | $T\Delta S°$ | $\Delta S°$ | $\Delta C_p$ |
|---|---|---|---|---|---|---|---|
| Trimethylacetic | (1) | 5.032 | +0.37 | −0.61 | −0.98 | −3.3 | +3 |
| Propionic | (1) | 4.875 | +0.16 | −0.12 | −0.28 | −0.9 | +1 |
| Hexoic | (1) | 4.857 | +0.14 | −0.59 | −0.73 | −2.5 | +2 |
| Isobutyric | (1) | 4.849 | +0.12 | −0.69 | −0.80 | −2.8 | +5 |
| Isohexoic | (1) | 4.845 | +0.12 | −0.61 | −0.73 | −2.5 | +4 |
| Valeric | (1) | 4.843 | +0.12 | −0.61 | −0.73 | −2.5 | +4 |
| Butyric | (2) | 4.818 | +0.08 | −0.61 | −0.69 | −2.3 | +2 |
| Isovaleric | (1) | 4.781 | +0.03 | −1.11 | −1.14 | −3.9 | +5 |
| Acetic | (3) | 4.756 | 0 | 0 | 0 | 0 | 0 |
| Diethylacetic | (1) | 4.736 | −0.03 | −1.92 | −1.89 | −6.4 | +8 |
| Succinic | (4) | 4.207 | −0.62 | +0.87 | +1.49 | +5.4 | +5 |
| Lactic | (5) | 3.860 | −1.23 | −0.06 | +1.17 | +3.9 | −3 |
| Glycollic | (6) | 3.831 | −1.27 | +0.21 | +1.06 | +4.9 | −3 |
| Formic | (7) | 3.752 | −1.38 | +0.07 | +1.45 | +4.8 | −4 |
| Iodoacetic | (8) | 3.182 | −2.16 | −1.31 | +0.85 | +2.8 | +4 |
| Bromoacetic | (8) | 2.902 | −2.53 | −1.13 | +1.40 | +4.7 | −1 |
| Chloroacetic | (8) | 2.868 | −2.58 | −1.01 | +1.57 | +5.2 | −9 |
| Fluoroacetic | (8) | 2.586 | −2.96 | −1.28 | +1.68 | +5.6 | +4 |
| Cyanoacetic | (9) | 2.470 | −3.12 | −0.78 | +2.34 | +7.8 | +1 |
| *p*-Hydroxybenzoic | (10) | 4.582 | −0.24 | +0.65 | +0.89 | +3.0 | −8 |
| Benzoic | (10) | 4.213 | −0.75 | +0.53 | +1.28 | +4.2 | −2 |
| *p*-Bromobenzoic | (10) | 4.002 | −1.03 | +0.22 | +1.25 | +4.2 | +6 |
| *p*-Chlorobenzoic | (10) | 3.986 | −1.06 | +0.34 | +1.40 | +4.7 | +14 |
| *m*-Chlorobenzoic | (10) | 3.827 | −1.27 | −0.07 | +1.20 | +4.0 | −4 |
| *m*-Bromobenzoic | (10) | 3.809 | −1.30 | +0.05 | +1.35 | +4.5 | +4 |
| *m*-Cyanobenzoic | (10) | 3.598 | −1.58 | +0.07 | +1.65 | +5.5 | −10 |
| *p*-Cyanobenzoic | (10) | 3.551 | −1.65 | +0.14 | +1.79 | +6.0 | −3 |
| *p*-Nitrobenzoic | (10) | 3.442 | −1.80 | +0.14 | +1.94 | +6.5 | +6 |
| $CO_2^-CO_2H$ | (11) | 4.266 | −0.67 | −1.55 | −0.88 | −2.9 | −22 |
| $CO_2^-CH_2CO_2H$ | (12) | 5.696 | +1.43 | −1.90 | −2.33 | −7.8 | −24 |
| $CO_2^-CH_2CH_2CO_2H$ | (4) | 5.638 | +1.20 | 0.00 | −1.20 | −4.0 | −15 |

1. D. H. Everett, D. A. Landsman, and B. R. W. Pinsent, *Proc. Roy. Soc.*, A, **215**, 403 (1952).
2. H. S. Harned and R. O. Sutherland, *J. Am. Chem. Soc.*, **56**, 2039 (1034); re-calculated by Everett, Landsman, and Pinsent (*loc. cit.*).
3. H. S. Harned and B. B. Owen, *Physical Chemistry of Electrolyte Solutions*, Reinhold, New York, 1943, Table 15-6-1A.
4. G. D. Pinching and R. G. Bates, *J. Res. Nat. Bur. Stand.*, **45**, 322, 444 (1950).
5. L. F. Nims and P. K. Smith, *J. Biol. Chem.*, **113**, 145 (1936).
6. L. F. Nims, *J. Am. Chem. Soc.*, **58**, 987 (1936).
7. H. S. Harned and N. D. Embree, *J. Am. Chem. Soc.*, **56**, 1042 (1934).
8. D. J. G. Ives and J. H. Pryor, *J. Chem. Soc.*, 2104 (1955).
9. F. S. Feates and D. J. G. Ives, *J. Chem. Soc.*, 2798 (1956).
10. G. Briegleb and A. Bieber, *Z. Elektrochem.*, **55**, 250 (1951).
11. G. D. Pinching and R. G. Bates, *J. Res. Nat. Bur. Stand.*, **40**, 405 (1948).
12. W. J. Hamer, J. O. Burton, and S. F. Acree, *J. Res. Nat. Bur. Stand.*, **24**, 269 (1940).

Table 5. *(Continued)*

*Amine cations.* Reaction $R_3NH^+ + NH_3 \rightleftharpoons R_3N + NH_4^+$
(pK refers to $[R_3N][H_3O^+]/[R_3NH^+]$)

| Acid | Reference | pK | kcal mol$^{-1}$ | | | cal mol$^{-1}$ deg$^{-1}$ | |
|---|---|---|---|---|---|---|---|
| | | | $\Delta G°$ | $\Delta H$ | $T\Delta S°$ | $\Delta S°$ | $\Delta C_p$ |
| Ammonium | (13) | 9.245 | 0 | 0 | 0 | 0 | 0 |
| Methylammonium | (14) | 10.624 | +1.87 | +0.69 | −1.18 | −3.9 | +8 |
| Ethylammonium | (15) | 10.631 | +1.91 | +1.18 | −0.83 | −2.8 | — |
| n-Propylammonium | (15) | 10.530 | +1.77 | +1.45 | −0.32 | −1.1 | — |
| n-Butylammonium | (15) | 10.597 | +1.85 | +1.67 | −0.18 | −0.9 | — |
| Hydroxymethylammonium | (16) | 9.498 | +0.34 | −0.32 | −0.64 | −2.1 | −1 |
| Dimethylammonium | (14) | 10.774 | +2.11 | −0.52 | −2.63 | −8.8 | +20 |
| Diethylammonium | (15) | 10.933 | +2.39 | +0.37 | −2.02 | −6.7 | — |
| Piperidinium | (17) | 11.123 | +2.53 | +0.40 | −2.13 | −7.0 | +21 |
| Trimethylammonium | (14) | 9.800 | +0.77 | −3.57 | −4.34 | −14.5 | +41 |
| Triethylammonium | (18) | 10.867 | +2.22 | −0.19 | −2.41 | −8.0 | — |
| Anilinium | (19) | 4.596 | −6.34 | −5.29 | +1.05 | +3.5 | 0 |
| o-Chloroanilinium | (20) | 2.634 | −9.02 | −6.40 | +2.62 | +8.7 | 0 |
| $NH_2(CH_2)_2NH_3^+$ | (21) | 9.928 | +0.94 | −0.58 | −1.52 | −5.1 | +10 |
| $NH_2(CH_2)_6NH_3^+$ | (21) | 10.930 | +2.30 | +1.51 | −0.79 | −2.6 | +8 |
| $\overset{+}{N}H_3(CH_2)_2\overset{+}{N}H_3$ | (21) | 6.848 | −3.27 | −1.53 | +1.74 | +5.8 | +18 |
| $\overset{+}{N}H_3(CH_2)_6\overset{+}{N}H_3$ | (21) | 9.830 | +0.80 | +1.42 | +0.62 | +2.1 | +8 |

*Values for acid dissociation of standard systems*

| | $\Delta G°$ | $\Delta H$ | $T\Delta S°$ | $\Delta S°$ | $\Delta C_p$ |
|---|---|---|---|---|---|
| $CH_3CO_2H \rightleftharpoons CH_3CO_2^- + H^+$ | 6.49 | −0.11 | −6.60 | −22.1 | −37 |
| $NH_4^+ \rightleftharpoons NH_3 + H^+$ | 12.61 | 12.40 | −0.21 | −0.7 | 0 |

13. R. G. Bates and G. D. Pinching, *J. Res. Nat. Bur. Stand.*, **42**, 419 (1949); *J. Am. Chem. Soc.*, **72**, 1393 (1950); D. H. Everett and D. A. Landsman, *Trans. Faraday Soc.*, **50**, 1221 (1954).
14. D. H. Everett and W. F. K. Wynne-Jones, *Proc. Roy. Soc.*, A, **177**, 499 (1941).
15. A. G. Evans and S. D. Hamann, *Trans. Faraday Soc.*, **47**, 34 (1951).
16. R. G. Bates and G. D. Pinching, *J. Res. Nat. Bur. Stand.*, **46**, 349 (1951).
17. R. G. Bates and N. E. Bower, *J. Res. Nat. Bur. Stand.*, **57**, 153 (1956).
18. J. E. Ablard, D. S. McKinney, and J. C. Warner, *J. Am. Chem. Soc.*, **62**, 2181 (1940).
19. K. J. Pedersen, *Kgl. Danske Vid. Selsk. Skr.*, **14**, 9 (1937).
20. K. J. Pedersen, *Kgl. Danske Vid. Selsk. Skr.*, **15**, 2 (1937).
21. D. H. Everett and B. R. W. Pinsent, *Proc. Roy. Soc.*, A, **215**, 416 (1952).

This is particularly the case for the amine cations when we are comparing secondary and tertiary amines with ammonia or a primary amine.

One method of attempting to allow for the solvent interaction is to use an electrostatic picture. By combining equations (8) with (12)–(15), we obtain for the electrostatic contributions to $\Delta S^\circ$ and $\Delta C_p$

$$(\Delta S^\circ)_e = \frac{Ne^2}{2}\left(\frac{1}{r_0} - \frac{1}{r}\right)\frac{d}{dT}\left(\frac{1}{\varepsilon}\right) \tag{17}$$

$$(\Delta C_p)_e = \frac{Ne^2}{2}\left(\frac{1}{r_0} - \frac{1}{r}\right)T\frac{d^2}{dT^2}\left(\frac{1}{\varepsilon}\right), \tag{18}$$

where $N$ is the Avogadro number. The exact value of $d^2\varepsilon/dT^2$ is uncertain, but if we use the table of values given by Harned and Owen[5] we find for water at 25°C

$$(\Delta S^\circ)_e = 9.70\left(\frac{1}{r_0'} - \frac{1}{r'}\right)\text{cal mol}^{-1}\text{ deg}^{-1} \tag{19}$$

$$(\Delta C_p)_e = 5.70\left(\frac{1}{r_0'} - \frac{1}{r'}\right)\text{cal mol}^{-1}\text{ deg}^{-1} \tag{20}$$

where $r_0'$ and $r'$ are measured in pm. Equations (19) and (20) might just be stretched to fit the experimental data for carboxylic acids, though it is necessary to assume $r < 100$ pm, which is hardly reasonable. They certainly fail to account for the large entropy and heat capacity changes found for the amine cations; for example, the reaction $NMe_3H^+ + NH_3 \rightleftharpoons NMe_3 + NH_4^+$ has $\Delta S^\circ = 14.5$, $\Delta C_p = +41$. Moreover, (19) and (20) predict $\Delta S^\circ = 1.7\Delta C_p$. Table 5 shows no sign of any such regularity, there being in fact some tendency (e.g., in the amine and fatty acid series) for $\Delta S^\circ$ and $\Delta C_p$ to vary in opposite directions.

The same difficulty arises when we consider processes such as $R_1CO_2H + R_2NH_2 \rightleftharpoons R_1CO_2^- + R_2NH_2^+$. The thermodynamic functions for 420 such reactions can be derived by combining the first and second parts of Table 5, and although $\Delta S^\circ$ and $\Delta C_p$ are both negative, in agreement with electrostatic calculation, their numerical values are frequently too high, and there is no sign of the predicted correlation between them ($\Delta S^\circ = 1.7\Delta C_p$). To take an extreme example, the reaction $MeCO_2H + NMe_3 \rightleftharpoons MeCO_2^- + NHMe_3^+$ has $\Delta S^\circ = -7$, $\Delta C_p = -78$.

There is thus ample evidence that an electrostatic picture is inadequate to account for the solvent–ion interaction. The problem is of course not restricted to acid-base equilibria but arises whenever we wish to interpret

[5] H. S. Harned and B. B. Owen, *Physical Chemistry of Electrolyte Solutions*, 3rd edn., Reinhold, New York, 1958, Table 5-1-3.

the entropies or heat capacities of ions in solution. The use of the dielectric constant is a crude macroscopic method of allowing for the orientation of the solvent molecules in the field of the ion, and there is no doubt that it is necessary to take a much more specfic picture of the molecular inter-actions involved. A first step is to attempt a detailed analysis of the motions of the first shell of co-ordinated water molecules, and then to apply the electrostatic equations to the solvent outside this shell. This approach has been used both for entropies[6] and for heat capacities[7] and is certainly an improvement on the Born equation. However, it is now clear that this approach does scant justice to the structure of water or to the effect of solutes upon this structure; for example, the dissolution of non-polar solutes (e.g., the inert gases or the hydrocarbons) in water causes a decrease of entropy which may even exceed the decrease caused by ions of comparable size.[8] The literature on the structure of water and aqueous solutions is copious and controversial, and we shall not attempt to summarize it here.[9] A popular but by no means universal view is that liquid water is best thought of as a mixture containing 'free' water molecules in mobile equilibrium with a structured component resembling ice; hence the frequent use of terms such as 'local icebergs', 'flickering clusters', and 'ice-likeness'. Apart from any more specific effects, the effect of adding a solute is to disturb the equilibrium between the free and structured components, with consequent changes in entropy and heat capacity. According to one picture, an ion of moderate size is surrounded by three successive regions of water: first, a co-ordination shell in which the water molecules are strongly oriented in the field of the ion, secondly a region which is *less* ordered than pure water (because of the competition between the two incompatible structures), and finally at greater distances a return to the comparative order of water itself. Such a picture could clearly lead to overall entropy changes of either sign, and ions have frequently been classified as *structure-making* or *structure-breaking*,[10]

[6] D. D. Eley and M. G. Evans, *Trans. Faraday Soc.*, **34**, 1093 (1938).

[7] D. H. Everett and C. A. Coulson, *Trans. Faraday Soc.*, **36**, 633 (1940).

[8] D. D. Eley, *Trans. Faraday Soc.*, **35**, 1281, 1421 (1939); H. S. Frank, *J. Chem. Phys.*, **13**, 507 (1945); R. E. Powell and W. M. Latimer, *J. Chem. Phys.*, **19**, 1139 (1951); W. F. Claussen and M. F. Polglase, *J. Am. Chem. Soc.*, **74**, 4817 (1952).

[9] Some key references are: *Disc. Faraday Soc.*, **24**, 133 (1957) (section on ion–solvent interaction); J. L. Kavanau, *Water and Solute-Water Interactions*, Holden-Day, San Francisco, 1964; *Hydrogen-bonded Solvent Systems* (ed. A. K. Covington and P. Jones), Taylor and Francis, London, 1968; D. Eisenberg and W. Kauzmann, *The Structure and Properties of Water*, Oxford, 1969.

[10] See particularly R. W. Gurney, *Ionic Processes in Solution*, McGraw-Hill, New York, 1953, Chapters 7 and 8.

though there is often disagreement about the classification of particular ions, depending upon what experimental criterion is used.

Considerations of this kind have very little predictive power, but they do at least provide a plausible reason for the wild variations in $\Delta H$, $\Delta S°$, and $\Delta C_p$ illustrated in Table 5. It might, however, appear to be a hopeless task to correlate protolytic equilibrium constants measured in polar solvents with quantitative (or even qualitative) considerations based on the molecular structure of the solutes, in view of the extreme complications introduced by solute–solvent interactions. Fortunately the position appears to be more satisfactory than might be expected, *provided that we confine ourselves to equilibrium constants (i.e., $\Delta G°$) at a single temperature*, and do not attempt to interpret the separate contributions to $\Delta H$ and $\Delta S°$. Similarly, in interpreting reaction velocities in terms of variations in the structures of the reacting molecules, it is usually more satisfactory to compare velocity constants at a single temperature rather than activation energies (i.e., to compare free energies of activation rather than enthalpies). The next few paragraphs will attempt to explain why this is so.

This raises a question of general importance: What is the correct basis for comparing the predictions of a molecular model with the results of thermodynamic or kinetic measurements? A molecular model, whether classical or quantal, essentially predicts energies (or effects on the energy) of molecules *in vacuo* at absolute zero, where there is of course no distinction between the enthalpy and the free energy. At a finite temperature these 'model energies' are overlaid by a large amount of thermal energy, distributed between many degrees of freedom, and when a solvent is present there will be a considerable contribution from the effect of the system being studied upon the kinetic and potential energy of the solvent molecules. The quantities $H$ and $G$ represent, in effect, different ways of averaging the molecular energies,* and it is not at all obvious which of them is more directly comparable with the model. In the present problem there might seem to be some justification for using the quantity $\Delta H_0$ in Equation (16), which has the formal appearance of an energy change at absolute zero and which is evaluated when fitting the experimental data to this equation. However, (16) implies that $\Delta C_p$ is independent of temperature, which may be a good approximation over a considerable

---

* If $\varepsilon_i$ represents the energy of an individual molecular level, then the two types of averaging correspond to

$$H = \sum_i \varepsilon_i e^{-\varepsilon_i/kT} \Big/ \sum_i e^{-\varepsilon_i/kT}, \qquad e^{-G/kT} = \sum_i e^{-\varepsilon_i/kT}$$

If all the systems are in the lowest state $\varepsilon_0$, then $H = G = \varepsilon_0$.

temperature range, but is certainly not valid down to absolute zero. Thus $\Delta H_0$ is not likely to have any more fundamental significance than $\Delta H$ or $\Delta G°$ at 25°C; and it is certainly much less accurately known because it is highly dependent upon the value taken for $\Delta C_p$ and in effect involves a long extrapolation from the experimental data. Similarly, no physical significance can be attached to the quantity $\Delta S_0°$ in Equation (16).

At a finite temperature, it seems intuitively likely that a change in structure will produce changes in $\Delta H$ and $\Delta S$ in the same direction; this is because any tightening of interactions in a system will decrease its energy, but will also decrease its entropy by restricting rotation and increasing vibrational frequencies. Since $\Delta G = H - T\Delta S$, the thermal contributions to $\Delta H$ and $T\Delta S$ will tend to compensate one another, so that at a finite temperature $\Delta G$ should approximate better to the model than does $\Delta H$. This can be put on a more formal basis, as follows.

Suppose that we start with an ideal situation, a vacuum at absolute zero, and proceed to complicate the situation by raising the temperature and introducing a solvent. Considering first the temperature change, for a given reaction, we have at absolute zero $\Delta H_0 = \Delta G_0$, $\Delta S_0 = 0$, and $\Delta C_p = 0$. At a finite temperature,

$$\Delta H = \Delta H_0 + \int_0^T \Delta C_p \, dT$$

$$\Delta G = \Delta H_0 + \int_0^T \Delta C_p \, dT - T \int_0^T \frac{\Delta C_p}{T} \, dT. \tag{21}$$

Suppose now that we modify the reaction slightly, for example, by introducing a substituent, and indicate the corresponding small changes in $\Delta H_0$, $\Delta C_p$, etc., by $\delta\Delta H_0$, $\delta\Delta C_p$, etc. Then

$$\delta\Delta H - \delta\Delta H_0 = \int_0^T \delta\Delta C_p \, dT \tag{22a}$$

$$\delta\Delta G - \delta\Delta H_0 = \int_0^T \delta\Delta C_p \, dT - T \int_0^T \frac{\delta\Delta C_p}{T} \, dT. \tag{22b}$$

There are now good grounds for believing that (22b) will most commonly be numerically smaller than (22a), i.e., that $\delta\Delta G$ approximates more closely to $\delta\Delta H_0$ than does $\delta\Delta H$. No general proof of this can be given, and indeed it will not be universally true, but we may illustrate the point by writing $\delta\Delta C_p$ as a power series, beginning with a term in $T^3$ so as to give the correct behaviour at absolute zero, i.e.,

$$\delta\Delta C_p = a_1 T^3 + a_2 T^4 + a_3 T^5 + \cdots. \tag{23}$$

This gives the result

$$\delta\Delta H - \delta\Delta H_0 = \frac{a_1}{4}T^4 + \frac{a_2}{5}T^5 + \frac{a_3}{6}T^6 + \cdots \tag{24a}$$

$$\delta\Delta G - \delta\Delta H_0 = -\frac{a_1}{3.4}T^4 - \frac{a_2}{4.5}T^5 - \frac{a_3}{5.6}T^6 - \cdots. \tag{24b}$$

We cannot make any general statements about the signs or magnitudes of $a_1, a_2, a_3, \cdots$, but it is clear that for random variations of these quantities there is a high probability that (24b) will be numerically smaller than (24a). This kind of consideration was first put forward by Evans and Polanyi,[11] using a somewhat different formulation.

The other major question is how far the quantities which we have termed $\delta\Delta H$ and $\delta\Delta G$ are affected by transfer to a solvent. The thermodynamic functions for a reaction in a solvent, $\delta\Delta H_s$, etc., can be expressed in terms of the corresponding quantities in the gas phase by the equations

$$\delta\Delta H_s - \delta\Delta H = \sum H_s \tag{25a}$$

$$\delta\Delta G_s - \delta\Delta G = \sum H_s - T\sum S_s \tag{25b}$$

where $H_s$ and $S_s$ are heats and entropies of solution and the summation is made over all the species involved in the reactions. There is good experimental evidence[12] that the heats and entropies of solution of a series of similar substances in a given solvent are approximately related by expressions of the form

$$TS_s = \alpha H_s + \beta \tag{26}$$

where $\alpha$ and $\beta$ are constants characteristic of the solvent, $\alpha$ being always positive and usually between 0.4 and unity. If Equation (26) applies to all the species concerned, it is clear that (25b) will be numerically smaller than (25a), i.e., the solvent will affect $\delta\Delta G$ less than it does $\delta\Delta H$, so that there is again some justification for using free energies rather than enthalpies for comparison with molecular models.

The argument in the last paragraph appeals to experiment for the relation between heats and entropies of solution of similar solutes in a given solvent, and it would be more satisfying to have a molecular

[11] M. G. Evans and M. Polanyi, *Trans. Faraday Soc.*, **32**, 1333 (1936); cf. also J. A. V. Butler, *Trans. Faraday Soc.*, **33**, 169 (1937); C. D. Ritchie and W. F. Sager, *Progr. Phys. Org. Chem.*, **2**, 323 (1964).
[12] J. A. V. Butler, *Trans. Faraday Soc.*, **33**, 229 (1937); R. P. Bell, *Trans. Faraday Soc.*, **33**, 496 (1937); I. M. Barclay and J. A. V. Butler, *Trans. Faraday Soc.*, **34**, 1445 (1938). For a summary see J. E. Leffler, *J. Org. Chem.*, **20**, 1202 (1955).

justification. Many interpretations based on particular models have been given, but the most satisfactory one rests on rather general considerations put forward by Ives and Marsden[1] with special reference to protolytic reactions in aqueous solution. Their arguments will be illustrated in terms of a hypothetical simple equilibrium $A \rightleftharpoons B$, in which only A is appreciably hydrated, so that the reaction may more properly be written

$$A(H_2O)_m \rightleftharpoons B + mH_2O. \tag{27}$$

Ives and Marsden now suppose that the observed thermodynamic functions for this equilibrium can be written as the sum of two contributions. The first relates to the species A and B themselves, and does not directly involve the water molecules; it may be termed the *reaction* or *internal* contribution. The second derives solely from changes in the water molecules as they pass from the solvation shell of A into the unmodified solvent, and is termed the *hydration* or *external* contribution.[13] We can therefore write

$$\Delta G = \Delta G_{int} + \Delta G_{ext}, \qquad \Delta H = \Delta H_{int} + \Delta H_{ext},$$
$$\Delta S = \Delta S_{int} + \Delta S_{ext} \tag{28}$$

where $\Delta G_{int}$, etc., refer to the reaction $A \rightleftharpoons B$, and $\Delta G_{ext}$, etc., to the change $mH_2O$ (in solvation shell) $\rightarrow mH_2O$ (in bulk water). Now, since the hydration shell is in equilibrium with bulk water (effectively the pure solvent in dilute solutions), we can write $\Delta G_{ext} = 0$, leading to the rather remarkable conclusion that $\Delta G = \Delta G_{int}$, i.e., the measured free energy change is the same as it would be in the absence of changes in hydration.* The same result obviously follows for more complex reactions in which several species are hydrated, such as protolytic reactions. On the other hand, there is no reason at all to suppose that $\Delta H_{ext}$ or $\Delta S_{ext}$ will be equal to zero; they are likely to be of considerable magnitude, but will be related by $\Delta H_{ext} = T\Delta S_{ext}$ because of the condition $\Delta G_{ext} = 0$, corresponding to $\alpha = 1$ in Equation (26). The observed values of $\Delta H$ and $\Delta S$ will therefore contain large contributions from solvation effects, and may in fact be a valuable source of information about such effects.

[13] The terms *internal* and *external* in this context were introduced by L. G. Hepler and F. O'Hara, *J. Phys. Chem.*, **65**, 811, 2107 (1965); *J. Am. Chem. Soc.*, **85**, 3089 (1963). These authors did not, however, carry the argument as far as Ives and Marsden.

* Ives and Marsden stress that $\Delta G_{int}$ does not refer to the hypothetical gas reaction $A \rightleftharpoons B$, but to the unhydrated species 'as they exist in the dissolved state'. It is difficult to give an exact significance to this phrase, but it presumably means that the internal energy states assumed for A and B must include any effects of the solvent upon them.

We have spent some time in justifying the use of $\Delta G°$ (e.g., dissociation constants at a single temperature) as a basis for relating acid-base equilibria to molecular structures, as will be attempted in the next chapter. The vulnerable point in the arguments of the last paragraph is the assumption of additivity expressed in Equations (28), but there seems little doubt that its consequences represent a good approximation to reality in many cases. For example, in the series $CMe_3CO_2H$, $CH_3CO_2H$, $HCO_2H$, $CH_2ICO_2H$, $CH_2BrCO_2H$, $CH_2ClCO_2H$, and $CH_2FCO_2H$ (cf. Table 5) the value of p$K$ decreases steadily, as would be predicted by electronic theory, while $\Delta H$ and $\Delta S°$ behave in an erratic manner and probably reflect solvation effects. We have seen in the last chapter that little is gained by attempting to eliminate the effect of the solvent either by using a non-polar medium or by extrapolating to infinite dielectric constant. It is possible that polar aprotic solvents such as dimethyl sulphoxide and dimethylformamide would be a better choice than water for elucidating structural problems, since they exhibit less specific solvation effects and offer fewer solubility problems. However, there are good practical reasons for adhering to dissociation constants in water, and these will be used in the following chapter.

Nevertheless, the compensation between the enthalpy and entropy of solvation will not always be complete, and caution is necessary in interpreting small differences in p$K$ at a single temperature. For example, the conventional dissociation equilibria of the carboxylic acids usually have small values of $\Delta H$, and $\Delta C_p$ is in the neighbourhood of $-40$ cal deg$^{-1}$ (cf. Table 5). This means that the value of $\Delta H$ frequently changes sign near room temperature, corresponding to a maximum in the dissociation constant. The experimental data can be represented closely by a parabolic expression of the form

$$\lg K = \lg K_m - p(T - T_m)^2 \qquad (29)$$

where $K_m$ is the maximum value of $K$ at the temperature $T_m$, $p$ is a constant which does not vary much from one acid to another, while $T_m$ varies considerably. Equation (29) is not exact and has no simple theoretical significance, but its form shows that the relative strengths of two acids will in general change with temperature, their order often being inverted either within the range studied experimentally, or not far outside it. This is illustrated for the fatty acid series by Table 6, taken from Everett, Landsman, and Pinsent.[14] There is obviously no particular

[14] D. H. Everett, D. A. Landsman, and B. R. W. Pinsent, *Proc. Roy. Soc.*, A, **215**, 403 (1952).

virtue in the standard temperature 25°C, and interpretations of the p$K$ variations at this temperature in terms of internal effects in the carboxylic acid or its ion might have to be reversed if a different temperature had been chosen. It is significant that the series considered here involves the substitution of alkyl groups. The inductive or hyperconjugative effect of these groups has been the subject of much controversy, but there is no doubt that they will have a considerable effect on the solvation of ions or molecules, and the compensation between the enthalpies and entropies of hydration is clearly not exact at all temperatures.

Table 6. INVERSIONS OF RELATIVE STRENGTHS IN THE FATTY ACID SERIES

$K_1 > K_2$ above inversion temperature

| (1) | (2) | Inversion temperature (°C) |
|-----|-----|----------------------------|
| Acetic | n-Hexoic | − 34 |
| Acetic | Isohexoic | − 25 |
| Acetic | Isovaleric | 16 |
| Acetic | Diethylacetic | 29 |
| Isovaleric | Diethylacetic | 51 |
| Propionic | n-Valeric | 58 |
| Propionic | Isobutyric | 46 |
| Propionic | n-Butyric | 90 |

The same problem arises to an even greater extent when alkyl substitution occurs very close to the acid-base centre, as in the alkylamines, since the separation into internal and external effects will be a poor approximation in this case. Thus in aqueous solution the sequence of acid strength is $NH_4^+ > Me_3NH^+ > MeNH_3^+ > Me_2NH_2^+$. The unexpected position of trimethylamine cannot be explained in terms of an internal effect of the methyl groups, but could arise if the hydration of the ions or molecules introduces a complicating factor. This explanation is supported by the observed entropy changes (Table 5) and also by measurements in aprotic solvents; we shall return to the problem in Chapter 10 in connection with basic catalysis by amines.

Since interaction with the solvent causes such complications, especially for alkyl substitution near the acid-base site, it is natural to ask whether any information can be obtained about the relative strengths of acids and bases in the gas phase. Measurements of thermal equilibria are not possible because of the very low equilibrium concentrations of ionic species, but many ions of interest can be generated by electron bombardment or electron transfer. It is then possible to use mass-spectrometric techniques to obtain semi-quantitative information about the energetics of proton-transfer reactions; for example, the fact that the reaction $MeO^- + EtOH \rightarrow MeOH + EtO^-$ can be observed in the forward but not in the reverse direction shows that it is exothermic, and hence that ethanol is a stronger acid that methanol in the gas phase. Particularly valuable is the *ion cyclotron double resonance* technique, in which a particular ionic species X can be selectively accelerated by irradiation with a radio-frequency corresponding to its cyclotron frequency. If it is found that the intensity of the peak due to a second ion Y is changed by this irradiation, this implies the occurrence of a process in which X reacts to produce Y.

Although the information obtained is essentially qualitative, many interesting results have been obtained during the last few years, of which only a few will be mentioned here. Several authors[15] agree that the proton affinities (basic strengths) of amines show the sequences $Me_3N > Me_2NH > MeNH_2 > NH_3; Et_3N > Et_2NH > EtNH_2 > NH_3;$ $Me_3CNH_2 > Me_3CCH_2NH_2 > Me_2CHNH_2 > MeCH_2CH_2NH_2 >$ $MeCH_2NH_2 > MeNH_2 > NH_3$. It is interesting that the *acid* strengths of the amines show similar sequences,[16] such as $Et_2NH > Me_2NH >$ $MeCH_2CH_2NH_2 > MeCH_2NH_2 > MeNH_2 > NH_3$, behaviour which is paralleled in the relative acid strengths of the alcohols,[17] $Me_3COH >$ $Me_2CHOH > MeCH_2OH > MeOH > H_2O$. Both acidic and basic strengths are thus increased by an increase in the size and number of alkyl substituents, which is most readily explained by supposing that both anions and cations are stabilized by the polarization of the alkyl groups by the charge on the oxygen or nitrogen atom, rather than in terms of the more popular inductive or hyperconjugation effects.

Results of this kind emphasize yet again that the behaviour of acids and bases in solution depends to a large extent on specific solute–solvent interactions. A striking example of this is the finding[17] that in the gas phase toluene is a stronger acid than water or methanol, though weaker than ethanol. Presumably its failure to behave as an acid in solution arises from the absence of solvation stabilization for an anion in which the charge is distributed over four carbon atoms.

[15] M. S. B. Munson, *J. Am. Chem. Soc.*, **87**, 2332 (1965); J. I. Braumann, J. M. Riveros, and L. K. Blair, *J. Am. Chem. Soc.*, **93**, 3914 (1971).
[16] J. I. Braumann and L. K. Blair, *J. Am. Chem. Soc.*, **93**, 3911 (1971).
[17] J. I. Braumann and L. K. Blair, *J. Am. Chem. Soc.*, **92**, 5986 (1970).

# 6 Acid-Base Strength and Molecular Structure

Many detailed discussions have been published on the effects of substituents on the strengths of organic acids and bases.[1] We have seen that caution is necessary in interpreting small differences in dissociation constants in terms of molecular models, and this chapter will deal only with a few of the more striking effects, with special reference to some which are of interest in reaction kinetics. We shall consider first the strengths of hydrides and oxyacids of different elements.

Table 7. APPROXIMATE p$K$ VALUES FOR SIMPLE HYDRIDES

| $CH_4$ | 46 | $NH_3$ | 35 | $OH_2$ | 16 | FH | 3 |
|--------|----|--------|----|--------|----|-----|-----|
| | | $PH_3$ | 27 | $SH_2$ | 7 | ClH | $-7$ |
| | | | | $SeH_2$ | 4 | BrH | $-9$ |
| | | | | $TeH_2$ | 3 | IH | $-10$ |

Table 7 gives approximate p$K$ values in water for a number of simple hydrides, some of which need special comment. $pK(CH_4) = 46$ derives from an argument due to Schwarzenbach,[2] who assumes that

$$pK(CH_4) - pK(\overset{+}{N}H_4) = pK(NH_3) - pK(\overset{+}{O}H_3),$$

the values for the last three being known experimentally. All that can be said with certainty is that methane is a much weaker acid than any of the others in Table 7. $pK(NH_3) = 35$ depends upon the ionic product of liquid ammonia, for which two concordant estimates give $[NH_4^+][NH_2^-] = 10^{-33}$ at 240 K. The first depends on a combination of entropy measure-

---

[1] For recent articles see, C. K. Ingold, *Structure and Mechanism in Organic Chemistry*, Bell, London, 1969, Ch. xiv; H. C. Brown, D. H. McDaniel, and O. Häfliger, article in *The Determination of Organic Structures by Physical Methods* (ed. E. A. Braude and F. C. Nachod), Academic Press, New York, 1955; G. W. Wheland, *The Theory of Resonance in Organic Chemistry*, Wiley, New York, 1955, Ch. vii.
[2] G. Schwarzenbach, *Z. Phys. Chem.*, **176A**, 133 (1936). He gives $pK(CH_4) = 34$, but employed an incorrect value for $pK(NH_3)$.

ments[3] with calorimetric values[4] for the heat of the reaction $NH_4^+ + NH_2^-$ $\rightarrow 2NH_3$. Since the arguments involved are rather complicated, it is satisfactory that recent measurements[5] of the e.m.f. of the cell

$$H_2Pt|NH_4Cl \text{ in } NH_3 \| KNH_2 \text{ in } NH_3|PtH_2$$

from 196 K to 237 K give exactly the same values both for the ionic product at 240 K and for the heat of ionization. Extrapolation to 298 K gives $10^{-28}$ for the ionic product, and hence $10^{-31}$ for the constant $[NH_4^+][NH_2^-]/[NH_3]^2$. This must be increased to allow for the change from liquid ammonia ($\varepsilon = 22$) to water ($\varepsilon = 78$), since we are basing our comparisons on the latter solvent, and our previous discussion (Chapter 4) on the effect of changing the solvent suggests a factor of about $10^5$, giving

$$[NH_4^+][NH_2^-]/[NH_3]^2 = 10^{-26}$$

for aqueous solutions at 25°C. In these solutions we have

$$[NH_3][H_3O^+]/[NH_4^+] = 10^{-9},$$

and hence finally

$$\frac{[NH_2^-][H_3O^+]}{[NH_3]} = \frac{[NH_4^+][NH_2^-]}{[NH_3]^2} \cdot \frac{[NH_3][H_3O^+]}{[NH_4^+]} = 10^{-35}.$$

The corresponding value $pK(PH_3) = 27$ was derived from kinetic measurements of deuterium exchange.[6] These gave values for the velocity constant of the reaction $PH_3 + OH^- \rightarrow PH_2^- + H_2O$, and the equilibrium constant was estimated by assuming that the reverse reaction took place on every collision.

In the series of hydrogen halides $pK(HF) = 3$ can be measured directly,[7] though some correction is necessary for the formation of bifluoride ion. The acid strength of HCl has been estimated in a number of ways, all of which involve the measured vapour pressure of HCl over concentrated aqueous solutions. This can be used to calculate the

[3] L. V. Coulter, J. R. Sinclair, A. G. Cole, and G. C. Roper, *J. Am. Chem. Soc.*, **81**, 2986 (1959).
[4] H. D. Mulder and F. C. Schmidt, *J. Am. Chem. Soc.*, **73**, 5575 (1951); W. L. Jolly, *Chem. Rev.*, **50**, 351 (1951); W. M. Latimer and W. L. Jolly, *J. Am. Chem. Soc.*, **75**, 4147 (1953).
[5] M. Renaud, *Canad. J. Chem.*, **47**, 4702 (1969). These measurements were carried out with very dilute solutions, $10^{-5}$–$10^{-4}$M, so that corrections for junction potentials, activity coefficients, and incomplete dissociation are minimized. They are to be preferred to the earlier results of V. A. Pleskov and A. M. Monoszon, *Acta Physicochim. U.R.S.S.*, **1**, 725 (1935).
[6] R. E. Weston and J. Bigeleisen, *J. Am. Chem. Soc.*, **76**, 3074 (1955).
[7] P. R. Patel, E. C. Moreno, and J. M. Patel, *J. Res. Nat. Bur. Stand.*, **75**, 205 (1971).

concentration of unddissociated HCl in solution provided that a value is assumed for the Henry's law constant of this species {i.e., its distribution coefficient between aqueous solutions and the vapour phase), and the problem is mainly one of estimating this constant. Wynne-Jones[8] assumed that Raoult's law applied to the system $H_2O + HCl$ (undissociated), and by using the vapour pressure of liquid HCl at room temperature arrived at $pK(HCl) = -7$. This value represents $K_c$ rather than the thermodynamic $K$, and Robinson[9] used the same data in conjunction with the measured activity coefficients to extrapolate to infinite dilution, obtaining $pK(HCl) = -6$ at 25°C; he also calculated $K$ over the temperature range 0–50°C and concluded that the reaction $HCl + H_2O + Cl^-$ was exothermic, with $\Delta H = -18 \, kcal \, mol^{-1}$. The application of Raoult's law is open to criticism, and a different assumption about the Henry's law constant was made by Ebert,[10] who regarded HCl as the first member of the series HCl, $CH_3Cl$, $C_2H_5Cl$, etc., and extrapolated the measured solubilities of the alkyl halides, thus obtaining $pK(HCl) = -7$. An alternative assumption is to suppose that the HCl molecule will behave in the same way as HCN, which is similar in size and polarity but is almost undissociated in aqueous solution; this leads also to $pK(HCl) = -7$. Thus, although each of the methods employed is open to some criticism, their agreement makes it unlikely that the value $K = 10^{+7}$ is in error by more than a power of ten.

There is less abundant information about HBr and HI, but if we apply Robinson's treatment to the vapour pressure data[11] we find $pK(HBr) = -8$, $pK(HI) = -9$, compared with $pK(HCl) = -6$, suggesting that the strengths of the three acids are approximately in the ratio $1:10^2:10^3$. Further, investigations in non-aqueous solvents often show that HBr is a considerably stronger acid than HCl.[12] Thus conductivity measurements in anhydrous acetic acid indicate a ratio of about 20 between the dissociation constants of these two acids, while in acetonitrile we have $pK(HBr) = 5.5$ and $pK(HCl) = 8.9$. It is also frequently found that HBr is a much more effective acid catalyst than HCl under conditions where both acids are undissociated.[13] It is of interest that studies

[8] W. F. K. Wynne-Jones, *J. Chem. Soc.*, 1064 (1930).
[9] R. A. Robinson, *Trans. Faraday Soc.*, **32**, 743 (1936); R. A. Robinson and R. G. Bates, *Analyt. Chem.*, **43**, 969 (1971).
[10] L. Ebert, *Naturwiss.*, **13**, 393 (1925).
[11] S. J. Bates and H. D. Kirschman, *J. Am. Chem. Soc.*, **41**, 1991 (1919).
[12] See Chapter 4 for references.
[13] A. Hantzsch, *Z. Phys. Chem.*, **134**, 406 (1928); R. P. Bell and R. le G. Burnett, *Trans. Faraday Soc.*, **35**, 324 (1939); P. B. D. de la Mare and P. W. Robertson, *J. Chem. Soc.*, 888 (1948).

of ion cyclotron resonance in the gas phase[14] show that the reaction $Cl^- + HBr \rightarrow HCl + Br^-$ can take place in the forward but not in the reverse direction, showing again that HBr is the stronger acid. The same technique gives qualitative confirmation of the relative strengths of many pairs of hydrides in Table 7, and enables us to add $SiH_4$ and $AsH_3$ in the expected sequence $AsH_3 > PH_3 > SiH_4$. There are a few unexpected results, notably that both $PH_3$ and $SiH_4$ are stronger acids than water in the gas phase, though the former of these is confirmed by the calculations given a little later in this chapter.

Table 7 shows two obvious trends, an increase of strength in passing from left to right along a period, and in going down a group. The former is what would be expected qualitatively in terms of the electronegativities of the elements, but the latter is less expected. For example, in the series HF, HCl, HBr, HI, the dipole moment and the 'ionic character' of the bond decreases. This might have been expected to lead to a decrease in acidity rather than the observed increase. We shall see that this is due to the decrease of bond energy in this series.

Consider first the reaction $HX \rightleftharpoons H^+ + X^-$ in the gas phase. This can be split up into a number of steps as follows:

$$HX \rightarrow H + X, \qquad \Delta H = D_e$$
$$H \rightarrow H^+ + e^-, \qquad \Delta H = I$$
$$X + e^- \rightarrow X^-, \qquad \Delta H = -E$$
$$HX \rightarrow H^+ + X^-, \qquad \Delta H_g = D_e + I - E.$$

$I$, the ionization energy of the hydrogen atom, has the same value $(313.3 \text{ kcal mol}^{-1})$ for all hydrides, but both $D_e$, the bond dissociation energy, and $E$, the electron affinity of the atom or radical X, will vary from one hydride to another. Values of the latter are now available for many of the hydrides in Table 7, so that it is possible to calculate $\Delta H_g$, the enthalpy of ionic dissociation in the gas phase. The results are given in Table 8.[15]

Table 8 agrees with experimental findings in the gas phase, and reproduces the main trends shown by Table 7 for dissociation in solution. As might be expected, the comparison with solution reactions reveals some anomalies. For example, $PH_3$ emerges as a stronger acid than HF; presumably the enhanced strength of HF in solution is due to the greater

[14] J. I. Braumann, J. R. Eyler, L. K. Blair, M. J. White, M. B. Comisarow, and K. C. Smyth, *J. Am. Chem. Soc.*, **93**, 6360 (1971).
[15] The numerical values are taken from a critical survey by Braumann *et al.*[14]

Table 8. GAS-PHASE IONIC DISSOCIATION OF SIMPLE HYDRIDES

(Energies in kcal mol$^{-1}$)

|                  | $D_e$ | $E$ | $\Delta H_g$ |        | $D_e$ | $E$ | $\Delta H_g$ |
|------------------|-------|-----|--------------|--------|-------|-----|--------------|
| $NH_3$           | 110   | 17  | 407          | $H_2S$ | 90    | 53  | 250          |
| $H_2O$           | 119   | 42  | 390          | HCl    | 103   | 83  | 333          |
| HF               | 136   | 80  | 370          | HBr    | 88    | 78  | 324          |
| $PH_3$           | 80    | 29  | 364          | HI     | 71    | 71  | 314          |

stabilization of fluoride ion (compared with $PH_2^-$) by solvation. Table 8 does show clearly that the increase of acid strengths in the series $NH_3$, $H_2O$, HF or $PH_3$, $H_2S$, HCl is mainly due to increase in electron affinity (partly compensated by an increase in bond dissociation energy); on the other hand, the increase in the series HF, HCl, HBr, HI depends essentially on the decrease in bond strength, the electron affinities varying only slightly.

In the case of the hydrogen halides the argument can be extended to give information about behaviour in aqueous solution,[16] as shown in Table 9. In the first place the standard entropies of reaction in the gas phase can be calculated from statistical mechanics, leading to $\Delta G_g^\circ$.

In proceeding from dissociation in the gas phase to dissociation in aqueous solution, we need to know the energies and entropies of hydration of the species $H^+$, $X^-$, and HX. These are accessible from experiment for $H^+ + X^-$ (though their division into separate contributions for $H^+$ and $X^-$ is an arbitrary one), but for the undissociated molecule HX values must be estimated from data for molecules of similar size and polarity.* The hydration terms are given in the second part of Table 9, and finally the calculated thermodynamic functions and pK values for dissociation in aqueous solution.

The agreement between the observed and calculated pK values in Table 9 is no doubt partly fortuitous, but it is worth emphasizing that the calculated values for HCl, HBr, and HI do not make use of the vapour pressure measurements which were used to estimate the 'observed' acid strengths; hence the present calculations serve to confirm these estimates, which involved some rather doubtful assumptions. It is clear from the table that the dissociation constants in water are governed by the interplay

[16] J. C. McCoubrey, *Trans. Faraday Soc.*, **51**, 743 (1955). In Table 9 some of McCoubrey's figures have been replaced by more recent ones.

* In computing the entropies of hydration it is important to take into account the difference in the standard states for the gas (1 atmosphere) and in solution (1 mol dm$^{-3}$).

Table 9. IONIC DISSOCIATION OF THE HYDROGEN HALIDES AT 25°C

All values of $\Delta H$, $T\Delta S°$, and $\Delta G°$ are given in kcal mol$^{-1}$, rounded off to the nearest unit, and relate to the process $HX \rightarrow H^+ + X^-$ unless otherwise stated.

|  | HF | HCl | HBr | HI |
|---|---|---|---|---|
| *Gas phase dissociation* |  |  |  |  |
| $\Delta H_g$ | 370 | 333 | 324 | 314 |
| $T\Delta S°_g$ | 7 | 7 | 7 | 7 |
| $\Delta G°_g$ | 363 | 326 | 317 | 307 |
|  |  |  |  |  |
| *Hydration terms* |  |  |  |  |
| $\Delta H$ (hydration of HX) | $-12$ | $-4$ | $-5$ | $-6$ |
| $\Delta H$ (hydration of $H^+ + X^-$) | $-382$ | $-349$ | $-341$ | $-330$ |
| $T\Delta S°$ (hydration of HX) | $-7$ | $-5$ | $-5$ | $-6$ |
| $T\Delta S°$ (hydration of $H^+ + X^-$) | $-20$ | $-16$ | $-15$ | $-14$ |
|  |  |  |  |  |
| *Dissociation in aqueous solution* |  |  |  |  |
| $\Delta H_{aq}$ | $-0$ | $-12$ | $-12$ | $-11$ |
| $T\Delta S°_{aq}$ | $-6$ | $-4$ | $-3$ | $-1$ |
| $\Delta G°_{aq}$ | 6 | $-8$ | $-9$ | $-10$ |
| p$K$ (calc.) | 4 | $-5$ | $-6$ | $-7$ |
| p$K$ (obs.) | 3 | $-7$ | $-9$ | $-10$ |

of a number of different factors, no one of which is dominant in determining the acid strengths as usually defined. Comparing $\Delta G°_{aq}$ with $\Delta G°_g$, we see that the considerable difference between HF and HCl is preserved, though it has been reduced from 36 to 13 kcal, largely because the hydration enthalpies of $F^-$ and $Cl^-$ operate so as to produce a difference in the opposite direction. In fact, if the ionic enthalpies were the only quantities involved, there would be an even smaller difference in the strengths of HF and HCl, the final value depending significantly both on the hydration entropies of the ions and on the hydration enthalpies of the undissociated molecules. Similar remarks apply, *a fortiori*, to the smaller differences between HCl, HBr, and HI. It is still roughly true to say that the anomalous weakness of HF in this series is associated with its high bond strength, but it appears little more than a coincidence that the other three acids preserve the same order in solution as in the gas phase. This is one of many instances in which dissociation constants in water appear to reflect molecular regularities more faithfully than might be anticipated.

There is of course a large *effect of charge* on acid strengths, as illustrated by the figures in Table 10. In each case an increase of positive charge or a decrease in negative charge is accompanied by a large increase in acid strength. This is what would be expected intuitively, and in some cases it

Table 10. EFFECT OF CHARGE UPON ACID STRENGTH

| | | | | | |
|---|---|---|---|---|---|
| $H_3O^+$ | $-1.7$ | $H_2O$ | 15.7 | $OH^-$ | 25 |
| $NH_4^+$ | 9.5 | $NH_3$ | 35 | | |
| | | $H_2S$ | 7.1 | $HS^-$ | 14.7 |

can be paralleled by information from gas-phase reactions. Thus we have
seen in Chapter 2 that $\Delta H$ for the reaction $H_3O^+(g) \rightarrow H_2O(g) + H^+(g)$
is 170 kcal mol$^{-1}$, which is much smaller than the value of 390 kcal mol$^{-1}$
given in Table 8 for $H_2O(g) \rightarrow OH^-(g) + H^+(g)$. The difference in free
energy changes in solution is very much smaller ($\Delta\Delta G^\circ = 24$ kcal mol$^{-1}$).
This is because, apart from $H^+$, which is common to both, the first reaction
involves the disappearance of an ion ($H_3O^+$), and the second reaction the
production of one ($OH^-$). Transfer from the gas phase to aqueous solution
will therefore favour the second reaction by the sum of the hydration
energies of $H_3O^+$ and $OH^-$, which is probably 180–200 kcal mol$^{-1}$, consistent with the observed quantities. Similar considerations could be
applied to other pairs, though the necessary quantitative information is
not often available.

Significant regularities exist in the strengths of the *inorganic oxyacids*,
as shown in Table 11. This has been pointed out by a number of authors,[17]
though their interpretations differ. As shown in the table, the acids fall
into four groups, according to the value of $n$ in the general formula
$XO_n(OH)_m$. The strength of the acid increases with increasing $n$ and does
not depend significantly upon the value of $m$.

Table 11. p$K$ VALUES FOR INORGANIC OXYACIDS

| $X(OH)_m$ | | $XO(OH)_m$ | | $XO_2(OH)_m$ | | $XO_3(OH)_m$ | |
|---|---|---|---|---|---|---|---|
| (very weak) | | (weak) | | (strong) | | (very strong) | |
| $Cl(OH)$ | 7.2 | $NO(OH)$ | 3.3 | $NO_2(OH)$ | $-1.4$ | $ClO_3(OH)$ | $(-10)$ |
| $Br(OH)$ | 8.7 | $ClO(OH)$ | 2.0 | $ClO_2(OH)$ | $-1$ | $MnO_3(OH)$ | — |
| $I(OH)$ | 10.0 | $CO(OH)_2$ | 3.9 | $IO_2(OH)$ | 0.8 | | |
| $B(OH)_3$ | 9.2 | $SO(OH)_2$ | 1.9 | $SO_2(OH)_2$ | $<0$ | | |
| $As(OH)_3$ | 9.2 | $SeO(OH)_2$ | 2.6 | $SeO_2(OH)_2$ | $<0$ | | |
| $Sb(OH)_3$ | 11.0 | $TeO(OH)_2$ | 2.7 | | | | |
| $Si(OH)_4$ | 10.0 | $PO(OH)_3$ | 2.1 | | | | |
| $Ge(OH)_4$ | 8.6 | $AsO(OH)_3$ | 2.3 | | | | |
| $Te(OH)_6$ | 8.8 | $IO(OH)_5$ | 1.6 | | | | |
| | | $HPO(OH)_2$ | 1.8 | | | | |
| | | $H_2PO(OH)$ | 2.0 | | | | |

[17] A. Kossiakoff and D. Harker, *J. Am. Chem. Soc.*, **60**, 2047 (1938); L. Pauling, *General Chemistry*, Freeman, San Francisco, 1947, p. 394; J. E. Ricci, *J. Am. Chem. Soc.*, **70**, 109 (1948); R. J. Gillespie, *J. Chem. Soc.*, 2537 (1950).

It is convenient first to consider some of the individual acids. A number of these fall into the appropriate group only if given the correct structural formula. Thus phosphorous and hypophosphorous acids cannot by $P(OH)_3$ and $HP(OH)_2$ respectively, and there is of course independent evidence that the formulae given in the table are the correct ones. Similarly telluric acid must be $Te(OH)_6$ [not $TeO_2(OH)_2$ or $TeO(OH)_4$], and periodic acid $IO(OH)_5$ [not $IO_3(OH)$ or $IO_2(OH)_3$]. It is remarkable that boric acid fits well into the class of very weak acids, since (as already mentioned in Chapter 2) there is good evidence that the borate ion in solution is $B(OH)_4^-$ rather than $^-OB(OH)_2$, so that the dissociation of boric acid involves the addition of $OH^-$ rather than the loss of $H^+$.[18]

The value $pK = 3.9$ given for carbonic acid is of course the true value for $H_2CO_3$, taking into account the fact that only 0.4% of the dissolved carbon dioxide is in this form. The conventional value for carbonic acid, $pK = 6.5$ (not distinguishing between $CO_2$ and $H_2CO_3$), would not fit into the table. It is interesting to note that the conventional value for sulphurous acid does fall into line with the other acids of formula $XO(OH)_m$, thus suggesting that sulphur dioxide in aqueous solution is much more strongly hydrated than is carbon dioxide. There is little direct evidence on this point, but it is in accord with the solubility of sulphur dioxide in water, which is much higher than that of most molecules of similar size and polarity.

The explanations given for the regularities in Table 11 all depend upon the number of equivalent oxygen atoms in the acid anion, which increases from one for the very weak acids $X(OH)_m$ to four for the very strong acids $XO_3(OH)_m$. The negative charge in the anion will be distributed equally between these oxygen atoms, but various views have been expressed as to why this stabilizes the ion and thus strengthens the acid. The first type of explanation is *electrostatic*, depending on the fact that the electrostatic energy of a charged system decreases when it is spread over a larger volume.* The second approach relies on *quantum theory*, and speaks of

---

[18] J. O. Edwards, G. C. Morrison, V. F. Ross, and J. W. Schulz, *J. Am. Chem. Soc.*, **77**, 266 (1955); R. P. Bell, J. O. Edwards, and R. B. Jones, in *The Chemistry of Boron and its Compounds* (ed. E. L. Muetterties), Wiley, New York, 1966, pp. 209–221.

*For example, the electrostatic free energy of a charge $e$ on a sphere of radius $r$ in a medium of dielectric constant $\varepsilon$ is $e^2/2\varepsilon r$. If the same charge is distributed between $(n+1)$ spheres of the same radius, as might be imagined in the anion derived from $XO_n(OH)_m$, the energy falls to $e^2/2(n+1)\varepsilon r$. This expression exaggerates the effect, since it neglects repulsion between the charges on the separate oxygen atoms. An alternative model, particularly appropriate for tetrahedral $XO_4^-$ ions is to suppose that the charge is located on a single sphere of volume $(n+1)$ times that of a single oxygen atom; this corresponds to an energy of $e^2/2(n+1)^{\frac{1}{3}}\varepsilon r$.

stabilization by resonance or mesomerism between several bond structures in the ion, or (equivalently) of the lowering of the molecular orbital energy by electron delocalization over several bonds. Although the quantal explanation ought to be the more fundamental one, it is difficult to assess its relevance to the present problem, since its application depends upon the view taken of the nature of the X—O bonds both in the acid molecule and in the anion. Thus for chlorous acid the dissociation process could be written in the following two ways:

(a) $\qquad$ $O{=}Cl{-}OH \rightarrow {}^{\frac{1}{2}-}O{-}Cl{-}O^{\frac{1}{2}-} + H^+$

(b) $\qquad$ $\bar{O}{-}\overset{+}{Cl}{-}OH \rightarrow \quad \bar{O}{-}\overset{+}{Cl}{-}\bar{O} \quad + H^+.$

In (a) the double bond and single bond in the acid are converted into two mesomeric bonds in the anion, giving stabilization by resonance, and the chlorine atom has 10 valency electrons throughout. In (b) the chlorine and oxygen are united by semipolar bonds, giving octet structures throughout, and there is no change in binding on ionization. Similar formulations apply to the other types of oxyacid, frequently with several intermediate possibilities, e.g., for perchloric acid,

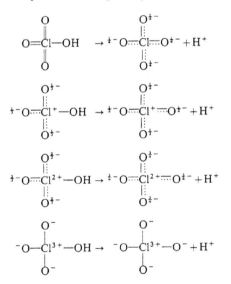

where the effect of resonance stabilization presumably decreases down the series, becoming zero in the last formulation. In principle any one of the structures for the acid might be transformed into any one of the ionic structures, so that on paper there are sixteen possibilities. The position could be resolved if there were firm experimental or theoretical evidence about the nature of the X—O bonds, but there is at present little agreement on this point, so that the electrostatic formulation might be preferred.*

Most of the work on structure and acidity has dealt with the effect of substituents upon the strength of organic acids and bases. We shall consider first a few of the simpler effects in aliphatic compounds and then the strengths of acids containing the group C—H, which are of importance in kinetic problems.

Much information about substituent effects relates to carboxylic acids, all of which are much stronger than alcohols. This recalls the behaviour of the inorganic oxyacids and can be related to the structure of the carboxylate anions,

An analogous situation exists in the amidines,

which are much stronger bases than the amines because the cation has the structure

and the same effect operates to an even greater extent in guanidine, $NH=C(NH_2)_2$, with the corresponding cation

$$C(=\overset{+}{N}H_2)_3.$$

---

*The concept of electron delocalization is of course necessary to account for the stability of species such as the allyl radical, which can be written as a resonance hybrid of $CH_2=CHCH_2$ and $CH_2CH=CH_2$. It also applies naturally to the dissociation of carboxylic acids mentioned in the next paragraph, since there would be no justification for writing this process $^-O\overset{+}{C}ROH \rightarrow {}^-O\overset{+}{C}RO^-$. A complete quantum treatment, allowing for interelectronic repulsion, would include what we have termed the electrostatic effect, so that the separation into quantal and electrostatic contributions is essentially an artefact depending upon our inability to solve the problem completely.

The simplest class of substituent effects is due to the presence in the molecule of a group (charged or dipolar) whose charge distribution is not appreciably changed during the interconversion of the acid-base pair. It is possible to picture the action of such a group kinetically, according to whether it helps or hinders the loss of a proton, but since we are considering equilibrium phenomena it is then also necessary to consider the effect of the group upon the reverse process. For quantitative purposes it is more satisfactory to examine the charge distribution in the initial and final states of the acid-base reaction, and thus to calculate the effect of the substituent upon the overall change in free energy.

The largest effects are encountered with groups bearing a net charge, and the best-known example deals with the *first and second dissociation constants of dicarboxylic acids*. In an acid $CO_2H(CH_2)_nCO_2H$ the free energy of the ion $CO_2^-(CH_2)_nCO_2^-$ will be increased by the mutual repulsion of the two negative charges, an effect which is absent both in the original molecule and in the singly charged ion $CO_2H(CH_2)_nCO_2^-$. This should cause the second dissociation constant to be smaller than the first one, and the effect should decrease with increasing distance between the two carboxyl groups, and hence with the number of $CH_2$

Table 12. THE SUCCESSIVE DISSOCIATIONS OF DICARBOXYLIC ACIDS

| Acid | $n(CH_2)$ | $pK_1$ | $pK_2$ | $\Delta pK$ | $r$ (calc.) pm | | | Model |
| | | | | | (a) | (b) | (c) | |
|---|---|---|---|---|---|---|---|---|
| Oxalic | 0 | 1.23 | 4.19 | 2.96 | 91 | 385 | 337 | 350– 444 |
| Malonic | 1 | 2.83 | 5.69 | 2.86 | 36 | 410 | 343 | 412– 487 |
| Succinic | 2 | 4.19 | 5.48 | 1.29 | 365 | 575 | 558 | 466– 666 |
| Adipic | 4 | 4.42 | 5.41 | 0.99 | 811 | 775 | 822 | 559– 902 |
| Azelaic | 7 | 4.55 | 5.41 | 0.86 | 119 | 985 | 1203 | 674–1242 |

groups. This is illustrated by the values in the first five columns of Table 12, which are taken from the compilation by Brown, McDaniel, and Häfliger.[1]* The first attempt to treat this problem quantitatively was due to Bjerrum,[19] who showed that on a simple electrostatic model the ratio of the successive dissociation constants should be given by

$$K_1/4K_2 = \exp(e^2/\varepsilon r k T) \qquad (30)$$

where $r$ is the distance between the two carboxyl groups. The factor 4 in

* Unless otherwise stated, all values of pK in the remainder of this chapter are taken from the same compilation.
[19] N. Bjerrum, *Z. Phys. Chem.*, **106**, 219 (1923).

Equation (30) represents the statistical effect, which occurs in various forms in both equilibrium and kinetic problems and is conveniently considered now. In the present instance it implies that $K_1 = 4K_2$ even when $r$ is so great that the electrostatic effect is negligible. This may be seen by considering the two processes

$$(1) \qquad \underset{\text{(i)}}{CO_2H(CH_2)_nCO_2H} \rightleftharpoons \underset{\text{(ii)}}{CO_2H(CH_2)_nCO_2^-} + H^+$$

$$(2) \qquad \underset{\text{(ii)}}{CO_2H(CH_2)_nCO_2^-} \rightleftharpoons \underset{\text{(iii)}}{CO_2^-(CH_2)_nCO_2^-} + H^+$$

where $n$ is so great that the tendency of the carboxyl group to lose a proton is the same in all the species concerned, while the same is true for the tendency of the group $-CO_2^-$ to accept a proton. Comparison of the species (i) and (ii) as acids reveals that (i) will dissociate twice as fast as (ii), because of its two equivalent $-CO_2H$ groups, and this would make $K_1/K_2 = 2$. Similarly, comparison of (ii) and (iii) as bases shows that (iii) will pick up protons twice as fast as (ii), because of its two equivalent $-CO_2^-$ groups, again giving $K_1/K_2 = 2$, and the total statistical effect will be $K_1/K_2 = 2 \times 2 = 4$. This naïve kinetic argument can be replaced by a more sophisticated one which invokes the symmetry numbers of the species (i), (ii), and (iii), which are respectively 2, 1, and 2, but the result is the same.

Returning to Bjerrum's equation, column (a) of Table 12 gives the values of $r$ obtained by substituting the experimental values of $K_1/K_2$ in (30), taking the dielectric constant $\varepsilon = 78$. They are of the right order of magnitude but are definitely too low for the earlier members of the series. This is best seen by comparison with the last column of the table, which contains the range of distances predicted by a molecular model, the variation corresponding to different conformations of the flexible chain. We have already seen that because of saturation effects $\varepsilon = 78$ is certainly too high a value to take for calculations on a molecular scale. An additional difficulty arises in dealing with interactions within the same molecule, since part of the space between the two charges is now occupied by the molecule itself, having an effective dielectric constant very much less than that of the solvent. Various methods have been proposed for improving Bjerrum's calculation so as to allow for these effects. Thus Kirkwood and Westheimer[20] use a model in which the

[20] J. G. Kirkwood and F. H. Westheimer, *J. Chem. Phys.*, **6**, 506 (1938); *J. Am. Chem. Soc.*, **61**, 555 (1939); F. H. Westheimer and J. G. Kirkwood, *J. Chem. Phys.*, **6**, 513 (1938).

charges are embedded in an ellipsoid of low dielectric constant, giving values of $r$ in column (b) of Table 12. Similarly, Gane and Ingold[21] attempt to allow individually for the polarizability of the different groups in the molecule and for the saturation of the solvent dielectric, leading to the values of $r$ in column (c) of the table. Both treatments (especially the former) lead to distances between the carboxyl groups which agree better with molecular models, but both methods of treatment involve parameters which are either adjustable or not known with certainty, so that exact quantitative agreement would be illusory.

The acid-base properties of the amino-acids offer many examples of the interactions between oppositely charged groups, since they frequently exist in the zwitterion form $\overset{+}{N}H_3\text{---}CO_2^-$. As with the dicarboxylic acids both of the groups concerned have acid-base properties, and a simpler situation exists when the positive charge resides on a group such as $-\overset{+}{N}Me_3$, which has no proton to lose. For example, $\overset{+}{N}Me_3CH_2CO_2H$ has $pK = 1.83$, as against $pK(CH_3CO_2H) = 4.75$, and this large increase of strength can be reasonably attributed to stabilization of the species $\overset{+}{N}Me_3 \cdot CH_2CO_2^-$ by the attraction between the opposite charges.

It would be interesting to consider a negatively charged substituent lacking the markedly basic properties of $-CO_2^-$, and it might appear that the sulphonate group $-SO_3^-$ would be suitable for this purpose, since the alkyl sulphonic acids are strong acids with $pK \sim 0$. The effect of this substituent, however, is at first sight surprising, since $SO_3^-CH_2CO_2H$ has $pK = 4.05$: i.e., it is about five times as strong as acetic acid, instead of being much weaker. An even larger effect is shown in the pair $pK(CH_3\overset{+}{N}H_3) = 10.64$, $pK(SO_3^-CH_2\overset{+}{N}H_3) = 5.75$, in which the substituent is closer to the acidic proton. Again the order of strength is the reverse of what might be expected.

This behaviour can be explained in terms of the electronic structure of the sulphonic acid group. As already discussed in connection with the strengths of the inorganic acids, the group $-SO_3^-$ can be written in a number of ways, such as

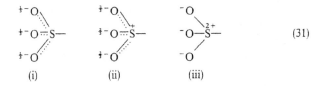

$$\text{(31)}$$

$$\text{(i)} \qquad\qquad \text{(ii)} \qquad\qquad \text{(iii)}$$

[21] R. Gane and C. K. Ingold, *J. Chem. Soc.*, 2153 (1931).

Substitution by structure (i) should certainly make an acid weaker, but this is not necessarily true for structures (ii) and (iii), since if the acidic proton is close to the $-SO_3^-$ group the effect of the positive charge on the sulphur atom may well outweigh that of the numerically greater negative charge on the oxygen atom.

The point may be illustrated by a simple calculation of the potential at a distance $r$ from the charge distributions shown below, corresponding to the charge distributions (ii) and (iii) just shown for the sulphonate group.

$$-2e \quad +e \qquad\qquad\qquad -3e \quad +2e$$

$$(32)$$

$$`a' \qquad `r' \qquad\qquad\qquad\qquad `a' \qquad `r'$$

$$\text{(ii)} \qquad\qquad\qquad\qquad\qquad \text{(iii)}$$

The two potentials are given by

$$V_{ii} = \frac{e}{\varepsilon}\left(\frac{1}{r} - \frac{2}{r+a}\right) = -\frac{e}{\varepsilon r}\phi_{ii}\left(\frac{r}{a}\right)$$

$$V_{iii} = \frac{e}{\varepsilon}\left(\frac{2}{r} - \frac{3}{r+a}\right) = -\frac{e}{\varepsilon r}\phi_{iii}\left(\frac{r}{a}\right) \qquad (33)$$

and the functions $\phi_{ii}(r/a)$ and $\phi_{iii}(r/a)$ are plotted in Figure 4, in which curve (i) represents a simple Coulomb potential, i.e.,

$$V_i = -\frac{e}{\varepsilon(r+a)} = -\frac{e}{\varepsilon r}\phi_i\left(\frac{r}{a}\right)$$

It will be seen that $\phi_{ii}$ and $\phi_{iii}$ are initially opposite in sign to $\phi_i$, and that their deviations from a simple Coulomb potential are considerable even at large values of $r/a$.

Table 13. p$K$ VALUES FOR ACIDS CONTAINING THE SULPHONATE GROUP

$pK(CH_3(CH_2)_nCO_2H) = 4.75–4.85$; $pK(CH_3(CH_2)_nNH_3^+) = 10.60–10.70$.

| | Value of $n$ | | | | | |
|---|---|---|---|---|---|---|
| | 1 | 2 | 3 | 4 | 5 | 10 |
| $pK(SO_3^-(CH_2)_nCO_2H)$ | 4.20 | 4.74 | 4.91 | 5.04 | — | 5.21 |
| $pK(SO_3^-(CH_2)_nNH_3^+)$ | 5.75 | 9.20 | 10.05 | 10.65 | 10.95 | 11.35 |

Figure 4 suggests that the group $-SO_3^-$ should have an acid strengthening effect when substituted close to the acidic site, as already mentioned,
but that this effect should decrease and eventually change sign as its

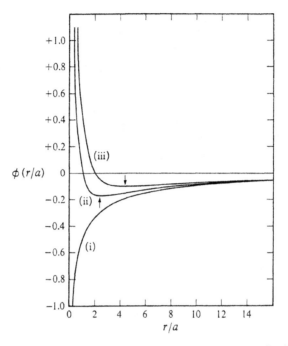

Fig. 4. Variation of potential with distance for different charge distributions.

distance is increased. This expectation is fulfilled by the observed $pK$
values for the series $SO_3^-(CH_2)_nCO_2H$ and $SO_3^-(CH_2)_nNH_3^+$, as shown
in Table 13.[22,23]

The continued rise in $pK$ between $n = 4$ and $n = 10$ is unexpected, and
must be due to other factors such as chain flexibility or solvation. These
can be eliminated by comparison with the series $CO_2^-(CH_2)_nCO_2H$ and
$CO_2^-(CH_2)_nNH_3^+$, in which the carboxylate group may be regarded as a
simple negative charge. Bell and Wright[22] concluded from this comparison, in conjunction with the model calculations of Kirkwood and
Westheimer,[20] that the $-SO_3^-$ group could be regarded as a negative
charge with a superimposed dipole of 3.7 D. This value is in good

[22] R. P. Bell and G. A. Wright, *Trans. Faraday Soc.*, **57**, 1377 (1961).
[23] P. Rumpf, *Bull. Soc. Chim. France*, 871 (1938).

agreement with the value 3.6D calculated on the basis of structure (31)(iii).[24]

There are several other pieces of evidence which confirm the presence of a large positive charge on the sulphur atom in the group $-SO_3^-$. Thus the alkaline hydrolysis of the ester $SO_3^- CH_2CO_2Et$ is faster than for ethyl acetate, while for acid hydrolysis the reverse is the case;[25] similarly, the base-catalysed iodination of $SO_3^- CH_2COCH_3$, the rate of which is determined by proton loss from the $CH_2$ group, is $10^3-10^4$ times faster than the corresponding reaction for acetone.[26] Correspondingly, the p$K$ value for dissociation of the $CH_2$ group in $SO_3^- CH_2COCH_3$ is 14, compared with 20 for acetone.[27] The group $-SO_3^-$ acts as a weakly *meta*-directing group in benzene substitution, in contrast to other negatively charged groups.[28] Finally, in proton magnetic resonance a $CH_2$ group adjacent to $-SO_3^-$ shows a chemical shift indicative of strong deshielding of the protons.[29] All these observations are consistent with structures such as (31)(ii) or (31)(iii), but not with (31)(i).

Curve (ii) in Figure 4 corresponds to a group such as $-SeO_2^-$ if we again adopt the structure

The figure suggests that this will act as a proton-repelling group only at very short distances, and this is borne out by the following values:[30]

$$pK(CH_3CO_2H) = 4.75, \qquad pK(SeO_2^- CH_2CO_2H) = 5.43,$$

$$pK(SeO_2^- CH_2CH_2CO_2H) = 5.99.$$

The effect of $-SeO_2^-$ in $SeO_2^- CH_2CO_2H$ is much smaller than that of $-CO_2^-$ in $CO_2^- CH_2CO_2H$ (cf. Table 12) and increases when we go to $SeO_2^-(CH_2)_2CO_2H$; this is qualitatively consistent with Figure 4.

[24] R. P. Bell and B. A. W. Coller, *Trans. Faraday Soc.*, **60**, 1087 (1964). This paper corrects an earlier suggestion (Bell and Wright[22]) that a moment of 3.7 D accords best with structure (31)(ii).
[25] R. P. Bell and D. A. Rawlinson, *J. Chem. Soc.*, 4387 (1958).
[26] R. P. Bell and G. A. Wright, *Trans. Faraday Soc.*, **57**, 1386 (1961).
[27] R. P. Bell, G. R. Hillier, J. W. Mansfield, and D. G. Street, *J. Chem. Soc.*, B, 827 (1967).
[28] C. K. Ingold, Ref. 1, p. 279.
[29] Unpublished observations by Dr. D. J. Barnes.
[30] H. J. Backer and W. van Dam, *Rec. Trav. Chim.*, **49**, 482 (1930).

Similar electrostatic explanations can be advanced to explain the effect of dipolar groups such as

$$\diagdown \underset{\diagup}{\overset{\delta+}{\underset{\phantom{x}}{C}}} \!\!-\!\! \overset{\delta-}{Cl}$$

in increasing the strength of acids. Quantitative calculations can again be made on the basis of models,[31] but the problem is more difficult than in the case of charged substituents, since the calculation demands a detailed knowledge of the charge distribution in the dipole, and the effect falls off more rapidly with distance, so that the results are sensitive to the assumptions made about molecular dimensions and local dielectric constants. It may be doubted, therefore, whether such calculations are of much value. One particular point may be mentioned, namely, the effect of several dipolar groups attached to the same carbon atom, for example, in the series $CH_3CO_2H$ (p$K$ = 4.75), $CH_2ClCO_2H$ (p$K$ = 2.76), $CHCl_2CO_2H$ (p$K$ = 1.30), and $CCl_3CO_2H$ (p$K$ = <0). It is sometimes argued that the effect of the group $-CCl_3$ should be about the same as that of $-CH_2Cl$, since the dipole moments of $CHCl_3$ and $CH_3Cl$ are approximately equal. However, this argument neglects two points. Firstly, the angle which the dipole makes with the C—C bond is different in the two cases, though it is difficult to judge the importance of this, since the $-CO_2H$ group is itself an angular group which can vary its orientation with respect to the C—C bond. Secondly (and more significantly), the resultant dipole is directly relevant only if we are considering the potential at a distance which is great compared with the charge separation in the dipole. The effect on a neighbouring group will be more closely related to the charge on the carbon atom, and there is no doubt that this will be greater in $-CCl_3$ than in $-CH_2Cl$. These dipolar effects can become very large, especially for fluorine substitution; for example, $CF_3CH(OH)_2$ has p$K$ = 10.21, and $(CF_3)_2CHOH$ p$K$ = 9.30, while the same alcohols without fluorine substitution have p$K$ values of 13.3 and 16.3 respectively.[32]

Many of the largest substituent effects upon acid-base properties arise when there is a large difference in electronic structure between the acid and its corresponding base, and these are commonly described as *resonance* or *mesomeric* effects. They have been treated in detail by many authors, especially for aromatic systems, and no general account will be

[31] See, e.g., Westheimer and Kirkwood[20]; C. Tanford, *J. Am. Chem. Soc.*, **79**, 5340 (1957); C. Tanford and J. G. Kirkwood, *J. Am. Chem. Soc.*, **79**, 5333 (1957).
[32] R. Stewart and M. M. Mocek, *Canad. J. Chem.*, **41**, 1161 (1963); W. J. Middleton and R. V. Lindsey, *J. Am. Chem. Soc.*, **86**, 4948 (1964).

attempted here. We shall, however, give some account of *carbon acids* (i.e., acids in which the proton is originally attached to carbon), since they are of particular interest in kinetic problems.

The paraffin hydrocarbons have no detectable acidic properties, and we have seen that $pK(CH_4)$ can be estimated to be about 46. The acidity is greatly increased by halogen substitution, and compounds $CHX_3$ (X = halogen) exchange hydrogen at a measurable rate in alkaline solution, as shown by experiments with deuterium or tritium,[33] but no estimates appear to have been made of their acid strength. In the absence of polar substituents, hydrocarbons can show measurable acidity when they contain certain aromatic systems, and a number of $pK$ values were obtained by McEwen,[34] using a stepwise procedure in which equilibria of the type $HX + Y^- \rightleftharpoons HY + X^-$ were investigated in ether solution. Later investigators have used spectrophotometric measurements in strongly basic or dipolar aprotic solvents, for example, cyclohexylamine[35] and dimethyl sulphoxide.[36]

There are difficulties in linking these measurements to $pK$ scales in aqueous solution, and the absolute values obtained are therefore subject to uncertainty, but there is no doubt that aromatic hydrocarbons containing extensive $\pi$-electron systems are much stronger acids than the paraffins. Typical values are $pK$(triphenylmethane) = 27–33, $pK$(fluorene) = 20–23, $pK$(indene) = 18–23. In these and similar cases we may consider that the anion is stabilized by a distribution of the negative charge over a number of atoms, which can be formally represented by writing a number of resonance structures. Thus for the anion of triphenylmethane we can write

and seven other similar structures, while for indene there are twelve structures such as

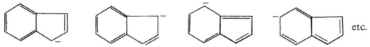

[33] J. Hine, R. C. Peck, and B. D. Oakes, *J. Am. Chem. Soc.*, **76**, 827 (1954); J. Hine and N. W. Burske, *J. Am. Chem. Soc.*, **78**, 3337 (1956); J. Hine, N. W. Burske, M. Hine, and P. B. Langford, *J. Am. Chem. Soc.*, **79**, 1406 (1957).
[34] W. K. McEwen, *J. Am. Chem. Soc.*, **58**, 1124 (1936).
[35] A. Streitwieser, J. I. Braumann, J. H. Hammons, and A. H. Pudjaatmaka, *J. Am. Chem. Soc.*, **87**, 384 (1965); A. Streitwieser, E. Ciuffarin, and J. H. Hammons, *J. Am. Chem. Soc.*, **89**, 63 (1967).
[36] E. C. Steiner and J. M. Gilbert, *J. Am. Chem. Soc.*, **87**, 382 (1965); C. D. Ritchie and R. E. Uschold, *J. Am. Chem. Soc.*, **89**, 1721, 2752, 2960 (1967).

It might seem strange that indene is so much stronger than triphenyl-methane, but this is because the unionized triphenylmethane molecule has eight 'normal' Kekulé structures, whereas indene has only the two,

Much higher acidities can be obtained with hydrocarbons designed so that there are very extensive possibilities of charge delocalization in the anion, as shown particularly by the work of Kuhn and his collaborators.[37] The paper quoted describes 15 hydrocarbons having pK values (referred to aqueous solution) between 15 and 8, while the hydrocarbon represented below has a pK less than 7, i.e. approaching that of a carboxylic acid; this is related to the fact that, in the anion formed by removing the proton marked with an arrow, the negative charge could be formally located on any one of 34 carbon atoms.

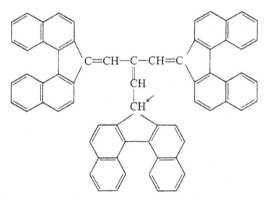

It has long been known that acetylenic hydrocarbons are much more acidic than the paraffins, since they form salts and will exchange hydrogen with deuterium and tritium in alkaline solution. A rough quantitative estimate gives $pK = 20-21$ for both acetylene and phenylacetylene.[38]

Much higher acidities occur in carbon acids where the negative charge can be transferred to an oxygen atom, which has a much higher electron affinity than carbon. This occurs in compounds containing the group

[37] R. Kuhn and D. Rewicki, *Annalen*, **706**, 250 (1967), and eight earlier papers.
[38] R. E. Dessy, Y. Okuzumi, and A. Chen, *J. Am. Chem. Soc.*, **84**, 2899 (1962).

which becomes

on losing a proton. Thus acetophenone, $C_6H_5COCH_3$, was found by McEwen[34] to have $pK = 19$, while it has been deduced from kinetic data[39] that $pK(\text{acetone}) = 20$. The $\beta$-diketones and $\beta$-keto-esters are even stronger acids, e.g.,

$$pK(CH_3COCH_2COCH_3) = 9, \quad pK(CH_3COCH_2CO_2Et) = 10,$$

since the charge of the anion can now be divided between two oxygen atoms. Thus the anion of acetylacetone can be written as

$$CH_3-\underset{\underset{O^{\frac{1}{2}-}}{\|}}{C}\cdots CH\cdots\underset{\underset{O^{\frac{1}{2}-}}{\|}}{C}-CH_3.$$

The above explanation of the acidity of ketones and similar substances implies that in the 'carbanions' formed the negative charge resides on oxygen atoms rather than on carbon. This has always appeared likely, in view of the magnitude of the effect and the high electronegativity of oxygen, but it is satisfactory that direct evidence is also forthcoming.[40] This arises from a study of chemical shifts in the proton magnetic resonance spectra of the anions of penta-1,4-diene and crotonaldehyde, which on our hypothesis would be formulated as

$$\overset{\frac{1}{2}-}{C}H_2\cdots CH\cdots CH\cdots CH\cdots\overset{\frac{1}{2}-}{C}H_2 \quad \text{and} \quad CH_2{=}CH-CH{=}CH-O^-$$
$$\quad\text{(b)}\qquad\qquad\text{(b)}\qquad\qquad\qquad\text{(c)}\qquad\qquad\qquad\text{(a)}$$

The observed shifts do in fact show that the densities of negative charge are in the order (a) > (b) > (c), in conformity with the above picture.

Groups other than carbonyl can act in the same way    notably the nitro group, where

$$\diagdown\!\!\!\!\!\diagup\!\!CH-NO_2$$

ionizes to give

[39] R. P. Bell, *Trans. Faraday Soc.*, **39**, 253 (1943).
[40] G. J. Heiszwolf and H. Kloosterziel, *Rec. Trav. Chim.*, **86**, 807 (1967).

Thus $pK(CH_3NO_2) = 10$. More than one group can be active in the same molecule, for example, $pK(CH_2NO_2COCH_3) = 5$, there now being three oxygen atoms to share the negative charge. The same groups can also enhance the acidity of

$$\diagdown_{\diagup}\!\!NH,$$

so that $pK(NH_2NO_2) = 6.5$ [cf. $pK(NH_3) = 35$], while the amides and the urethanes are both more acidic than the amines.

The loss of a proton from carbon is also strongly promoted by the groups —CN and —$SO_2R$, as shown by the following $pK$ values:

| $CH_3CN$ | ca. 25 | $CH_2(CN)_2$ | 11.2 | $CH(CN)_3$ | −5 |
|---|---|---|---|---|---|
| $CH_3SO_2CH_3$ | ca. 23 | $(CH_3SO_2)_2CH_2$ | 12.7 | $(CH_3SO_2)_3CH$ | <0 |

These effects are commonly attributed to stabilization of the anion by a mesomeric shift of the negative charge on the nitrogen or oxygen atoms, but it seems more likely that the effect is essentially a polar one, since the same groups have a powerful strengthening effect on carboxylic acids where no redistribution of charge is possible, e.g., $pK(CH_2CNCO_2H) = 2.47$, $pK(CH_3SO_2CO_2H) = 2.36$, compared with 4.75 for acetic acid. The interpretation of the acidity of the sulphones is beset by the same valency problems as we have already discussed in connection with the strengths of inorganic oxyacids and the substituent effect of the sulphonate group. There has been much controversy, on the basis of stereochemical and other evidence, as to whether an $\alpha$-sulphocarbanion such as $[RSO_2CH_2]^-$ has a planar or a pyramidal configuration, but the question is far from being resolved.[41] However, a recent theoretical study[42] of $[HSO_2CH_2]^-$ claims that its most stable configuration is definitely pyramidal, with no appreciable conjugation with the $3d$-orbitals of sulphur; this of course implies that the acid-strengthening effect of the $RSO_2$ group is essentially dipolar in character. We shall see later (Chapter 10) that the kinetic behaviour of cyanocarbon and sulphocarbon acids suggests that their ionization involves little bond reorganization, which is consistent with this view.

An interesting aspect of the effect of molecular structure appears in the acid-base properties of *electronically excited states*. Information of this

[41] See, e.g., D. J. Cram, R. D. Trepka, and P. St. Janiak, *J. Am. Chem. Soc.*, **88**, 2749 (1966); M. J. Gresser, *Quart. Rep. Sulphur Chem.*, Suppl. 4, p. 29 (1969); B. S. Thyagarajan, *ibid.*, p. 115.
[42] S. Wolfe, A. Rauk, and I. G. Czmidia, *J. Am. Chem. Soc.*, **91**, 1567 (1969).

kind is now available for a variety of aromatic species, as described in several review articles.[43-45]

The simplest method in principle depends upon the *Förster cycle*,[46] and relates to the first excited singlet state, which is normally involved in the absorption band of longest wavelength, and in fluorescence. Figure 5 illustrates the energy levels for an acid HA, its dissociation products $A^- + H^+$, and the corresponding excited state species HA* and $A^-* + H^+$.

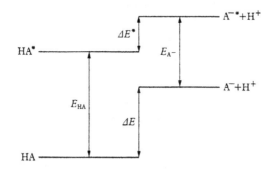

Fig. 5. Electronic energy levels for an acid and its anion.

It is clear from Figure 5 that $\Delta E^* - \Delta E = E_{A^-} - E_{HA}$, and since $\Delta E$ and $\Delta E^*$ relate to the dissociation of the acid in the ground state and the excited state respectively, we can write to a good approximation

$$pK^* - pK = (\Delta E^* - \Delta E)/2.3RT$$

$$= (E_{A^-} - E_{HA})/2.3RT = (\mathbf{h}\nu_{A^-} - \mathbf{h}\nu_{HA})/2.3RT \qquad (34)$$

where $\nu_{HA}$ and $\nu_{A^-}$ are the frequencies corresponding to the transitions $HA \rightarrow HA^*$ and $A^- \rightarrow A^-*$, each species being in its lowest vibrational state.* The values of $\nu_{HA}$ and $\nu_{A^-}$ can in principle be obtained from the absorption or fluorescence spectra of HA and $A^-$; however, it may be difficult to eliminate the effect of vibrational energy, and a common procedure is to use the mean of the frequencies corresponding to the maximum intensities of the absorption and fluorescence bands.

[43] A. Weller, *Progr. Reaction Kinetics*, **1**, 187 (1961).
[44] B. L. van Duuren, *Chem. Rev.*, **63**, 325 (1963).
[45] E. Vander Donckt, *Progr. Reaction Kinetics*, **5**, 273 (1970).
[46] T. Förster, *Z. Elektrochem.*, **54**, 42 (1950).
*It is usually stated that Equation (34) depends on the assumption that HA and HA* have equal entropies of ionization. However, we have seen in Chapter 5 that molecular energies are usually more closely related to $\Delta G$ than to $\Delta H$, so that Equation (34) is likely to be valid independent of any equality of experimental entropies of ionization.

Another method, also applicable to the first excited singlet state, depends upon the variation of the fluorescence spectrum with pH, and is best illustrated by an example. The acridinium ion, $AcrH^+$, has $pK = 5.45$ in its normal state. It exhibits a green fluorescence in acid solution, attributable to the cation, and this remains unchanged with increasing pH up to about pH = 10, although the acridine has been converted almost entirely into the free base. This is explicable if the $pK$ of the excited state is much greater than 5, so that in the pH range 7–10 the process is

(i)   $Acr + h\nu \qquad \rightarrow Acr^*$
(ii)  $Acr^* + H_2O \qquad \rightarrow Acr^*H^+ + OH^-$ (fast)
(iii) $Acr^*H^+ \qquad \rightarrow AcrH^+ + h\nu'$
(iv)  $AcrH^+ + OH^- \rightarrow Acr + H_2O$

As the pH is increased further, reaction (ii) becomes less complete, and the green fluorescence is gradually replaced by blue fluorescence, attributable to the reverse of process (i). A quantitative analysis of this behaviour leads to $pK = 10.6$ for the excited state, in good agreement with the value obtained from the frequency difference between the two fluorescent spectra. A simple interpretation assumes that process (ii) is much faster than the reverse of (i), i.e., that the acidic and basic forms of the excited state are in equilibrium. Since the lifetimes of the excited states are very short (ca. $10^{-8}$ s) this is rarely the case, and a closer analysis of the problem gives information about the rates of the fast proton-transfers involved. If the object of the investigation is to determine $pK^*$ these processes can be speeded up by the use of suitable buffer solutions.

These methods are not directly applicable to triplet states, but these can

Table 14. p$K$ VALUES FOR EXCITED STATES

| Acid | Ground | 1st excited singlet | Triplet |
|---|---|---|---|
| 2-Naphthol | 9.5 | 3.1 | 7.9 |

| Acid | Ground | 1st excited singlet | Triplet |
|---|---|---|---|
| 2-Naphthol | 9.5 | 3.1 | 7.9 |
| 2-Naphthoic ($C_{10}H_7CO_2H$) | 4.2 | 6.6 | 4.1 |
| $C_{10}H_7CO_2H_2^+$ | −6.9 | 1.5 | — |
| 2-Naphthylammonium | 4.1 | −2 | 3.2 |
| Acridinium | 5.5 | 10.6 | 5.6 |

often be obtained in considerable concentrations by high-intensity light flashes, and their absorption spectra studied as a function of pH.[47] Owing to the longer lifetime of triplet states there is no difficulty about attaining acid-base equilibrium in suitable buffer solutions.

Because of various uncertainties the quantitative agreement between different methods and different workers is often poor,[48] but the general features are clear. Some typical results are given in Table 14.

The effect of excitation to the singlet is throughout considerable, and usually greater than can be produced by any chemical substitution; this of course reflects the profound changes of electron distribution caused by excitation. A quantitative understanding of these effects must rest on a theoretical calculation of electron distributions in excited states, and some progress has been made in this direction.[49] However, a qualitative valence-bond picture is also illuminating. For example, a crude picture of the first excited singlet state of 2-naphthol is

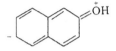

and this will naturally be a much stronger acid than the ground state. (Participation of this kind of resonance state is often invoked to explain why phenols are stronger acids than alcohols.) Similarly, the excited state of 2-naphthoic acid can be written as

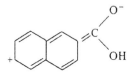

and this will be a weaker acid and a much stronger base than the ground state, as shown in Table 14.

Table 14 also shows that triplet states have acid-base properties very similar to those of the ground state. This can be rationalized if we remember that triplet states can be pictured as having unpaired electrons

[47] G. Jackson and G. Porter, *Proc. Roy. Soc.*, A, **260**, 13 (1961).
[48] For a critique, see N. Arnal, M. Deumié, and P. Viallet, *J. Chim. Phys.*, **66**, 421 (1969).
[49] See, e.g., J. Bertran, O. Chalvet, and R. Daudel, *Theor. Chim. Acta*, **14**, 1 (1969).

on different atoms, so that the triplet states corresponding to the two examples above can be written as

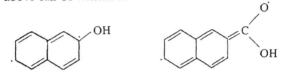

Excitation of this type does not involve any appreciable charge separation, and the effect on acid-base properties is thus small.

# 7 The Direct Study of Rates of Simple Proton-Transfer Reactions

Everyday laboratory experience suggests that, with very few exceptions, reactions between acids and bases are extremely fast, since no time lag is observable in the dissociation of acids or bases, buffer action, hydrolysis, etc. In fact, for many purposes proton-transfer reactions involving simple acids and bases are fast enough to be treated as equilibrium processes. However, there are two reasons why the rates of these processes are of interest. In the first place modern techniques have made it possible to measure the rates of extremely fast reactions, with half-times down to about $10^{-9}$ second, and hence to obtain information about the mechanism of such reactions. In the second place, when proton-transfer reactions are coupled with other chemical processes they may lead to slow observable changes, in particular to the catalysis of reactions by acids and bases. The latter type of approach is historically the older, but it is more logical to consider first the direct observation of reactions between simple acids and bases, as will be done in this chapter. Some general features of the experimental results will be described, but detailed consideration of the relations between rates, equilibria, and structures will be deferred until Chapter 10, so as to include the information obtained less directly from studies of acid-base catalysis, described in Chapters 8 and 9.

Since the first edition of this book was published in 1959, a number of books and review articles have appeared both on the experimental methods for studying fast reactions in solution,[1-7] and on their applica-

[1] Conference on Fast Reactions, *Z. Elektrochem.*, **64**, 1–204 (1960).
[2] *Technique of Organic Chemistry* (ed. E. L. Friess, E. S. Lewis, and A. Weissberger), Vol. VIII B (Investigation of Rates and Mechanisms of Reactions), Interscience, 1963. This volume contains articles on most of the experimental techniques.
[3] K. Tamm, *Handbuch de Physik*, Vol. XI (1), Springer, Heidelberg, 1961, p. 202 (Ultrasonic methods).
[4] E. F. Caldin, *Fast Reactions in Solution*, Blackwell, Oxford, 1964.
[5] G. Czerlinski, *Chemical Relaxation*, Dekker, New York, 1966.
[6] *Fast Reactions and Primary Processes in Reaction Kinetics* (5th Nobel Symposium) (ed. S. Claesson), Interscience, 1967.
[7] J. N. Hague, *Fast Reactions*, Wiley, New York and London, 1971.

tion to proton transfers.[8-11] This chapter will therefore contain only a very general account of the available experimental techniques, and few detailed references. The scope of the available methods is conveniently summarized in Table 15.

Table 15. METHODS FOR STUDYING FAST REACTIONS IN SOLUTION

| Method | Characteristic time range, seconds |
|---|---|
| Stopped flow | $> 10^{-3}$ |
| Temperature jump | $1-10^{-6}$ |
| Pressure jump | $> 10^{-5}$ |
| Electric field impulse | $10^{-4}-10^{-8}$ |
| Dielectric absorption | $10^{-6}-10^{-9}$ |
| Ultrasonics | $10^{-3}-10^{-9}$ |
| Electrochemical methods | $1-10^{-6}$ |
| N.m.r. line broadening ($^1$H) | $1-10^{-3}$ |
| N.m.r. line broadening ($^{17}$O) | $10^{-3}-10^{-7}$ |
| Optical line broadening | $10^{-10}-10^{-14}$ |

Since proton-transfers are essentially bimolecular processes, the velocity constants which correspond to the time ranges in Table 15 will depend upon the reactant concentrations which can conveniently be used ($k \sim 1/at$, where $a$ is the larger of the concentrations of the two reacting species) and will therefore vary with the nature of the system and the instrumental set-up employed. However, more than half of the methods listed in Table 15 can, under favourable circumstances be used to measure second-order velocity constants up to $10^{10}-10^{11}$ dm$^3$ mol$^{-1}$ s$^{-1}$, which will be shown later to represent the maximum attainable value for solution reactions.

In a reversible acid-base reaction $A_1 + B_2 \rightleftharpoons B_1 + A_2$, the equilibrium constant is given by the ratio of the forward and reverse rate constants, $K = k_f/k_r$. If the equilibrium is very far on the left-hand side $k_f \ll k_r$: $k_f$ may in this case be quite small, since the activation energy of the forward reaction must be at least equal to its endothermicity. It might be thought, therefore, that a reaction involving a very weak acid or a very weak base (or both) would always be slow enough to be followed by conventional means. However, the time taken to reach equilibrium will still

[8] M. Eigen, *Angew. Chem.*, **75**, 489 (1963); *Internat. Edn.*, **3**, 1 (1964); *Pure Appl. Chem.*, **6**, 97 (1963).
[9] M. Eigen, W. Kruse, G. Maas, and L. de Maeyer, *Progr. Reaction Kinetics*, **2**, 285 (1964).
[10] E. Grunwald. *Progr. Phys. Org. Chem.*, **3**, 317 (1965).
[11] W. J. Albery, *Progr. Reaction Kinetics*, **4**, 353 (1966).

be very short, since it depends upon the values of both $k_f$ and $k_r$. In the simple case of first-order kinetics in both directions (e.g., when the concentrations of $A_2$ and $B_2$ are much larger than those of $A_1$ and $B_1$) the approach to equilibrium is characterized by a first-order constant equal to the sum of those for the forward and reverse reactions, and, in general, if $k_f$ and $k_r$ have very different magnitudes the approach to equilibrium is governed by the larger of them.*

If the concentration of one of the reacting species can be monitored by measuring an optical or electrical property (e.g., optical absorbance, optical rotation, fluorescence, or electrical conductivity) it is no great problem to devise electronic equipment for following these changes with a time resolution of down to $10^{-9}$ second. The limiting factor thus becomes the time required to mix the reacting solutions, which in ordinary laboratory practice will rarely be less than a few seconds. This can be reduced to about $0.1\,s$ by efficient stirring, and to $10^{-3}\,s$ by allowing jets of the two solutions to impinge in a carefully designed mixing chamber, as in the much used *stopped flow* technique. However, even with reactant concentrations as low as $10^{-3}$M a half-time of $1\,ms$ corresponds to a velocity constant of only $10^6\,dm^3\,mol^{-1}\,s^{-1}$.

Much faster reactions can be studied by the use of *relaxation methods*, developed largely by Eigen and his collaborators, which avoid any mixing problems, though they can only be applied to systems in which detectable quantities of both reactants and products are present at equilibrium. In its simplest form the relaxation method consists of perturbing the equilibrium position by a rapid change in some external parameter (usually temperature, pressure, or electric field) and observing by optical or electrical means the rate at which the system adjusts to the new equilibrium position. A typical example is the application of the *temperature jump* method to the rate of dissociation of a weak acid HX, characterized by the scheme

$$HX \underset{k_r}{\overset{k_d}{\rightleftharpoons}} H^+ + X^-$$

where $k_d$ and $k_r$ refer to the dissociation and recombination reactions respectively. Provided that the enthalpy of dissociation is not zero, the degree of dissociation $\alpha$ will depend upon temperature, and after a sudden temperature rise the degree of dissociation will readjust itself to a new

---

*It is sometimes possible to measure $k_f$ separately, for example, if the reaction products are removed by isotopic exchange or by some other chemical reaction. We shall meet examples of this procedure later, especially in connection with acid-base catalysis, but for the present only a simple approach to equilibrium will be considered.

value. If the perturbation is a small one (as is usually the case in practice) the readjustment of the equilibrium follows a simple exponential course, which can be characterized either by a first-order velocity constant $k$, or by a *relaxation time* $\tau = 1/k$, related to the half-time by $t_{\frac{1}{2}} = \tau \ln 2 = 0.69\tau$. This will be demonstrated for the above reaction scheme.

Let the initial equilibrium concentrations of HX, $X^-$, and $H^+$ be $\bar{a}$, $\bar{b}$, and $\bar{h}$ respectively, and their concentrations at time $t$ after the temperature jump be $\bar{a} + \delta_a$, $\bar{b} + \delta_b$, and $\bar{h} + \delta_h$. The rate of disappearance of HX is then

$$-d[\text{HX}]/dt = -d\delta_a/dt = k_d(\bar{a} + \delta_a) - k_r(\bar{h} + \delta_h)(\bar{b} + \delta_b).$$

Conservation of X and charge gives $\delta_h = \delta_b = -\delta_a$, and at equilibrium we have $k_d\bar{a} = k_r\bar{h}\bar{b}$, so that if we neglect the second-order term in $\delta_a^2$ the equation simplifies to

$$-d\delta_a/dt = \{k_d + k_r(\bar{h} + \bar{b})\}\delta_a \qquad (35)$$

which represents a first-order change with a velocity constant $k_d + k_r(\bar{h} + \bar{b})$; the same constant governs the appearance or disappearance of the species $H^+$ and $X^-$. The separate values of $k_d$ and $k_r$ can be obtained either by using the measured equilibrium constant, which is equal to $k_d/k_r$, or by measuring the relaxation time as a function of the concentrations of $H^+$ and $X^-$.*

The temperature jump, which may amount to several degrees, is commonly produced by Joule heating, produced by applying a high potential (several kV) for a very short time. This method applies only to conducting solutions, and for non-conducting systems dielectric absorption of microwave pulses can be used,[12] while absorption of light from a pulsed laser, often by a suitable dye, applies to both types of system.[13] All these methods will produce a temperature rise within a few microseconds, and the temperature jump technique has been widely used for investigating acid-base reactions. Similar principles apply to the *pressure jump* method, in which the equilibrium is shifted both by the pressure itself (because

---

*The above simple treatment assumes that there is only one route by which the equilibrium $HX \rightleftharpoons H^+ + X^-$ can be set up. This is not necessarily the case in an amphoteric solvent; for example, in aqueous solution an alternative route is $HX + OH^- \rightleftharpoons X^- + H_2O$ coupled with $H_2O \rightleftharpoons H^+ + OH^-$. When this is taken into account it is found that the system has two relaxation times, each of which depends in principle on all the six velocity constants involved. However, it is often possible to choose conditions under which a simpler treatment is valid, and we shall not pursue the subject here. For a full discussion see the article by M. Eigen and L. de Maeyer in Ref. 2.

[12] E. F. Caldin and J. E. Crooks, *J. Sci. Instr.*, **44**, 449 (1967); K. J. Ivin, J. J. McGarvey, and E. L. Simmons, *Trans. Faraday Soc.*, **67**, 97 (1971).

[13] See, e.g., H. Koffer, *Ber. Bunsengesell. Phys. Chem.*, **75**, 1245 (1971).

most acid-base reactions involve a small volume change) and by the temperature change produced by adiabatic compression; however, this technique is usually less convenient to use in practice.

It is also possible to obtain kinetic information on acid-base systems by rapid variations of the applied electric field, but since the application of high fields for long periods causes complications due to heating and electrolysis, short *electric impulses* are used instead. The degree of dissociation of a weak electrolyte is increased to a calculable extent by the application of a strong electric field; this phenomenon is known as the dissociation field effect, or the second Wien effect. If the field is changed suddenly, the degree of dissociation will not change immediately to its new value, and Figure 6 shows how the degree of dissociation $\alpha$ will change

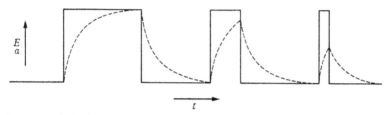

Fig. 6. Change of degree of dissociation for square-wave impulses of different durations.

during a square-wave electric impulse of varying duration. Since the conductivity depends upon the degree of dissociation, the high-field conductivity of the system will vary with the duration of the impulse, and this variation can be used to determine the rate of dissociation. In practice it is often convenient to use a damped harmonic impulse instead of a square one, and the conductivity is measured by comparison with a strong electrolyte so as to eliminate the part of the Wien effect which is due to interionic attraction. In order to obtain an appreciable change in the degree of dissociation, fields of the order of $10^5$ volt cm$^{-1}$ must be used, with an impulse time of $10^{-5}$–$10^{-7}$ s, which is of the order of the relaxation time.

Similar principles apply when a *high-frequency alternating field* is applied to a solution of a weak electrolyte. This is illustrated in Figure 7, where the upper curve represents the variation of the field strength with time, and the lower three curves the corresponding variation of the degree of dissociation $\alpha$ for different reaction velocities. Curve (a) represents a low velocity, so that $\alpha$ remains throughout at the value corresponding to zero field, whereas for curve (b) the velocity is so high that $\alpha$ follows the

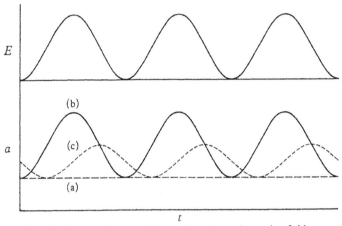

Fig. 7. Variation of degree of dissociation in an alternating field.

instantaneous field strength and varies in phase with it. The case of interest here is when the half-time of the reaction is of the same order of magnitude as the periodic time of the field; we then obtain a curve such as (c), which has a smaller amplitude than (b) and is now out of phase with the applied field. As first pointed out by Pearson,[14] the conductivity of the solution should vary with frequency in this region, but the effect is very small (because of the low fields employed) and is difficult to detect experimentally. An alternative approach is to measure the dielectric loss, which arises from the phase difference between $\alpha$ and the applied field; this makes the process partially irreversible and leads to the dissipation of electrical energy as heat. However, early claims to have measured the contribution to dielectric loss arising from the dissociation of boric acid[15] have been shown to be unsound,[16] and this method does not appear to have been applied successfully to proton-transfer reactions, though it has been used for dimerization processes accompanied by a large change in polarity.[17]

Exactly the same principles apply to the study of *ultrasonic absorption* in acid-base systems, except that the quantity displacing the equilibrium is now the oscillating pressure of the sound wave rather than the electric field. The effect of pressure on the equilibrium constant is given by

$$\partial \ln K / \partial p = \Delta \overline{V} / RT$$

[14] R. G. Pearson, *Disc. Faraday Soc.*, **17**, 187 (1954).
[15] W. R. Gilkerson, *J. Chem. Phys.*, **27**, 914 (1957).
[16] R. P. Bell and R. R. Robinson, *Trans. Faraday Soc.*, **58**, 2358 (1962).
[17] K. Bergmann, M. Eigen, and L. de Maeyer, *Ber. Bunsengesell. Phys. Chem.*, **67**, 819 (1963).

where $\Delta V$ is the volume change in the reaction, and since almost all reactions involve some volume change the method is widely applicable. Since the changes in a sound wave are essentially adiabatic, the equilibrium is also perturbed by a periodic temperature fluctuation, though this effect is small for aqueous solutions. In principle it would be equally informative to measure the dispersion of sound (variations of velocity with frequency), the heat produced, or the absorption of sound as a function of frequency. In practice the last is the most convenient, and a variety of techniques are used to cover a wide range of frequencies. The appearance of a maximum at a frequency $f_c$ on the plot of absorption against frequency indicates the presence of a rate process with a relaxation time equal to $1/2\pi f_c$. The advantage of this method lies in the wide range of relaxation times ($10^{-3}$–$10^{-9}$ s) which can be studied. It suffers from the disadvantage that sound absorption can also be associated with a number of other rate processes, such as the formation and dissociation of hydrogen bonds or ion pairs, being in fact widely used for investigating these processes. In applying the method to acid-base reactions care is therefore needed in identifying the process corresponding to an observed absorption maximum.

In the relaxation methods described so far, the equilibrium is disturbed by changing a general parameter ($\tau$, $p$, field strength) and observations are then made on either transient or steady-state properties of the solution. In *electrochemical methods*[18] the perturbation is introduced by removing one of the species by an electrode reaction, usually reduction at a mercury or platinum cathode. The quantity measured is either the current or the electrode potential, and either transient or steady-state observations can be made, many of the techniques involving some form of *polarography*. Under ordinary conditions the limiting current observed at a dropping or rotating cathode (corresponding to the flat part of the polarographic curve) is controlled by the rate at which the reducible species can diffuse to the electrode and is therefore determined by the concentration of this species. However, if a small quantity of a reducible species is in chemical equilibrium with a second species which is not reducible, then under suitable conditions the observed current will be controlled by the rate of the chemical process producing the reducible species. This was first shown for aqueous solutions of formaldehyde,[19] in which

[18] For a general account see H. Strehlow, Ref. 2; P. Delahay, *New Instrumental Methods in Electrochemistry*, Interscience, New York, 1954; H. Schmidt and M. von Stackelberg, *Die neuartigen polarographischen Methoden*, Verlag Chemie, Weinheim, 1962.
[19] R. Brdicka, *Coll. Trav. Chim. Czech.*, **12**, 213 (1947).

the equilibrium $CH_2O + H_2O \rightleftharpoons CH_2(OH)_2$ is far over to the right. Of the two species $CH_2O$ and $CH_2(OH)_2$ only the former is reducible, so that under suitable conditions of drop rate, concentration, and so forth the observed current is a direct measure of the rate of the process $CH_2(OH)_2 \rightarrow CH_2O + H_2O$. It is difficult to get an exact mathematical solution for the kinetic problems involved, but an approximate absolute value for the velocity constant can easily be obtained and fairly accurate relative values for a series of similar reactions. In acid-base systems it is often found that only one member of an acid-base pair is reducible at the cathode; for example, undissociated pyruvic acid, $CH_3COCO_2H$, is reducible, while the anion $CH_3COCO_2^-$ is not. In this system it is thus possible to get information about the rate of the reaction $CH_3COCO_2^- + H^+ \rightarrow CH_3COCO_2H$, for which $k_2$ is about $10^9$ $dm^3 mol^{-1} s^{-1}$.

This procedure is limited to the rather small class of reducible acids and bases, but the scope of the polarographic method can be greatly extended by using a somewhat different principle. The cathodic reduction of azobenzene takes place according to the scheme

$$C_6H_5N = NC_6H_5 + 2H^+ + 2e^- \rightarrow C_6H_5NH - NHC_6H_5,$$

the hydrogen ions being commonly supplied by a buffer system $A \rightleftharpoons B + H^+$. Under suitable conditions the rate of reduction, and hence the polarographic current, is determined by the rate at which hydrogen ions are supplied by the reaction $A \rightarrow B + H^+$, and since the equilibrium constant is known the velocity constant of the reverse reaction $B + H^+ \rightarrow A$ can also be determined. Other reducible substances, such as *p*-nitroaniline, can be used. Alternatively, the equilibrium $HX \rightleftharpoons H^+ + X^-$ can be perturbed directly by the reduction of hydrogen ions at a rotating platinum or palladium cathode, though there are complications due to the reduction of water molecules at high cathode potentials.

By the use of a variety of polarographic and allied techniques, reaction half-times down to about $10^{-4}$ s can be studied, and a recent technique, known as 'high level Faradaic rectification'[20] extends this range to about $10^{-6}$ s. However, there are some problems in interpreting kinetic results for proton-transfer reactions derived from electrochemical studies. In the first place, the relation between the observed quantities and chemical rate constants is a complicated one, and its derivation usually involves physical or mathematical approximations and corrections. Secondly, the informa-

[20] H. W. Nürnberg, *Fortschr. Chem. Forsch.*, **8**, 241 (1967).

tion obtained relates to a layer close to the electrode which may be as thin as $10^3$ pm for the faster reactions investigated. It may be doubted whether the normal laws of diffusion can be applied under these conditions, and it is possible that adsorbed molecules may contribute to the reaction rate. Moreover, the electrical potential gradient in these layers may be extremely high, leading to a large increase in the rates of dissociation. These uncertainties probably account for the fact that the velocity constants derived from electrochemical measurements do not always agree with those measured by other methods; this applies particularly to some of the earlier investigations, which did not take into account the problems mentioned above, and reported velocity constants are sometimes too high to be physically reasonable. However, there is no doubt that carefully designed electrochemical experiments can yield reliable results, especially for the relative rates of a series of similar proton-transfer reactions. A minor difficulty stems from the fact that it is frequently convenient to make measurements in the presence of a high concentration of inert electrolyte (e.g., 1M KCl), so that the velocity constants obtained are not directly comparable with those at low ionic strengths.

The final class of methods which we shall consider may be termed *lifetime methods*, and depend upon the observation of *spectroscopic line shapes*, so that they do not involve either mixing problems or perturbation of equilibria. In particular they can be used to determine the rate of a symmetrical exchange reaction such as $XH + X^- \rightleftharpoons X^- + XH$, which is not observable by any other means. These methods depend upon the *uncertainty principle*, according to which a state having a lifetime $\tau$ has an uncertainty in its energy $\delta\varepsilon$ given by $\delta\varepsilon \sim \mathbf{h}/2\pi\tau$. If this state is involved in a spectroscopic transition involving radiation of frequency $v$, there will be a corresponding uncertainty $\delta v$ in the frequency, given by $\delta v = \delta\varepsilon/\mathbf{h} \sim 1/2\pi\tau$; thus a reduction in lifetime, by chemical reaction or otherwise, will lead to a broadening of the spectroscopic line.

Since the lifetime of a state may be shortened by factors other than chemical reaction, notably by intermolecular or intramolecular energy transfer, information about chemical reactions is obtained by observing the effect of factors such as temperature, reactant concentrations, or pH upon the line shape. Optical spectra are rarely suitable for this purpose since they will only be broadened by extremely fast reactions, but the broadening of the Raman lines of the trifluoroacetate ion in presence of trifluoroacetic acid has been attributed to proton exchange in ion pairs.[21]

[21] M. M. Kreevoy and C. A. Mead, *J. Am. Chem. Soc.*, **84**, 4596 (1962); *Disc. Faraday Soc.*, **39**, 166 (1965).

There are, however, difficulties in the quantitative interpretation of the results,[22] and the method has not been widely used.

Much more information can be obtained by a study of *nuclear magnetic resonance spectra*, and *proton magnetic resonance* is particularly appropriate for our present purpose.[23] Reaction times in the range $1-10^{-3}$ s are appropriate to proton resonance, while considerably faster reactions can be studied when the $^{17}O$ nucleus is observed.

If a solution contains two sets of protons in different environments, the observed spectrum depends upon the rate at which the two sets can interchange. If this interchange is slow compared with the frequency separation, two distinct peaks will be observed in the spectrum, but if it is fast there will be only a single peak at an intermediate frequency. At some intermediate frequency there will be a gradual change from one type of spectrum to the other, and in this frequency range the mean lifetime of the proton in one of its situations is of the same order of magnitude as $1/v$, where $v$ is the frequency separation. Similarly, the splitting of the lines into a number of components (which is due to the interactions between the observed nucleus and other magnetic nuclei) is lost if the mean lifetime is much smaller than $1/v$, and will be restored if either the lifetime or the frequency is increased. In principle, therefore, the velocity constant of the interchange process can be determined by observing the changes in the spectrum when the reaction velocity is progressively varied by changing some property of the solution such as the concentration or the pH.

The theory of the relation between the line shape and the lifetimes is rather complicated, and the best results are obtained by a computer fitting of the observed and calculated curves over their whole range.[24] Less accurate velocity constants can be derived by the commoner procedures of measuring the width of a line at half-height, or determining the conditions under which two lines coalesce to form a single broad line.

Some of the earliest applications to acid-base kinetics are due to Ogg.[25] In very pure liquid ammonia the proton-resonance line shows a triplet structure, due to interaction with the nucleus $^{14}N$, but if very small amounts of the ions $NH_2^-$ or $NH_4^+$ are introduced (the latter by adding a trace of water) the structure disappears. This is undoubtedly due to the

[22] See, various authors, *Disc. Faraday Soc.*, **39**, 172–182 (1965).

[23] For a general account, see H. Strehlow, Ref. 2, p. 865; J. A. Pople, W. G. Schneider, and H. J. Bernstein, *High Resolution Nuclear Magnetic Resonance*. McGraw-Hill, New York, 1959, especially Ch. 10; E. F. Caldin, Ref. 1, Ch. 11.

[24] See, e.g., B. G. Cox, F. G. Riddell, and D. A. R. Williams, *J. Chem. Soc.*, B, 859 (1970); B. G. Cox, *J. Chem. Soc.*, B, 1780 (1970).

[25] R. A. Ogg, *J. Chem. Phys.*, **22**, 560 (1954); *Disc. Faraday Soc.*, **17**, 215 (1954).

exchange reactions $NH_3 + NH_2^- \rightleftharpoons NH_2^- + NH_3$ and $NH_3 + NH_4^+ \rightleftharpoons NH_4^+ + NH_3$, but only a very rough estimate of the velocity constants could be obtained, partly because the very low ionic concentrations were not well defined. Similarly, an acid aqueous solution of an ammonium salt gave a triplet due to $^{14}NH_4^+$ and a singlet due to $H_2O$, but in a neutral solution these were merged into a single peak by the occurrence of exchange reactions involving $NH_3$ or $OH^-$. Similar results were obtained by observing the nuclear resonance of nitrogen rather than that of the proton. More recent work with liquid ammonia solutions[26] has provided quantitative information about the rates of these exchange processes.

The most extensive measurements of this kind have been made by Grunwald, Loewenstein, Meiboom, and their collaborators, particularly on solutions of amines and carboxylic acids in water and methanol.[27] Their results for methylamine are well suited to illustrate the kind of information which can be obtained. All the solutions used contained the methylamine essentially in the form of the cation $CH_3\overset{+}{N}H_3$. When the solution is sufficiently acid, the proton exchange is not fast enough to affect the spectrum, which consists of three distinct components, (1) a quadruplet from $CH_3$, the splitting being due to the protons of the $\overset{+}{N}H_3$ group, (2) a triplet from $\overset{+}{N}H_3$, the splitting being due to the $^{14}N$ nucleus, and (3) a sharp single line from $H_2O$.* If the pH of the solution is increased, proton exchange becomes important, and the spectrum changes in a number of ways. With increasing pH the $CH_3$ quadruplet first broadens and then coalesces into a singlet which finally sharpens; similarly, the $\overset{+}{N}H_3$ triplet first broadens and then disappears, while the $H_2O$ singlet first broadens and then sharpens again, simultaneously shifting slightly in the direction of the $\overset{+}{N}H_3$ frequency. Since the hydrogens of the $CH_3$ group are certainly not exchanging rapidly under these experimental conditions, all these effects must be due to protolytic exchange on the $\overset{+}{N}H_3$ group, also involving the $H_2O$ protons. The sharp singlet which at high pH replaces both the $\overset{+}{N}H_3$ triplet and the $H_2O$ singlet actually represents an average for both kinds of protons, the exchange between them being now too fast for two lines to be distinguishable. Information about the rate processes involved is best obtained from

---

[26] T. J. Swift, S. B. Marks, and W. G. Sayre. *J. Chem. Phys.*, **44**, 2796 (1966): T. J. Swift and H. H. Lo, *J. Am. Chem. Soc.*, **88**, 2994 (1966): D. R. Clutter and T. J. Swift, *J. Am. Chem. Soc.*, **90**, 601 (1960).
[27] For some collected results, see A. Loewenstein and T. M. Conner, *Ber. Bunsengesell. Phys. Chem.*, **67**, 280 (1963).
* It is not immediately obvious why the $CH_3$ quadruplet is not further split by the $^{14}N$ nucleus or the $NH_3$ triplet by the $CH_3$ protons, but a quantitative theoretical treatment shows that these further splittings should not be detectable.

the broadening which takes place at intermediate acidities in all three components of the spectrum. Thus the broadening of either the $CH_3$ quadruplet or the $\overset{+}{N}H_3$ triplet gives a measure of the mean life of the $\overset{+}{N}H_3$ protons. Similarly the broadening of the $H_2O$ singlet measures how long a proton remains on oxygen before it is transferred to nitrogen. By studying the way in which the three kinds of broadening depend upon the pH and upon the concentration of the methylammonium ions, Grunwald, Loewenstein, and Meiboom were able to discriminate between various exchange mechanisms and to obtain approximate velocity constants for some of them. The mechanisms considered (after excluding some improbable ones) were as follows:

(1)    $CH_3\overset{+}{N}H_3 + H_2O \rightarrow CH_3NH_2 + H_3O^+$

(2)    $CH_3\overset{+}{N}H_3 + OH^- \rightarrow CH_3NH_2 + H_2O$

(3a)   $CH_3\overset{+}{N}H_3 + CH_3NH_2 \rightarrow CH_3NH_2 + CH_3\overset{+}{N}H_3$

(3b)   $CH_3\overset{+}{N}H_3 + \underset{\underset{H}{|}}{O}\!-\!H + CH_3NH_2 \rightarrow CH_3NH_2 + H\!-\!\underset{\underset{H}{|}}{O} + CH_3\overset{+}{N}H_3.$

Reaction (1) was not detected for methylamine or dimethylamine (though it was for trimethylamine) and must there have had a velocity constant less than about $4 \times 10^{-3}$ $dm^3$ $mol^{-1}$ $s^{-1}$; this is consistent with evidence from rates of isotopic exchange. Reaction (2) was not detectable for any of the amines studied, but it would not be significant at the acidities concerned even if its velocity constant were as high as $10^{10}$–$10^{11}$ $dm^3$ $mol^{-1}$ $s^{-1}$. The observed exchange rate could therefore be attributed entirely to reactions (3a) and (3b), both of which show the same dependence on pH and concentration. At least part of the exchange must also involve a water molecule, since the $H_2O$ singlet broadens at the same time as the other components of the spectrum, and a quantitative treatment shows that (3a) and (3b) are of roughly equal importance, each having a velocity constant of about $3 \times 10^8$. For dimethylamine and trimethylamine the reactions involving a water molecule are increasingly important, the analogue of (3a) being undetectable for trimethylamine. Measurements of proton magnetic resonance obviously offer a valuable method for detecting the participation of a solvent molecule in proton-transfer reactions, which may be more usual than is commonly supposed.

Before discussing any of the numerical values obtained for proton-transfer rates by the above methods, it is convenient to consider some general problems concerned with fast reactions in solution.[28] In a gas, molecular collisions occur singly, a repeated collision being a rare event, so that even a very efficient reaction will cause very little disturbance of the spatial distribution of the molecules. In solution, on the other hand, once two solute species have diffused together they will be surrounded by a cage of solvent molecules and are likely to collide with one another many times before they diffuse apart again. Such a group of collisions is termed an *encounter*, and two solute species in the same solvent cage constitute an *encounter complex*; the number of collisions per encounter and the lifetime of an encounter complex (neglecting for the moment any chemical reaction) depend upon the diffusion coefficients, and hence upon the viscosity of the solvent. Under these conditions an efficient reaction between two solvent species can cause a considerable depletion in their equilibrium spatial distribution, and in particular in the number of encounter complexes.

The application of this picture to a proton-transfer reaction $HX + Y \rightleftharpoons X + HY$ (where the charges are omitted for the sake of generality) leads to the kinetic scheme

$$HX + Y \underset{k_{ba}}{\overset{k_{ab}}{\rightleftharpoons}} XH \cdots Y \underset{k_{cb}}{\overset{k_{bc}}{\rightleftharpoons}} X \cdots HY \underset{k_{dc}}{\overset{k_{cd}}{\rightleftharpoons}} X + HY \qquad (36)$$

$$\text{(a)} \qquad\qquad \text{(b)} \qquad\qquad \text{(c)} \qquad\qquad \text{(d)}$$

where (b) and (c) represent the encounter complexes formed from the reactants and the products respectively. The broken lines indicate only that the two species are present in the same solvent cage, though in some instances a hydrogen bond may be formed. For dilute solutions in water or similar solvents the concentrations of the encounter complexes are very small compared with those of the other species, and the application of the steady-state approximation leads to the following general expressions for the observed forward and reverse velocity constants:

$$k_f = k_{ab} k_{bc} k_{cd}/S, \qquad k_r = k_{dc} k_{cb} k_{ba}/S \qquad (37)$$

where

$$S = (k_{ba} + k_{bc})k_{cd} + k_{ba} k_{cb}$$
$$= (k_{cd} + k_{cb})k_{ba} + k_{cd} k_{bc}. \qquad (38)$$

These expressions can be considerably simplified if, as is usually the case, at least one of the following conditions holds:

$$k_{ba} \ll k_{bc}, \qquad k_{cd} \ll k_{cb} \qquad (39)$$

[28] This section follows closely the treatment given by M. Eigen, *Z. Phys. Chem.* (Frankfurt), **1**, 176 (1954); *Angew. Chem. Internat. Edn.*, **3**, 1 (1964).

when we have

$$k_f = \frac{k_{ab}}{1 + K_{bc} k_{ba}/k_{cd}}, \qquad k_r = \frac{k_{dc}}{1 + k_{cd}/(k_{ba} K_{bd})} \qquad (40)$$

in which $K_{bc} = k_{bc}/k_{cb}$ is the equilibrium constant for proton transfer in the encounter complex. Since, as we shall see, $k_{ba}$ and $k_{cd}$ are usually of similar magnitude, $\lg K_{bc}$ is close to the difference between the p$K$ values for the two acids HX and HY.

The situation is formally simpler for reactions involving only one acid-base pair other than the solvent, since no diffusion process is involved in bringing a solvent molecule up to a solute species. Thus the reaction schemes for the dissociation of an uncharged acid HX and for its reaction with hydroxide ion can be written

$$H^+(aq) + X^- \rightleftharpoons H^+(aq) \cdots X^- \rightleftharpoons HX$$

$$HO^-(aq) + HX \rightleftharpoons HO^-(aq) \cdots HX \rightleftharpoons H_2O + X^- \qquad (41)$$

$$(1) \qquad\qquad (2) \qquad\qquad (3)$$

and the observed velocity constants for either reaction are

$$k_f = k_{12} k_{23}/(k_{21} + k_{23}), \qquad k_r = k_{32} k_{21}/(k_{21} + k_{23}). \qquad (42)$$

The condition for a diffusion-controlled reaction is now $k_{23} \gg k_{21}$, giving

$$k_f = k_{12}, \qquad k_r = k_{21} k_{32}/k_{23}. \qquad (43)$$

$k_{12}$, $k_{ab}$, and $k_{dc}$, which govern the rate at which the reactants diffuse together, represent the maximum observable velocity constants for the appropriate reactions. The approximate magnitude of either of them can be deduced theoretically from a simplified picture of the diffusion process. For two spherical molecules which react when they approach to a distance $R_{ij}$, it was first shown by Smoluchowski[29] that if the forces between the molecules are neglected, the diffusion-controlled velocity constant $k_D$ is given by

$$k_D = \frac{4\pi N_0 (D_i + D_j) R_{ij}}{1000} \, dm^3 \, mol^{-1} \, s^{-1} \qquad (44)$$

[29] A. Smoluchowski, *Z. Phys. Chem.*, **92**, 129 (1917).

where $N_0$ is the Avogadro number, $D_i$ and $D_j$ the diffusion coefficients in c.g.s. units, and $R_{ij}$ is in centimetres.*

Equation (44) can be transformed further by making a few more assumptions. Writing $R_{ij} = r_i + r_j$, and introducing the Stokes–Einstein expression for the diffusion constant of spheres in a continuous medium of viscosity $\eta$

$$D = kT/6\pi\eta r \tag{45}$$

we obtain

$$k_D = \frac{8RT}{3000\eta}\left\{\frac{(r_i+r_j)^2}{4r_ir_j}\right\} \sim \frac{8RT}{3000\eta}\ \mathrm{dm^3\,mol^{-1}\,s^{-1}} \tag{46}$$

where the last approximation is valid if $r_i$ and $r_j$ have similar values. The last expression suggests that the velocity constant should depend only upon the temperature and the viscosity of the solvent. The reason for the insensitivity of the rate to the size of the solute molecules is that the slower diffusion of a larger molecule will be compensated by the larger target area which it offers. Although Equation (46) involves more assumptions than (44), it may be equally useful in practice, since the distance $R_{ij}$ is not known with any certainty and the use of macroscopic diffusion coefficients on a molecular scale is open to question.

So far no account has been taken of the forces between the two species constituting the encounter complex. It can be shown[30,31] that the effect of a potential energy of interaction $E_{ij}$ at a distance $R_{ij}$ is to multiply the diffusion-controlled velocity constant by a factor $q$, given by

$$q = \frac{E_{ij}}{kT}\left\{\exp\left(\frac{E_{ij}}{kT}\right) - 1\right\}^{-1} \tag{47}$$

---

* A more accurate treatment shows that expression (44) for $k_D$ should be divided by a factor $1+\{4\pi N_0(D_i+D_j)R_{ij}/1000k\}$ where $k$ is the velocity constant (in $\mathrm{dm^3\,mol^{-1}\,s^{-1}}$) which would be observed if the equilibrium spatial distribution of the reactants were not disturbed by the reaction. However, this correction is a small one, and since the whole treatment is approximate it is usually omitted. It should be mentioned that the diffusion model also gives a value for the rate of dissociation of the encounter complex; since the equilibrium constant between two species and their encounter complex cannot be affected by rates of diffusion, the first-order rate constant for dissociation must also contain diffusion constants, and in our present nomenclature is given by $(3/R_{ij}^2)(D_i+D_j)$. For a critical account of diffusion control in reactions in solution, see R. M. Noyes, *Progr. Reaction Kinetics*, **1**, 129 (1961); J. M. Schurr, *Biophys. J.*, **10**, 700 (1971); K. S. Schmitz and J. M. Schurr, *J. Phys. Chem.*, **76**, 534 (1972).

[30] L. Onsager, *J. Chem. Phys.*, **2**, 599 (1934).
[31] P. Debye, *Trans. Electrochem. Soc.*, **82**, 265 (1942).

The case of most interest in the present context is when both the reactants are ions; for dilute solutions $E_{ij}$ is then given approximately by the simple Coulombic expression

$$E_{ij} = z_i z_j e^2 / \varepsilon R_{ij} \tag{48}$$

where $z_i$ and $z_j$ are the algebraic values of the charges, $e$ the electronic charge, and $\varepsilon$ the dielectric constant. If we insert the appropriate values for water at 298 K, we find that the effect is not very large for singly charged ions, the factor $q$ being between 2 and 3 for values of $R_{ij}$ between 200 and 300 pm.

If we insert in Equation (46) the value of the viscosity of water at 298 K we obtain $k_d = 0.7 \times 10^{10} \, \mathrm{dm^3 \, mol^{-1} \, s^{-1}}$, and similar values are obtained by inserting experimental values of diffusion coefficients and plausible values of $R_{ij}$ in Equation (44). This is therefore the velocity constant which might be anticipated for any thermodynamically favourable (or 'downhill') reaction between two solute species which is not retarded by the need for activation energy or other requirements. In view of the crude nature of the model, this prediction is not likely to be valid to much better than a power of ten, so that $k_D \sim 10^{10} \, \mathrm{dm^3 \, mol^{-1} \, s^{-1}}$ is a more realistic prediction. Equations (47) and (48) predict that reactions between oppositely charged ions will be somewhat faster than those in which at least one reactant is uncharged, and those between ions of like charge somewhat slower.

We shall now consider how far the rates of simple proton-transfer reactions fulfil these predictions. It should first be noted that velocity constants of the magnitude with which we are concerned are not known with any accuracy, and that the discrepancies between the results obtained by different methods are frequently far outside the limits of error estimated by the investigators. This is illustrated by Table 16 which collects rate constants reported for the reaction between hydrogen ions and acetate ions in aqueous solution. Although some of the discrepancies may be due to differences in experimental conditions (e.g., electro-chemical methods often involve the use of high concentrations of inert electrolyte*), there are certainly considerable differences which must stem from experimental or theoretical uncertainties in the techniques used. Caution is therefore necessary in interpreting minor variations in velocity constants of this magnitude, especially if they have been obtained by different methods.

---

*Equation (48) is only valid for dilute solutions, since it neglects the effect of the ionic atmosphere: there are therefore theoretical grounds for expecting the value of $k_D$ to vary with ionic strength.

Table 16. RATE OF COMBINATION OF HYDROGEN IONS AND ACETATE IONS
(Aqueous solutions at 293–298 K)

| $10^{-10} k$, dm$^3$ mol$^{-1}$ s$^{-1}$ | Method | Reference |
|---|---|---|
| 4.5 | Field impulse | (a) |
| 5.1 | Field impulse | (b) |
| 1.1 | Field impulse | (c) |
| 2.7 | Chronopotentiometry, 2M KCl | (d) |
| 2.6 | Rotating disc electrode, 1M KCl | (e) |
| 5.2 | Rotating disc electrode | (f) |
| 3.8 | Faradaic rectification, 1M LiBr | (g) |
| 5.1 | Fluorescence quenching | (h) |

(a) M. Eigen and J. Schoen, *Z. Elektrochem.*, **59**, 483 (1955); (b) M. Eigen and E. M. Eyring, *J. Am. Chem. Soc.*, **84**, 3254 (1962); (c) B. R. Staples, D. J. Turner, and G. Atkinson, *J. Chem. Instr.*, **2**, 127 (1969); (d) J. Giner and W. Vielstich, *Z. Elektrochem.*, **64**, 128 (1960); (e) W. Vielstich and D. Jahn, *Z. Elektrochem.*, **64**, 43 (1960); (f) W. J. Albery and R. P. Bell, *Proc. Chem. Soc.*, 169 (1963); (g) H. Nürnberg, H. W. Dürbeck, and G. Wolff, *Z. Phys. Chem.* (Frankfurt), **52**, 144 (1967); (h) A. Weller, *Z. Phys. Chem.* (Frankfurt), **3**, 238 (1955).

The results for a selection of simple acid-base reaction are given in Tables 17–19, taken from compilations by Eigen.[1,8,9] As indicated in the last paragraph, somewhat different values could be quoted from other authors, and many similar values could be added,[32] but the general picture would not be affected.

Considering first the reactions involving hydrogen or hydroxide ions (Tables 17 and 18) we see that all the velocity constants are in the range $10^{10}$–$10^{11}$ dm$^3$ mol$^{-1}$ s$^{-1}$ and that they show no trend with the p$K$ of the

Table 17. RATE CONSTANTS FOR REACTIONS OF HYDROGEN
ION WITH BASES IN AQUEOUS SOLUTION
$k$ in dm$^3$ mol$^{-1}$ s$^{-1}$, $T \sim 298$ K

| Corresponding acid | p$K$ | lg $k$ |
|---|---|---|
| $H_2O$ | 15.75 | 11.1 |
| $H_2S$ | 7.24 | 10.8 |
| $NH_4^+$ | 9.25 | 10.6 |
| $H_2CO_3$ | 3.77 | 10.7 |
| $CH_3CO_2H$ | 4.75 | 10.7 |
| $C_6H_5CO_2H$ | 4.17 | 10.5 |
| $m$-Nitrophenol | 7.15 | 10.6 |
| $(CH_3)_3NH^+$ | 9.79 | 10.4 |
| $C_3N_2H_3^+$ (Imidazole) | 6.95 | 10.2 |

[32] See, e.g., the results for seventeen carboxylic acids given by Nürnberg, Ref. 20.

Table 18. RATE CONSTANTS FOR REACTIONS OF HYDROXIDE
ION WITH ACIDS IN AQUEOUS SOLUTION

$k$ in $dm^3 \, mol^{-1} \, s^{-1}$, $T \sim 298 \, K$

| Acid | p$K$ | lg $k$ |
|---|---|---|
| $H_3O^+$ | $-1.75$ | 11.1 |
| $NH_4^+$ | 9.25 | 10.6 |
| $(CH_3)_3NH^+$ | 9.79 | 10.3 |
| $NH_3^+CH_2CO_2^-$ | 9.78 | 10.1 |
| $C_3N_2H_5^+$ (Imidazole) | 6.95 | 10.3 |
| $C_6H_5OH$ | 9.98 | 10.1 |

Table 19. RATE CONSTANTS FOR $A_1 + B_2 \rightleftharpoons B_1 + A_2$ IN AQUEOUS SOLUTION

$k$ in $dm^3 \, mol^{-1} \, s^{-1}$, $T \sim 298 \, K$

| Reaction | p$K_1 -$p$K_2$ | lg $k_f$ | lg $k_r$ |
|---|---|---|---|
| *Acetic acid* | | | |
| +dichloroacetate* | $+3.27$ | 6.8 | 10.0 |
| +monochloroacetate* | $+1.90$ | 7.4 | 9.3 |
| +*m*-chloroaniline | $+1.44$ | 8.0 | 9.5 |
| +formate* | $+1.00$ | 8.0 | 9.0 |
| +aniline | $+0.16$ | 8.4 | 8.5 |
| +propionate* | $-0.12$ | 8.6 | 8.5 |
| +malonate* | $-0.95$ | 8.7 | 7.8 |
| +cacodylate* | $-1.44$ | 9.2 | 7.8 |
| +imidazole | $-2.28$ | 9.1 | 6.8 |
| +$HPO_4^{2-}$* | $-2.45$ | 9.7 | 7.2 |
| +hydrazine | $-3.36$ | 9.3 | 6.0 |
| *Phenol* | | | |
| +imidazole | $+3.07$ | 6.9 | 10.0 |
| +hydrazine | $+1.85$ | 7.8 | 9.6 |
| +ammonia | $+0.70$ | 8.5 | 9.2 |
| +n-propylamine | $-0.62$ | 8.8 | 8.2 |
| +piperidine | $-1.14$ | 8.9 | 7.8 |

*These values are from M. L. Ahrens and G. Maass, *Angew. Chem. Internat. Edn.*, 7, 818 (1968).

reacting species. For these species Equation (46) is hardly appropriate, since the abnormally high mobility of hydrogen and hydroxide ions is due to a chain process involving the handing on of a proton by the processes $H_2O + H_3O^+ \rightarrow H_3O^+ + H_2O$ and $HO^- + H_2O \rightarrow H_2O + HO^-$, and cannot be represented by using a small value of $r$ in Equation (45).

However, if the observed diffusion coefficients together with a reasonable value of $R_{ij}$ are inserted in Equation (44) we do in fact obtain values of $k_D$ close to those observed, so that all these reactions may be regarded as diffusion-controlled, involving no chemical activation energy.

The velocity constant given for the reaction $H^+ + OH^- \rightharpoonup H_2O$, $1.4 \times 10^{11} \, dm^3 \, mol^{-1} \, s^{-1}$, is one of the highest for any bimolecular process in solution, being even greater than that of the reaction $H^+ + e^-(aq)$, which is $2 \times 10^{10} \, dm^3 \, mol^{-1} \, s^{-1}$. If the observed velocity constant and diffusion coefficients are inserted in Equation (44), allowing for the electrostatic effect by means of (47) and (48), we find $R_{ij} \sim 800 \, pm$. This is considerably greater than the sum of the radii of $OH^-$ and $H_3O^+$, and has been interpreted to mean that proton-transfer can take place through intervening water molecules (probably two) contained in the species $H_3O(H_2O)_3^+$ and $OH(H_2O)_3^-$ referred to in Chapter 2. However, two recent investigations by electric field pulse[33] and pulse radiolysis[34] methods suggest that the earlier values[35] for the velocity constant may be too high by about a factor of 2, so that this conclusion may need to be revised.

Because of the great acidic and basic strength of $H_3O^+$ and $OH^-$ respectively, all the reactions in Tables 17 and 18 are highly favourable thermodynamically, the smallest equilibrium constant involved being about $10^6$. The reverse reactions are therefore relatively slow, although, as previously explained, the approach to equilibrium will be governed mainly by the larger of the two rate constants. Reactions between two solute species, such as those listed in Table 19, may involve much smaller free-energy changes. The diffusion coefficients of these species are considerably smaller than those of $H_3O^+$ and $OH^-$, and rather smaller rates are to be expected even for diffusion-controlled reactions; a reasonable average rate constant for such reactions is $7 \times 10^9 \, dm^3 \, mol^{-1} \, s^{-1}$, as predicted by Equation (46), no account being taken of electrostatic effects. Moreover, Equation (40) predicts a further small reduction in the observed rates (by a factor of up to 2) when the equilibrium constant is close to unity. The broken lines in Figure 8 represent the predictions of Equation (40) when $k_{ab} = k_{dc} = 7 \times 10^9 \, dm^3 \, mol^{-1} \, s^{-1}$, and $k_{ba} = k_{cd}$. Figure 8 also shows the experimental results from Table 19. These lie on curves of the expected general shape, though when $\Delta pK$ is less than about 2 the

[33] G. Brière and F. Gaspard, *J. Chim. Phys.*, **64**, 403 (1967).
[34] G. C. Barker and D. C. Sammon, *Nature*, **213**, 65 (1967).
[35] M. Eigen and L. de Maeyer, *Z. Elektrochem.*, **59**, 986 (1955); G. Ertl and H. Gerischer, *Z. Elektrochem.*, **65**, 629 (1961); **66**, 560 (1962).

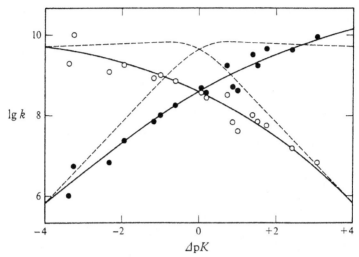

Fig. 8. Velocity constants for proton-transfer reactions of oxygen and nitrogen acids in aqueous solution. (Values from Table 19.)

observed rates are significantly less than those predicted.* This presumably implies that even thermodynamically favourable reactions in this range require a small activation energy other than that involved in the diffusion process. However, the following generalization is widely applicable. *For thermodynamically favourable proton transfers between oxygen or nitrogen atoms in aqueous solution, the velocity constant is approximately* $10^{10}$ *dm³ mol⁻¹ s⁻¹, and is smaller by a factor* $10^{\Delta pK}$ *in the reverse (unfavourable) direction.*

Acids and bases which conform to this generalization have been termed 'normal' by Eigen. There are a number of factors which in special cases can lead to a reduction of these 'normal' rates, even for reactions which are highly favourable thermodynamically, and a few examples will now be given. We have already seen that because of *electrostatic interaction* there should be a reduction in rate for reactions between species bearing charges of the same sign. This is illustrated[36] by the reaction of hydroxide ions with the species $HPO_4^{2-}$, $HP_2O_7^{3-}$, and $HP_3O_{10}^{4-}$, for which the velocity constants are respectively $2 \times 10^9$, $4.7 \times 10^8$, and $2.1 \times 10^8$

---

*It is not possible to observe directly the rates of proton transfers between solute species in aqueous solution when $\Delta pK$ is very large, because processes involving solvent species then become dominant. Information of this kind can, however, be obtained indirectly from studies of acid-base catalysis, and will be considered in subsequent chapters.

[36] M. Eigen and W. Kruse, *Z. Naturforsch.*, **18b**, 857 (1963).

$dm^3 mol^{-1} s^{-1}$. *Internal hydrogen bonding* can also retard reaction, as in the reaction of salicylate ion with hydrogen ions,[37] which has $k = 6.4 \times 10^9$ compared with $(3.8 \pm 1.0) \times 10^{10}$ for eleven other carboxylate ions; similarly, the reaction of $N,N$-dimethylanthranilic acid with hydroxide ions[36] has $k = 1.1 \times 10^7 dm^3 mol^{-1} s^{-1}$. *Steric hindrance* is not often prominent in proton-transfer reactions, but is probably the reason why the rate constants for $CH_3CO_2^- + H^+$ and $(CH_3)_3CCO_2^- + H^+$ are respectively $5.2 \times 10^{10}$ and $1.5 \times 10^{10} dm^3 mol^{-1} s^{-1}$.[38]

Much more striking deviations from normal behaviour are found in reactions involving *carbon acids*, i.e., acids in which the mobile proton is attached to a carbon atom. Much of the information on this point originally came from studies of acid-base catalysis, and will be described later in this book, but there is now ample evidence from direct rate measurements. In fact, some of these reactions are so slow that they can be studied by conventional methods, without resort to any of the special techniques described in this chapter. Thus the slow reactions between nitroparaffins and hydroxide ions have been familiar ever since they were first reported by Hantzsch,[39] and led to the introduction of the term

Table 20. RATE CONSTANTS FOR PROTON TRANSFERS INVOLVING CH ACIDS IN AQUEOUS SOLUTION

$k$ in $dm^3 mol^{-1} s^{-1}$, $T \sim 298$ K

| Reaction | $\Delta pK$ | $\lg k$ | Reference |
|---|---|---|---|
| $CH_3NO_2 + OH^-$ | 5.5 | 1.4 | (a) |
| $(CH_3)_2CHNO_2 + OH^-$ | 8.0 | $-0.5$ | (a) |
| $[CH_2NO_2]^- + H^+$ | 12.0 | 2.8 | (a) |
| $[CH_2NO_2]^- + CH_3CO_2H$ | 5.5 | 1.3 | (b) |
| $CH_2NO_2CO_2C_2H_5 + OH^-$ | 10.0 | 5.2 | (c) |
| $CH_2NO_2CO_2C_2H_5 + C_6H_5O^-$ | 4.2 | 4.5 | (c) |
| $[CHNO_2CO_2C_2H_5]^- + H^+$ | 7.6 | 4.0 | (c) |
| $(CH_3CO)_2CH_2 + OH^-$ | 6.9 | 4.6 | (d) |
| $[(CH_3CO)_2CH]^- + H^+$ | 10.6 | 7.2 | (d) |
| $[(CH_3CO)_2CH]^- + CH_3CO_2H$ | 4.1 | 4.7 | (d) |
| $[(CH_3CO)_2CH]^- + C_6H_5NH_3^+$ | 4.3 | 5.1 | (d) |
| $[(CH_3CO)_3C]^- + H^+$ | 9.5 | 6.0 | (c) |

(a) R. P. Bell and D. M. Goodall, *Proc. Roy. Soc.*, A, **294**, 273 (1966); (b) D. M. Goodall and F. A. Long, *J. Am. Chem. Soc.*, **90**, 238 (1968); (c) R. P. Bell and D. J. Barnes, *Proc. Roy. Soc.*, A, **318**, 421 (1970); (d) M. L. Ahrens, M. Eigen, W. Kruse, and G. Maass, *Ber. Bunsengesell. Phys. Chem.*, **74**, 380 (1970).

[37] H. Nürnberg, Ref. 20.
[38] W. J. Albery and R. P. Bell, *Proc. Chem. Soc.*, 169 (1963).
[39] A. Hantzsch, *Ber.*, **32**, 575 (1899).

*pseudo-acid*, which is still sometimes used to describe carbon acids in general.

Some examples of velocity constants obtained by direct measurement are given in Table 20. All the reactions listed are thermodynamically highly favourable, with $\Delta pK > 4$, but the velocity constants are nevertheless far below the value of $10^{10} \, dm^3 \, mol^{-1} \, s^{-1}$ which characterizes diffusion control; a considerable 'chemical' activation energy must therefore be involved. This has been variously attributed to the inability of carbon to take part in hydrogen-bond formation, or to the considerable structural reorganization which often accompanies the interconversion of the acidic and basic forms of carbon acids. We shall return to this point in Chapter 10, when we shall see that in fact some carbon acids behave normally in the same sense as oxygen and nitrogen acids.

# 8 The Indirect Study of Rates of Proton-Transfer

We have so far paid no attention to any processes which may occur after a proton has been transferred from an acid to a base, with the production of two new species. These new species may have electron distributions differing considerably from those of the original acid and base, and may therefore undergo further reactions such as decomposition, rearrangement, reaction with the solvent, or with some other species present in solution. It is frequently the case that in the reaction $A_1 + B_2 \rightarrow B_1 + A_2$ one product (say $B_1$) does undergo further reaction to a product X while the other ($A_2$) does not, so that $B_2$ can easily be regenerated. Under these circumstances we speak of *acid-base catalysis*, and the reaction just given would be described as the conversion of the *substrate* $A_1$ into X, catalysed by the base $B_2$. Reactions catalysed by acids or bases are of course very common, especially in organic chemistry, and a few examples will be discussed in detail in the next chapter. In the present chapter we shall be concerned with the general kinetic consequences of such behaviour, and in particular how the rates of proton-transfer reactions can be inferred indirectly from studies of acid-base catalysis and related observations. It will then be possible in subsequent chapters to discuss the interpretation of these rates, using information obtained by both direct and indirect methods.

The early study of catalysis by acids and bases was concerned chiefly with the use of catalysed reactions for investigating general problems of physical chemistry. For example, the first correct formulation of the kinetic laws of a first-order reaction was made by Wilhelmy in 1850 in connection with his measurements of the catalytic inversion of cane sugar by acids.[1] Catalytic reactions also played an important part in the foundation of the classical theory of electrolytic dissociation towards the end of the nineteenth century, and kinetic measurements (notably on the

---

[1] L. F. Wilhelmy, *Pogg. Ann.*, **81**, 413, 499 (1850).

hydrolysis of esters) were used widely for investigating the state of electrolyte solutions.

The classical theory of acid-base catalysis assumed that hydrogen and hydroxide ions are the only effective catalysts, that the reaction velocity is proportional to the concentration of the catalysing ion, and that the degrees of dissociation of the catalysing ions were given directly by their electrolytic conductivities. These assumptions gave a good general description of the facts, but a number of discrepancies arose when they were applied quantitatively. The next phase in the study of acid-base catalysis, especially associated with the name of J. N. Brönsted, dealt mainly with the clearing up of these discrepancies, partly by the application of modern views on electrolyte solutions and partly by the deduction from experiment of new laws governing catalytic phenomena. The most important of the latter were the concept of general acid-base catalysis and the establishment of relations between catalytic power and acid-base strength. In this way the systematics of acid-base catalysis were largely established in the decade 1920–1930, and little has been added later to the fundamental aspects of this subject. This part of the story is well known[2] and will not be repeated here, though comment is needed on some particular points.

The subject of *salt effects* arises in all kinetic work involving electrolytes, and it was a thorough investigation of these effects which served to clarify many of the early misunderstandings about acid-base catalysis. Since the catalyst is commonly ionic, the same problems often arise even when no other electrolyte has been added to the system. Salt effects are commonly classified as *primary* and *secondary*. Secondary salt effects are not essentially kinetic in nature but arise from the effect of ionic environment on ionic equilibria. In acid-base catalysis these are commonly the dissociation equilibria of weak acids or weak bases which are acting as catalysts; a change in the ionic environment affects the concentration of hydrogen or hydroxide ions in the solution and hence the velocity of a reaction catalysed by these ions. This type of effect can be investigated by thermodynamic measurements on the catalyst solution, without any addition of reactant, and at sufficiently low ionic concentrations it can be accounted for by the theory of interionic attraction.[3]

Primary salt effects arise in the actual kinetic step of the reaction and appear in their simplest form when all the electrolytes present in the solution are completely dissociated, so that there are no equilibria which

---

[2] For an account of the historical development of the subject, see R. P. Bell, *Acid-Base Catalysis*, Oxford, 1941.
[3] See Bell, Ref. 2, Ch. 2.

can be displaced. If we are considering the effect of environment upon the velocity $v$ of a reaction between two species X and Y, the fundamental equation is

$$v = k[X][Y]f_X f_Y / f_\ddagger, \tag{49}$$

where $k$ is independent of environment and the symbol $\ddagger$ refers to the transition state of the reaction. This expression was originally advanced by Brönsted[4] on not very clear theoretical grounds but would now be regarded as a consequence of the transition-state theory of reaction velocities. Equation (49) leads to useful predictions when both X and Y bear charges, since the charge on the transition state will then differ from either, and both theory and experiment show that in dilute solutions the activity coefficients of ions depend primarily upon their charges. However, in most instances of acid-base catalysis the substrate is uncharged; thus for hydrogen ion catalysis of a substrate S, Equation (49) becomes

$$v = k[S][H^+]f_S f_{H^+} / f_\ddagger, \tag{50}$$

where the transition state now has a single positive charge. In this equation $f_S$ is accessible experimentally, but there is no way of measuring the ratio $f_{H^+}/f_\ddagger$ except by the kinetic experiments themselves, and the interionic theory only makes the prediction that it should differ little from unity in dilute solution.

For many purposes this uncertainty does not matter, since the primary salt effect can often be neglected in dilute solutions or eliminated by extrapolation to zero ionic strength. Salt effects sometimes occur, however, which are much greater than would be expected, and which are specific to the nature of the ion added. They are particularly large when multiply charged ions can interact with reactants or transition states bearing charges of the opposite sign, and are commonly interpreted in terms of *ion association*.[5] In addition, some ions (particularly metallic cations with multiple charges) may exert a large positive catalytic effect on account of particular electronic interactions, especially the formation of chelate complexes which stabilize the transition state relative to the initial state. This type of catalysis has been particularly investigated for decarboxyla-

[4] J. N. Brönsted, *Z. Phys. Chem.*, **102**, 169 (1922).
[5] For general accounts, see, e.g., C. W. Davies, *Ion Association*, Butterworths, London, 1962; G. N. Nancollas, *Interactions in Electrolyte Solutions*, Elsevier, Amsterdam, 1966; A. D. Pethybridge and J. E. Prue, in *Inorganic Reaction Mechanisms* (ed. J. O. Edwards), Wiley-Interscience, New York, 1972.

tion reactions,[6] but it is also present in a number of reactions catalysed by acids and bases.[7] It is certainly not appropriate to describe these phenomena as 'salt effects', and since they involve the acceptance of electron pairs by the metallic cations they could be classified as 'acid catalysts' in the Lewis sense of the term. However, we shall not pursue this subject further in a book devoted to the chemical behaviour of the proton.

It will be seen that there are many phenomena which are not adequately described in terms of the original concept of salt effects, though in aqueous solutions there is still a wide area in which primary salt effects can be neglected and secondary effects satisfactorily dealt with by the interionic theory. This is commonly true in solutions of ionic strength less than 0.1 not containing multiply charged ions, though an exception must be made for the ions $Tl^+$ and $Ag^+$. The position is otherwise in non-aqueous solvents of lower dielectric constant, where electrostatic interactions are much greater. We have already seen in Chapter 4 the importance of ion-pair formation for acid-base equilibria in non-aqueous solvents, and work by Winstein and his collaborators[8] has shown that they also play an important part in the solvolysis reactions of many organic compounds, leading to large and specific salt effects. Somewhat similar behaviour has been observed by Eastham[9] in the base-catalysed mutarotation of tetramethyl- and tetraacetyl-glucose in pyridine and nitromethane. The catalytic effect of uncharged bases alone is very low, but it is greatly enhanced by the addition of a wide variety of salts. For example, 0.02M $LiClO_4$ increases the catalytic effect of pyridine by a factor of ten, though the magnitude of the effect varies widely from one salt to another. In the absence of salt the mechanism of the reaction (which involves the intermediate formation of the aldehydic form of glucose) would be written schematically as

[6] E.g., R. Steinberger and F. H. Westheimer, *J. Am. Chem. Soc.*, **73**, 429 (1951); K. J. Pedersen, *Acta Chem. Scand.*, **6**, 285 (1952); J. E. Prue, *J. Chem. Soc.*, 2331 (1952); E. Gelles and A. Salama, *J. Chem. Soc.*, 3689 (1958); J. V. Rund and R. A. Plane, *J. Am. Chem. Soc.*, **86**, 367 (1964); J. V. Rund and K. G. Claus, *Inorg. Chem.*, **7**, 860 (1968); D. W. Larson and M. W. Lister, *Canad. J. Chem.*, **46**, 823 (1968); K. G. Claus and J. V. Rund, *Inorg. Chem.*, **8**, 59 (1969); R. W. Hay and K. N. Leong, *J. Chem. Soc.*, A, 3639 (1971).
[7] K. J. Pedersen, *Acta Chem. Scand.*, **2**, 252, 385 (1948); H. Kroll, *J. Am. Chem. Soc.*, **74**, 2036 (1952); L. Meriwether and F. H. Westheimer, *J. Am. Chem. Soc.*, **78**, 5119 (1956); H. L. Conley and R. B. Martin, *J. Phys. Chem.*, **69**, 2923 (1965); R. W. Hay and L. J. Porter, *J. Chem. Soc.*, A, 127 (1969); R. W. Hay and P. J. Morris, *J. Chem. Soc.*, A, 1524 (1971); J. E. Meany, *J. Phys. Chem.*, **73**, 3421 (1969).
[8] A. H. Fainberg and S. Winstein, *J. Am. Chem. Soc.*, **78**, 328, 2763, 2767, 2777, 2780, 2784 (1956); S. Winstein, S. Smith, and D. Darwish, *J. Am. Chem. Soc.*, **81**, 5511 (1959).
[9] A. M. Eastham, E. L. Blackall, and G. A. Latremouille, *J. Am. Chem.Soc.*, **77**, 2182 (1955); E. L. Blackall and A. M. Eastham, *J. Am. Chem. Soc.*, **77**, 2184 (1955).

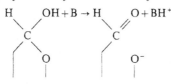

so that the transition state will involve a considerable separation of positive and negative charge. Eastham considers that an ion pair $M^+X^-$ can stabilize the transition state electrostatically by a configuration such as

where the effectiveness of an ion pair will obviously depend upon its individual geometry and charge distribution. Eastham has termed these effects 'electrolyte catalysis', and they are likely to occur fairly frequently in solvents of low dielectric constant.

The second modification which Brönsted introduced into the classical theory was the idea of *general acid-base catalysis*, and this is now generally accepted essentially in its original form. It was closely bound up with the definition of acids and bases, discussed in Chapter 2. As soon as it was realized that the hydrogen and hydroxide ions in water were not unique but were members of the general class of proton donors and proton acceptors respectively, it became natural to expect that they would not have a monopoly of catalytic power. This was soon found to be the case for a number of reactions, notably the iodination of acetone,[10] the decomposition of nitramide,[11] and the mutarotation of glucose.[12] The experimental evidence for general acid-base catalysis in aqueous solution rests essentially on the form of the observed rate laws in solutions containing weak acids and bases; for example, if a reaction exhibits general catalysis by both acids and bases, its velocity in an acetate buffer solution will be given by an expression of the form

$$v = v_0 + k_H[H_3O^+] + k_{OH}[OH^-] + k_{HOAc}[HOAc] + k_{OAc}[OAc^-] \quad (51)$$

where the so-called 'spontaneous' rate $v_0$ is really due to catalysis by water molecules. This kind of behaviour is found not only in catalysed

[10] H. M. Dawson and F. Powis, *J. Chem. Soc.*, 2135 (1913), and later papers.
[11] J. N. Brönsted and K. J. Pedersen, *Z. Phys. Chem.*, **108**, 185 (1923).
[12] T. M. Lowry and G. F. Smith, *J. Chem. Soc.*, 2539 (1927); J. N. Brönsted and E. A. Guggenheim, *J. Am. Chem. Soc.*, **49**, 2554 (1927).

reactions but in any other experimental study of the rates of acid-base reactions. Thus several of the investigations of fast processes described in the last chapter revealed terms in the rate equation which did not involve either hydrogen ion or hydroxide ions. In principle all the constants in Equation (51) can be determined experimentally by making measurements in buffer solutions of varying ratios and concentrations, but of course this will be difficult in practice if some of the terms contribute little to the observed velocity, and under these conditions the small uncertainties due to salt effects may make it impossible to decide whether some of these terms are present at all. These difficulties become especially great in solvents of lower dielectric constant, such as the alcohols, where both primary and secondary salt effects are much greater than in water and our knowledge of acid-base equilibria is less complete.

The position is simpler in principle in non-dissociating solvents such as the hydrocarbons, where the solvent itself has no catalytic power and there are no analogues of the hydrogen and hydroxide ion. For example, in a solution of acetic acid in benzene the only species which can be catalytically active is the acetic acid molecule, and simpler rate laws might be expected. In practice, however, there are frequently complications due to association in this type of solvent, and it will be shown later that the actual kinetic processes may also be more complex.

Early workers on acid-base catalysis had no clear ideas about the *mechanism* of catalysis, which was regarded as some kind of 'influence'. These early views envisaged reactions which could take place in the absence of a catalyst but were facilitated by its presence; however, evidence gradually accumulated to show that most of the reactions subject to acid-base catalysis cannot take place at all in the complete absence of catalysts, apparently spontaneous reactions being usually due to catalysis by solvent molecules or by some adventitious acidic or basic impurity. This indicates that the catalyst takes part in some fundamental way in the reaction mechanism, and the present view is that acid-base catalysis always involves *an acid-base reaction between catalyst and substrate.*

We turn now to the *kinetic analysis* of acid-base catalysis, following a number of previous treatments[13] It is customary to classify these reaction mechanisms according to the number of proton transfers involved. Of course, in any truly catalytic reaction the catalyst must be finally regenerated, so that every initial proton transfer must be reversed; however,

[13] K. J. Pedersen, *J. Phys. Chem.*, **38**, 581 (1934); *Trans Faraday Soc.*, **34**, 237 (1938); A. Skrabal, *Z. Elektrochem.*, **46**, 146 (1940); Bell, Ref. 2, Ch. 6; *Adv. Catalysis*, **4**, 151 (1952).

from the kinetic point of view we are not interested in fast reactions which occur after the rate-determining step, and the number of kinetically significant proton transfers is usually either one or two.

We shall treat first *base-catalysed reactions involving a single proton transfer*, which probably constitute the most important class of reactions, and initially only aqueous solutions will be considered. For a substrate SH the sequence of reactions is

$$\text{(a)} \quad SH + \sum_i B_i \underset{k_{-1}}{\overset{k_1}{\rightleftharpoons}} S^- + \sum_i A_i$$

$$\text{(b)} \quad S^- \xrightarrow{k_2} X \tag{52}$$

where reaction (b), giving the products X, does not involve an acid-base reaction, though it may involve some other reagents present in the solution.* For the sake of simplicity we shall suppose that reaction (b) is irreversible, or is made effectively so by removing the product X as quickly as it is formed. If it is assumed that the concentrations of the acid-base pairs $A_i$–$B_i$ are held constant during a given reaction then the first stage of the reaction can be characterized by the two first-order velocity constants $k_1$ and $k_{-1}$, the values of which will depend on the concentrations of acids and bases present in the system. $k_1$ can be expressed in the form $k_1 = \sum \pi_i [B_i]$, where the $\pi_i$ are characteristic velocity constants for proton transfers to the different bases, and $k_{-1}$ is related to $k_1$ through $K_{SH}$, the conventional acid strength of the substrate. If we consider the hypothetical equilibrium between SH and $S^-$, we have

$$k_1/k_{-1} = [S^-]_e/[SH]_e = K_{SH}/[H^+] = K_{SH}[OH^-]/K_w. \tag{53}^\ddagger$$

Similarly, the second stage of the reaction is characterized by a first-order velocity constant $k_2$, which may well depend upon the concentration of other reactants. For example, in halogenation reactions showing acid-base catalysis it will involve the concentration of halogenating species in the solution.

The general solution for the kinetic scheme (52) is

$$1 - \frac{[X]}{a} = \frac{\rho_2}{\rho_2 - \rho_1} e^{-\rho_1 t} - \frac{\rho_1}{\rho_2 - \rho_1} e^{-\rho_2 t} \tag{54}$$

---

* These reagents may be acids or bases, but it is implied that their reaction in (52b) shall not involve the transfer of a proton. For example, the reaction may consist of solvolysis by a water molecule, which could not be brought about by any other acidic or basic species.
‡ It should be noted that (53) does not imply that the equilibrium between SH and $S^-$ is actually established under experimental conditions.

where $a$ is the initial concentration of substrate, $[X]$ the concentration of product after time $t$, and $\rho_1$ and $\rho_2$ the roots of the equation $\rho^2 - (k_1 + k_{-1} + k_2)\rho + k_1 k_2 = 0$. This does not represent a reaction of any simple order. Further, the reaction velocity is not a simple function of the base concentrations (including $[OH^-]$) and the effects of different catalysts are not additive. In practice, however, simpler behaviour is usually found, corresponding to special cases for the relative values of $k_1$, $k_{-1}$, and $k_2$. These will now be treated separately.

(a) $k_1 \ll k_{-1}$. This corresponds to the commonest case in which SH is a very weak acid and the hydroxide ion concentration is not great enough to convert an appreciable proportion of it into $S^-$, even if equilibrium is reached. Equation (54) reduces to a single exponential term, corresponding to a first-order reaction with velocity constant $k$. Since $[S^-]$ is now small throughout, the same result can be obtained by the usual steady-state approximation. The observed first-order velocity constant $k$ is given by

$$\frac{1}{k} = \frac{1}{k_1} + \frac{k_{-1}}{k_1 k_2} = \frac{1}{\sum \pi_i [B_i]} + \frac{\sum \pi_i' [A_i]}{k_2 \sum \pi_i [B_i]}$$

$$= \frac{1}{\sum \pi_i [B_i]} + \frac{K_w}{k_2 K_{SH} [OH^-]}. \tag{55}$$

Equation (55) corresponds qualitatively to general base catalysis, but the reaction velocity is not a linear function of catalyst concentrations, and the effect of different catalysts is not additive.

(b) $k_1 \ll k_{-1}$ and $k_2 \gg k_{-1}$. Equation (55) becomes

$$k = k_1 = \sum \pi_i [B_i]. \tag{56}$$

This corresponds to general base catalysis, with the reaction velocity a linear function of the catalyst concentrations; the observed velocity is determined directly by the proton transfers from the substrate to the bases.

(c) $k_1 \ll k_{-1}$ and $k_2 \ll k_{-1}$. Equation (55) becomes

$$k = k_1 k_2 / k_{-1} = k_2 K_{SH} [OH^-] / K_w, \tag{57}$$

corresponding to specific catalysis by hydroxide ions. The first stage of the reaction is now effectively at equilibrium, and the reaction rate depends only on $[OH^-]$, although the proton transfers taking place involve all the bases in solution.

If we now drop the restriction $k_1 \ll k_{-1}$, there are still two cases giving simple expressions. The assumption $k_1 \sim k_{-1}$ means that the substrate is a strong enough acid and the hydroxide ion concentration high enough so that an appreciable proportion of SH would be converted into $S^-$ if equilibrium were reached.

(d) $k_1 \sim k_{-1}$ and $k_2 \gg k_{-1}$. $[S^-]$ still remains small, since it is removed by the rapid second stage of the reaction, and we arrive at the same result as (56), $k = \sum \pi_i [B_i]$.

(e) $k_1 \sim k_{-1}$ and $k_2 \ll k_{-1}$. The first stage of the reaction is effectively in equilibrium throughout, and an appreciable proportion of substrate is actually present as $S^-$. Equation (54) gives approximately

$$\rho_1 = k_1 k_2 / (k_{-1} + k_1), \qquad \rho_2 = k_{-1} + k_1,$$

and since $\rho_2 \gg \rho_1$ the second term of (54) is negligible except for very small values of $t$. This again represents first-order kinetics, with a velocity constant

$$k = \frac{k_1 k_2}{k_1 + k_{-1}} = \frac{k_2 K_{SH}[OH^-]}{K_w + K_{SH}[OH^-]}. \tag{58}$$

Equation (58) corresponds qualitatively to specific catalysis by hydroxide ions, but the velocity is not proportional to $[OH^-]$; in fact, it should reach a limiting value at high concentrations of alkali corresponding to complete conversion of SH into $S^-$.*

Equations exactly analogous to (52)–(58) can be derived for *acid catalysis involving a single proton transfer*, in which the first step is the protonation of the substrate SH to give $SH_2^+$, which then reacts further to give products. There is again a distinction between general acid catalysis, in which the protonation step is rate-determining, and specific catalysis by hydrogen ions, in which the protonation step is effectively at equilibrium, followed by a subsequent rate-determining reaction; more complicated situations analogous to (55) and (58) are also possible. Examples of different types of behaviour, for both acid and base catalysis, will be given in the next chapter.

We must now consider the kinetic analysis of catalysed reactions involving *two proton transfers*. These include the so-called *prototropic isomerizations*, notably keto–enol tautomerism. The simplest mechanisms for catalysis by acids and bases are as follows:

---

*The conversion of SH into $S^-$ will of course remove $OH^-$ from the solution, and our initial supposition that the acid-base concentrations do not change during a reaction will be satisfied only if the concentration of catalyst is considerably greater than that of substrate.

*Acid catalysis*

$$\overset{\diagdown}{\underset{\diagup}{\phantom{C}}}CH-\overset{\diagup}{\underset{\phantom{C}}{C}}\!\!=\!\!O + A \rightleftharpoons \overset{\diagdown}{\underset{\diagup}{\phantom{C}}}CH-\overset{\diagup}{\underset{\phantom{C}}{C}}\!\!=\!\!OH^+ + B \rightleftharpoons \overset{\diagdown}{\underset{\diagup}{\phantom{C}}}C\!\!=\!\!\overset{\diagup}{\underset{\phantom{C}}{C}}\!\!-\!OH + A \qquad (59)$$

*Base catalysis*

$$\overset{\diagdown}{\underset{\diagup}{\phantom{C}}}CH-\overset{\diagup}{\underset{\phantom{C}}{C}}\!\!=\!\!O + B \rightleftharpoons \overset{\diagdown}{\underset{\diagup}{\phantom{C}}}C\!\!=\!\!\overset{\diagup}{\underset{\phantom{C}}{C}}\!\!-\!O^- + A \rightleftharpoons \overset{\diagdown}{\underset{\diagup}{\phantom{C}}}C\!\!=\!\!\overset{\diagup}{\underset{\phantom{C}}{C}}\!\!-\!OH + B \qquad (60)$$

and similar schemes can be written for analogous transformations such
as lactam–lactim, nitroso–isonitroso, nitro–*aci*-nitro, and three-carbon
tautomerism

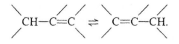

For simplicity we shall write the two isomers schematically as HS and
SH, and if the overall reaction is made irreversible by removing the
product as quickly as it is formed, the kinetic scheme for acid catalysis
becomes

$$HS + \sum A_i \underset{k_{-1}}{\overset{k_1}{\rightleftharpoons}} HSH^+ + \sum B_i \overset{k_2}{\longrightarrow} SH + \sum A_i, \qquad (61)$$

Provided that the acid-base composition of the solution remains constant
during a reaction, the kinetics can again be described by three first-order
constants $k_1$, $k_{-1}$, and $k_2$, all of which will now depend upon the concentrations of the acids and bases present. A kinetic analysis can be carried
out just as for reactions involving a single proton transfer, with the
following results:

(a) $k_1 \sim k_{-1} \sim k_2$. Complicated kinetics result (cf. Equation (53));
there is no experimental evidence for this case.

(b) $k_1 \ll k_{-1} \ll k_2$. The rate is now determined by the first step in (61),
and only the first proton transfer is kinetically significant. Hence we have

$$k = k_1 = \sum \pi_i [A_i] \qquad (62)$$

i.e., general acid catalysis.

(c) $k_1 \ll k_{-1} \gg k_2$. The first step of (61) is now effectively at equilibrium,
and the rate is given by

$$v = [HSH^+] \sum \pi_i' [B_i] = [HS] \sum \pi_i' [B_i] [HSH^+] [HS].$$

The observed first-order constant is therefore

$$k = \sum \pi_i'[B_i][HSH^+]/[HS]$$
$$= \sum \pi_i'[B_i][H^+]/K_{HSH} = \sum \pi_i'[A_i]K_i/K_{HSH} \tag{63}$$

where $K_i$ and $K_{HSH}$ are the acid strengths of $A_i$ and $HSH^+$. This result is experimentally indistinguishable from (62), so that general acid catalysis is again observed although the rate-determining step is actually the transfer of a proton from basic catalysts to the ion $HSH^+$. This is in contrast to reactions involving a single proton transfer, where the same kinetic scheme led to specific catalysis by hydrogen ions. The identical form of (62) and (63) is of course due to the fact that in both cases the transition state is composed of one molecule of HS and one molecule of $A_i$. This does not mean, however, that the two mechanisms are identical, since the structures of the two transition states are different. For example, if a keto–enol change is catalysed by an acid HX, the transition states for cases (b) and (c) are

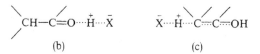

(b)                               (c)

(d) $k_1 \ll k_{-1} \sim k_2$. The concentration of $HSH^+$ is still small throughout, so that the steady-state treatment can be applied, giving

$$\frac{1}{k} = \frac{1}{k_1} + \frac{k_{-1}}{k_1 k_2} = \frac{1}{\sum \pi_i[A_i]} + \frac{\sum \pi_i''[B_i]}{\sum \pi_i[A_i]\sum \pi_i'[B_i]}$$
$$= \frac{1}{\sum \pi_i[A_i]} + \frac{1}{K_{HSH}\sum K_i\pi_i'[A_i]} \tag{64}$$

where $k_1 = \sum \pi_i[A_i]$, $k_2 = \sum \pi_i'[B_i]$, $k_{-1} = \sum \pi_i''[B_i]$. If only a single acid-base pair is effective, Equation (64) reduces to

$$k = \pi\pi'[A]/(\pi' + \pi'') \tag{65}$$

which represents general acid catalysis with linear dependence of velocity upon concentration. If catalysis involves more than one acid-base pair, including those derived from the solvent, then the reaction will show the qualitative characteristics of general acid catalysis but in general will not follow the usual quantitative laws. In particular, the velocity will not be a linear function of catalyst concentration, and the effect of several catalysts present simultaneously will not be additive. A simple result is, however, still obtained if we can write

$$\pi_1'/\pi_1'' = \pi_2'/\pi_2'' = \cdots = \pi_i'/\pi_i'' = \cdots = \beta \tag{66}$$

when (63) becomes

$$k = \frac{\beta}{1+\beta} \sum \pi_i [A_i]. \tag{67}$$

Although there is no reason why (66) should be generally valid, it is likely to hold approximately for catalysts of similar structure, and this may be why the more complex behaviour represented by (64) has not been observed in prototropic reactions.

(e) $k_1 \sim k_{-1} \ll k_2$. If equilibrium were attained in the first stage of the reaction, a considerable proportion of HS would be converted into $HSH^+$; however, since $k_2 \gg k_{-1}$ the concentration of $HSH^+$ remains very small and the steady-state treatment can again be applied, giving results identical with (d).

(f) $k_1 \sim k_{-1} \gg k_2$. The solution will now contain an appreciable proportion of $HSH^+$, and it is necessary to distinguish between the rate of disappearance of HS and the rate of appearance of SH. In terms of the latter we find

$$k = \frac{k_1 k_2}{k_1 + k_{-1}} = \frac{\sum \pi_i [A_i] \sum \pi_i' [B_i]}{\sum \pi_i [A_i] + \sum \pi_i'' [B_i]} = \frac{\sum \pi_i' K_i [A_i]}{[H^+] + K_{HSH}} \tag{68}$$

This predicts general acid catalysis, but also an inverse dependence on hydrogen ion concentration. In practice the latter is likely to be important only at high acidities, and under these conditions general catalysis by acids other than the hydrogen ion will be difficult to detect; moreover, the simple equilibrium expressions used in the above derivation will break down. The situation is thus a complex one. Equations analogous to (62)–(68) can obviously be derived for base-catalysed prototropy.

It will be seen that under conditions of general catalysis by acids and bases information can be obtained about the rate of proton transfer in one direction between the catalyst and the substrate; this contrasts with the direct methods described in Chapter 7 in which the observed changes always involve both the forward and the reverse processes, being dominated by the faster of the two rates. Observations on general acid-base catalysis can therefore be used to study a slow forward proton-transfer even when the reverse reaction is very fast. The same applies to isotope exchange and racemization reactions, which have been used extensively for this purpose.

Considering first *isotope exchange*, if an isotopically labelled substrate SL (where L is either deuterium or tritium) ionizes in a large excess of solvent $H_2O$, the ion $S^-$ will revert to SH in almost every case, so that the rate of conversion of SL into SH is a measure of its rate of ionization. Similarly, a protonated species $SLH^+$ can revert either to SL or SH, so that in acid solution the rate of exchange is related to the rate of protonation. (A statistical factor is necessary here to allow for the fact that either SH or SL may be re-formed, and it is usually necessary to allow for the different rates of reaction of the different isotopes.) An alternative procedure is to label the solvent with deuterium or tritium, and to observe the rate at which these isotopes are incorporated into the substrate.

Similar remarks apply to the *base-catalysed racemization* of optically active compound $R_1R_2R_3CH$. The removal of the proton will produce an anion whose geometry may depend upon the nature of the groups which impart an appreciable acidity to the C—H group. For ketones and similar substances the structure of the ion will correspond closely to

so that the left-hand carbon atom will be in the same plane as the three groups attached to it, and if the compound $R_2R_3CHCOR_1$ is regenerated the two optical isomers will be formed in equal amounts. In other instances the carbon atom may bear a considerable negative charge, corresponding to a pyramidal carbanion, but such ions would be expected to invert their configuration readily, so that racemization will again take place.* A two-stage reaction scheme such as (52) (or the analogous scheme for acid catalysis) is not really appropriate to isotope exchange or racemization,

* The discussion in the last two paragraphs implies that the intermediate anions and cations have a sufficiently independent existence to undergo isotopic exchange, racemization, or further reaction (e.g., with halogens) in preference to reverting to their original state. This in turn implies that for a given reaction exhibiting general catalysis by acids and bases the rates of these various processes should be equal. This is frequently found to be the case, but in some instances, especially in solvents of low dielectric constant, matters may be complicated by ion pair formation between the primary products, and different rates may be observed for the different processes. For a further discussion see D. J. Cram, *Fundamentals of Carbanion Chemistry*, Academic Press, New York, 1965; C. K. Ingold, *Structure and Mechanism in Organic Chemistry*, 2nd edn., Bell, London, 1969, pp. 831–834.

since in these processes the reaction leading to the observed phenomena is chemically identical with the reverse of the initial acid-base reaction, differing from it only isotopically or stereochemically. There is therefore no state of affairs corresponding to specific catalysis by hydrogen ions or hydroxide ions, and the rates observed always refer to proton transfers between the catalyst and the substrate.

Returning to prototropic reactions, for example keto–enol tautomerism, in the base-catalysed reaction the formation of the anion will lead to racemization and isotope exchange without involving the enol at all. Similarly, it should react with halogens even more rapidly than the enol itself. Observed rates of base-catalysed racemization, isotope exchange, or halogenation are therefore rates of ionization rather than rates of enolization. As Hammett has expressed it, 'There is no reason to suppose that the formation of an electrically neutral enol form represents anything more than an unimportant by-path into which a portion of the reacting substance may transiently stray'.[14] These reactions can therefore be regarded as involving a single proton transfer, since all subsequent events are kinetically unimportant, and general base catalysis will be observed as in Equation (56). On the other hand, in acid catalysis according to scheme (59) the hydrogen atom attached to carbon is not removed until the second step of the reaction is accomplished, so that the rate of racemization or isotope exchange will be equal to the rate of formation of enol and independent of its reversion to the keto form or to the cation*

Similarly, the cation will be even less reactive than the keto form towards halogens or other electrophilic reagents; hence, provided that the reaction between halogen and enol is fast enough, the rate of halogenation will also be equal to the rate of enol formation and independent of the concentration or nature of the halogenating agent, as is usually found.

The kinetic analyses given so far in this chapter have been developed for aqueous solutions, in which the catalysts are in equilibrium with the species $H_2O$, $H_3O^+$, and $OH^-$, and there is no association of ions to give ion pairs. The results would also apply to any analogous amphiprotic

[14] L. P. Hammett, *Physical Organic Chemistry*, McGraw-Hill, New York, 1940, p. 231.
* Expressions (61) to (68) all apply to the rate of *formation* of enol, any reverse reaction being neglected.

solvent, such as an alcohol, though there is little systematic experimental work on catalysis in such solvents. However, the situation is different if we go to an *aprotic solvent*, i.e., one which does not take any part in acid-base equilibria. This can be illustrated by considering catalysis by a single uncharged acid dissolved in such a solvent, for example, a solution of trichloroacetic acid in benzene.

If the reaction involves only a single proton transfer, the kinetic scheme is as before

$$S + A \underset{k_{-1}}{\overset{k_1}{\rightleftharpoons}} SH^+ + B^-, \qquad SH^+ \overset{k_2}{\longrightarrow} X \qquad (69)$$

assuming that the solution is sufficiently dilute for ion pairing to be neglected. If $k_{-1} \ll k_2$, the rate will be determined by $k_1$ and will be equal to $\pi_A[A]$; $\pi_A$ is characteristic of A, and the behaviour is typical of general acid catalysis. If $k_{-1} \gg k_2$, the first stage is at equilibrium, but the concentrations of A and $B^-$ are no longer related to the concentration of any 'hydrogen ion', and there is no analogue of the specific hydrogen ion catalysis encountered in water. If none of the base $B^-$ has been added to the system, we have $[SH^+] = [B^-] \propto [S]^{\frac{1}{2}}[A]^{\frac{1}{2}}$, so that the reaction would not follow any simple kinetic order.

In most aprotic solvents it is more realistic to regard the ions as being present in the form of ion pairs, so that the reaction scheme for a single proton transfer becomes $S + A \rightleftharpoons SH^+B^- \rightarrow X$. If the first step is rate-determining, we find general acid catalysis with simple kinetics, as before. When the second step is rate-determining, we still have simple kinetics, since the reaction velocity is $k_2 K[S][A]$, where $K$ is the equilibrium constant for the first stage. However, the value of $k_2$ may well now depend upon the nature of B, even if no proton transfer is involved in the second stage.

Similar considerations apply to reactions involving two proton transfers in aprotic solvents. If the first step is rate-determining, the problem reduces to a single proton transfer. When the second step is rate-determining, the assumption of free ions leads to the same prediction as in aqueous solution, since the composition of the transition state is still equivalent to one molecule of substrate plus one molecule of acid, and this remains true even if the ions $HSH^+$ and $B^-$ are associated in a pair. However, the ion pair initially formed in the first proton transfer may well have a configuration which is unfavourable for the second proton transfer, which involves the removal of a proton from a different atom. This raises the possibility that the second proton transfer may involve another basic species. This could be a second molecule of substrate, giving (for a keto–enol transformation) a transition state such as

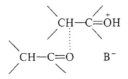

or a second ion pair, with the transition state

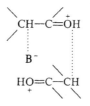

Either of these possibilities would of course lead to kinetic orders higher than the first in substrate or catalyst.

The above considerations suggest that the kinetics of catalysis in aprotic solvents may be more complex than in aqueous solution, in spite of the apparent chemical simplicity of the system. This general conclusion is borne out by the existing experimental material, which is only incompletely understood. In particular, the reaction velocity is rarely directly proportional to the catalyst concentration, the apparent order being sometimes greater and sometimes less than unity. This may be related to the kinetic predictions made in the last paragraph, but an additional complication arises in that the catalysts mainly used in these investigations were carboxylic acids, which are associated to a varying extent in aprotic solvents.[15]

Returning to the general problem of prototropic reactions, the discussion so far has assumed that the two proton transfers take place consecutively, as in Equations (59)–(61). There is another possibility, the

[15] For kinetic studies of acid-base catalysis in aprotic solvents see: J. N. Brönsted and R. P. Bell, *J. Am. Chem. Soc.*, **53**, 2478 (1931); R. P. Bell, *Proc. Roy. Soc.*, A, **143**, 377 (1934); R. P. Bell and J. F. Brown, *J. Chem. Soc.*, 1520 (1936); R. P. Bell and O. M. Lidwell, *J. Chem. Soc.*, 1096 (1936); R. P. Bell and E. F. Caldin, *J. Chem. Soc.*, 382 (1938); R. P. Bell, O. M. Lidwell, and J. Wright, *J. Chem. Soc.*, 1861 (1938); R. P. Bell, O. M. Lidwell, and M. W. Vaughan-Jackson, *J. Chem. Soc.*, 1792 (1936); R. P. Bell and P. V. Danckwerts, *J. Chem. Soc.*, 1724 (1939); R. P. Bell and J. A. Sherred, *J. Chem. Soc.*, 1202 (1940); R. P. Bell and S. M. Rybicka, *J. Chem. Soc.*, 24 (1947); R. P. Bell and A. D. S. Tantram, *J. Chem. Soc.*, 370 (1948); R. P. Bell and B. G. Skinner, *J. Chem. Soc.*, 2955 (1952).

so-called *concerted mechanism*, in which the transition state contains *two* catalyst molecules, one acid and one base.* This can be written schematically as

$$B_1 + HS + A_2 \rightleftharpoons A_1 + SH + B_2 \tag{70}$$

or more specifically

$$B_1 + HX\!-\!Y\!=\!Z + A_2 \rightleftharpoons A_1 + X\!=\!Y\!-\!ZH + B_2 \tag{71}$$

where $A_1\text{–}B_1$ and $A_2\text{–}B_2$ may be either the same or different acid-base pairs. The concerted mechanism makes no distinction between acid and base catalysis, while the mechanism so far considered (which will be distinguished by the name *consecutive mechanism*) postulates two separate routes:

$$\text{(a) } HS + A \rightleftharpoons HSH^+ + B \rightleftharpoons SH + A$$
$$\text{(b) } HS + B \rightleftharpoons S^- + A \rightleftharpoons SH + B. \tag{72}$$

On the other hand, Equations (70) and (71) do not imply a ternary collision, since just the same kinetic result is achieved if the substrate first associates with the acid and is subsequently attacked by the base, or vice versa. The real distinction between (70) and (72) is that in the latter the intermediate stage involves the separation of two entities rather than an association, and the transition state contains only one catalytic species.

If the formation of ion pairs can be neglected, as is usually the case in aqueous solution, there appears to be a ready means of distinguishing between the consecutive and concerted mechanisms. Provided that the concentration of intermediates is low, scheme (70) leads to an observed velocity constant of the form

$$k = \sum_i \sum_j k_{ij}[A_i][B_j] \tag{73}$$

where $k_{ij}$ is a constant characteristic of each acid-base combination. The number of constants can often be reduced if we make the reasonable assumption that the relative effectiveness of any two acids is independent

*Other names have been given to this mechanism, notably *ternary*, *synchronous*, and *push-pull*. The first indicates the number of species in the transition state and has been widely used. However, Ingold (*Structure and Mechanism in Organic Chemistry*, Cornell U.P., Ithaca, New York, 1953, p. 551) has used the description *bimolecular* for the same mechanisms, referring to the number of catalyst molecules involved, so that confusion can easily arise. The second and third might imply either that the acid and the base arrive simultaneously to the substrate molecule or that the two proton transfers take place simultaneously, neither of which need be the case. We have therefore preferred *concerted* as least likely to cause misunderstanding.

of the base with which they are coupled, and similarly for any two bases. This leads to

$$k = \left(\sum_i r_i[A_i]\right)\left(\sum_j r_j[B_j]\right) \tag{74}$$

which when expanded still consists of a sum of terms each involving the product of two concentrations. On the other hand, we have seen that under the same conditions scheme (72) leads to the rate equation

$$k = \sum k_i[A_i] + \sum k_j[B_j] \tag{75}$$

independent of the relative rates of the two successive steps.

Although (75) differs radically in form from (73) and (74), it is not easy to distinguish the two possibilities experimentally in aqueous solution. This is because of the presence of a large and constant concentration of water molecules, which can act either as acid or as base so that any term in (73) involving $k[X][H_2O]$ is indistinguishable from the corresponding term $k[X]$ in (75). A real distinction becomes possible in principle in the case of terms involving two solute species; for example, in an acetate buffer solution the concerted mechanism would predict a term in $[HOAc][OAc^-]$. One reaction where such a product term has been established in aqueous solution is the iodination of acetone, as first reported by Dawson and Spivey.[16] Their data certainly require such a term, but it might be objected that they refer to solutions containing high and variable concentrations of electrolyte ($> 0.75$M) in which complications might arise from salt effects. However, more recent work[17] at a constant ionic strength of 0.2 confirms the necessity for a product term somewhat larger than that found by Dawson and Spivey. The rate expression obtained by Bell and Jones is

$$10^6 v = 5 \times 10^{-4}[H_2O] + 1600[H^+] + 1.5 \times 10^7[OH^-]$$
$$+ 5.0[HOAc] + 15[OAc^-] + 20[HOAc][OAc^-] \tag{76}$$

where $v$ is the rate of disappearance of halogen in moles per litre per minute, referred to a ketone concentration of one mole per litre, and the last term contributes up to 20% of the observed velocity. Less extensive measurements with trimethylacetate and glycollate buffers indicated the presence of product terms of similar magnitude.

It is reasonable to attribute these product terms to a concerted process, but this of course leaves open the question of whether the remaining

[16] H. M. Dawson and E. Spivey, *J. Chem. Soc.*, 2180 (1930). For another example, see B. E. Banks, *J. Chem. Soc.*, 63 (1962).
[17] R. P. Bell and P. Jones, *J. Chem. Soc.*, 88 (1953).

terms in (76) represent concerted processes involving a molecule of water. An apparently strong argument against the concerted interpretation has been put forward by Pedersen.[18] This interpretation would allot the terms in Equation (76) according to the following scheme:

| Velocity | $5 \times 10^{-4}$ | 1600 | $1.5 \times 10^7$ | 5 | 15 | 20 |
|---|---|---|---|---|---|---|
| Acid | $H_2O$ | $H_3O^+$ | $H_2O$ | HOAc | $H_2O$ | HOAc |
| Base | $H_2O$ | $H_2O$ | $OH^-$ | $H_2O$ | $OAc^-$ | OAc |

This means that when the acid is $H_2O$ changing the base from $H_2O$ to $OAc^-$ increases the velocity by a factor of $3 \times 10^4$, but when the acid is HOAc the same change of base produces only a fourfold increase. Similarly, a change of the acid from $H_2O$ to HOAc increases the velocity by $10^4$ when the base is $H_2O$, but only by 30% when the base is $OAc^-$. These large discrepancies seem quite unreasonable, and on this basis Pedersen rejected the general application of the concerted mechanism.

This argument has been widely quoted, but it has been shown to be invalid by Swain[19] who pointed out that the interpretation of several of the terms in (76) is not unambiguous. For example, the term $15[OAc^-]$ might conceal either $k[OAc^-][H_2O]$ or $k[HOAc][OH^-]$, which are kinetically indistinguishable; the observed term will in general be a sum of the two. The scheme given above must therefore be replaced by the following

| Velocity | $5 \times 10^{-4}$ | | 1600 | $1.5 \times 10^7$ | 5 | | 15 | | 20 |
|---|---|---|---|---|---|---|---|---|---|
| Acid | $H_2O$ | $H_3O^+$ | $H_3O^+$ | $H_2O$ | HOAc | $H_3O^+$ | $H_2O$ | HOAc | HOAc |
| Base | $H_2O$ | $OH^-$ | $H_2O$ | $OH^-$ | $H_2O$ | $OAc^-$ | $OAc^-$ | $OH^-$ | $OAc^-$ |

Swain showed that on this basis the discrepancies pointed out by Pedersen no longer arose and that the experimental coefficients could in fact be represented approximately by an expression having the form of (74). If we use the numerical values of (76) rather than the original ones of Dawson and Spivey, such an expression is*

$$10^6 v = 10^{-5}([H_2O] + 10^2[HOAc] + 1.7 \times 10^6[H^+])([H_2O]$$
$$+ 1.4 \times 10^4[OAc^-] + 3 \times 10^{10}[OH^-])$$
$$= 7 \times 10^{-4}[H_2O] + 1000[H^+] + 1.7 \times 10^7[OH^-]$$
$$+ 5.2[HOAc] + 7.7[OAc^-] + 14[HOAc][OAc^-] \qquad (77)$$

[18] K. J. Pedersen, *J. Phys. Chem.*, **38**, 590 (1934).
[19] C. G. Swain, *J. Am. Chem. Soc.*, **72**, 4578 (1950).
*The numerical values differ somewhat from those in the original paper of Bell and Jones, which contains a misprint and some arithmetical errors; however, the conclusions remain unchanged.

using values for the ionic product of water and the dissociation constant of acetic acid appropriate to an ionic strength of 0.2. The coefficients in (77) agree with those in (76) within a factor of two, and no better agreement could be expected in view of the simplifying assumption made in obtaining (74) from (73). Further, Swain showed that the application of the same treatment to the mutarotation of glucose predicted no observable product term in $[HOAc][OAc^-]$, in agreement with experiment, and the same is true for the depolymerization of dihydroxyacetone,[20] which proceeds by a similar mechanism. On these grounds Swain concluded that the concerted mechanism was of major importance in the iodination of acetone and the mutarotation of glucose, and probably for other reactions showing catalysis by both acids and bases in aqueous solution.

However, it now seems certain that this is not so. In the first place, the agreement between (76) and (77) is deceptive. The factor $10^2[HOAc]$ in the first line of (77) contributes very little to the coefficients of $[HOAc]$ and $[OAc^-]$ in the second line but enters directly into the coefficient of $[HOAc][OAc^-]$. This factor can therefore be adjusted within wide limits to fit the observed value of the product term. In fact, any coefficient between zero and about 1000 could be accommodated by the concerted scheme. The approximate agreement found with the observed value of 20 is thus illusory.

Moreover, a general treatment in terms of equation (74) reveals some interesting general ambiguities.* Suppose that a catalysed reaction is taking place entirely by a concerted mechanism in an aqueous solution containing one acid-base pair $HA$–$A^-$ in addition to the solvent species. Equation (74) then becomes

$$k = a(1 + r_1[H^+] + r_3[HA])(1 + r_2[OH^-] + r_4[A^-])  \qquad (78)$$

while the experimental observations would be recorded in the form

$$k = k_0 + k_1[H^+] + k_2[OH^-] + k_3[HA] + k_4[A^-] + k_5[HA][A^-].  \qquad (79)$$

Comparison of (78) and (79) gives the relations

$$k_0 = a(1 + r_1 r_2 K_w), \qquad k_1 = ar_1$$
$$k_2 = ar_2, \qquad k_3 = a(r_3 + r_1 r_4 K)  \qquad (80)$$
$$k_4 = a(r_4 + r_2 r_3 K_w K^{-1}), \qquad k_5 = ar_3 r_4$$

where $K_w$ is the ionic product of water and $K$ the dissociation constant

[20] R. P. Bell and E. C. Baughan, *J. Chem. Soc.*, 1947 (1937).
*The author is indebted to Professor E. L. King for pointing this out.

of HA. Each of the expressions for $k_0$, $k_3$, and $k_4$ in (80) consists of the sum of two terms, of which the first will be termed *direct*, since they correspond to concerted processes in which the only solvent species present is $H_2O$, while the second are termed *indirect*, since they correspond to the combinations $H^+$, $OH^-$, $H^+$, $A^-$, and $OH^-$, HA.

The first three equations in (80) are sufficient to determine $a$, $r_1$, and $r_2$ in terms of observed rate coefficients. In particular, $a$ is determined by the quadratic equation

$$a^2 - k_0 a + k_1 k_2 K_w = 0 \qquad (81)$$

which must have either two real positive roots, or two imaginary ones.*

In the case of real roots there is no physical reason for preferring one value to another, so that we have two alternative values $a_+$ and $a_-$, given by

$$2a_\pm = k_0 \pm (k_0^2 - 4k_1 k_2 K_w)^{\frac{1}{2}}. \qquad (82)$$

For each of these values we can calculate $R$, the ratio of the indirect to the direct contributions to $k_0$, giving

$$R_+ = k_1 k_2 K_w / a_+^2, \qquad R_- = k_1 k_2 K_w / a_-^2, \qquad (83)$$

and since $a_+ a_- = k_1 k_2 K_w$, this leads to $R_+ R_- = 1$. Thus if any set of experimental data are interpreted to show that $k_0$ corresponds to a predominantly indirect mechanism, they can equally well be interpreted to demonstrate the opposite conclusion.

The same ambiguity appears in the coefficients $r_1$, $r_2$, $r_3$, and $r_4$, which are given by the first four equations of (80) by inserting $a_+$ or $a_-$ together with the observed rate coefficients. It is again found that the corresponding ratios of the direct to the indirect contributions to $k_3$ and $k_4$ obey the relation $R_+ R_- = 1$. Finally, the rate constant $k_5$ is given by $a_+ r_3^+ r_4^+$ or $a_- r_3^- r_4^-$, and tedious but straightforward algebra shows that these two values are identical; thus the observation of a rate term proportional to $[HA][A^-]$ is of no assistance in discriminating between the two possibilities.

Finally, some general arguments have been advanced by Bell and Jones[17] on the consequences of the concerted mechanism for any reaction which shows general catalysis both by acids and by bases in aqueous solution. These arguments are too long to reproduce here, but they lead to

---

*In fact, the values of $k_0$, $k_1$, and $k_2$ in (76) lead to imaginary roots for $a$, but this is not in itself a reason for rejecting a concerted mechanism for this reaction since the replacement of (73) by (74) involves an assumption which is not necessarily correct.

the following conclusion for a typical buffer solution composed of an acid HX and its anion X: *either* the product term $k[HX][X^-]$ must contribute a large part of the observed velocity in dilute buffer solutions *or* the relation $\alpha + \beta = 1$ must hold for the exponents of the Brönsted relations.* Neither of these results is found in practice, and we must conclude that the concerted mechanism has no general application to those reactions which are known to exhibit general acid-base catalysis in aqueous solution, although it may account for the small term in $[HOAc][OAc^-]$ observed in the prototropy of acetone. This term has alternatively been attributed,[21] as in the original work of Dawson and Spivey,[16] to catalysis by the hydrogen-bonded species $AcO\cdots HOAc$, for which there is some evidence,[22] but it is necessary to attribute to this complex a catalytic activity considerably greater than that of either $AcO^-$ or $HOAc$; this seems improbable, since hydrogen bonding should reduce both the acidic and the basic strength of the separate species. Other attempts to detect kinetic product terms such as $[HOAc][OAc^-]$, and the bearing of hydrogen isotope effects upon concerted mechanisms, will be described in Chapters 9 and 12.

The position may well be different in non-aqueous solvents. In the interconversion of isomeric methyleneazomethines

$$R_1R_2CHN{=}CR_3R_4 \rightleftharpoons R_1R_2C{=}NCHR_3R_4$$

in sodium ethoxide dissolved in dioxan–ethanol the initial rates of interconversion, racemization, and deuterium uptake are the same, and this was interpreted to mean that the reaction must take place by a concerted mechanism, without the formation of a free anion.[23] However, later work on the same type of reaction[24] has shown that the rates of interconversion and hydrogen exchange may differ considerably, and has provided an alternative explanation for the original observations.

In solvents of low dielectric constant the formation of free ions becomes less likely, so that a concerted mechanism may be favoured even for

---

* These are the exponents in the equations $k_A = G_A K_A^{\alpha}$ and $k_B = G_B(1/K_A)^{\beta}$ which relate catalytic power to acid-base strength and will be dealt with in the next chapter.

[21] F. J. C. Rossotti, *Nature*, **188**, 936 (1970).

[22] D. L. Martin and F. J. C. Rossotti, *Proc. Chem. Soc.*, 60 (1959); H. N. Farrer and F. J. C. Rossotti, *Acta Chem. Scand.*, **17**, 1824 (1963).

[23] C. K. Ingold and C. L. Wilson, *J. Chem. Soc.*, 1403 (1933); 93 (1934); S. K. Hsü, C. K. Ingold, and C. L. Wilson, *J. Chem. Soc.*, 1774 (1935); E. de Salas and C. L. Wilson, *J. Chem. Soc.*, 319 (1938); S. K. Hsü, C. K. Ingold, C. G. Raisin, E. de Salas, and C. L. Wilson, *J. Chim. Phys.*, **45**, 232 (1948); R. Perez Ossorio and E. D. Hughes, *J. Chem. Soc.*, 426 (1952). For a summary see Ingold, *Structure and Mechanism in Organic Chemistry*, 2nd edn., Bell, London, 1969, p. 835.

[24] D. J. Cram and R. D. Guthrie, *J. Am. Chem. Soc.*, **87**, 397 (1965); **88**, 5760 (1966).

reactions which do not follow it in water or similar solvents. In fact, the concerted mechanism was first put forward by Lowry[25] on the basis of observations on the mutarotation of tetramethylglucose in media of low dielectric constant. This reaction was very slow in dry pyridine (possessing no acid properties) or in dry cresol (possessing hardly any basic properties), but was rapid in a mixture of the two solvents or in either solvent when moist, suggesting that both an acid and a base must take part in the reaction.[26] It is difficult to attach any quantitative significance to this result, in view of the drastic changes of medium involved, but the same point is made more strongly by the measurements of Swain and Brown[27] on the same reaction in dilute benzene solutions of amines and phenols. They found that the reaction was kinetically of the third order, the velocity being proportional to the product of the concentrations of phenol, amine, and tetramethylglucose, as expected from Equation (73). It is also of interest that 2-hydroxypyridine is a powerful specific catalyst for the mutarotation; at a concentration of 0.001M it is 7000 times as effective as a mixture of 0.001M pyridine and 0.001M phenol, though it is only one ten-thousandth as strong a base as pyridine and one-hundredth as strong an acid as phenol. Since, further, the velocity is proportional to the first power of the concentration of 2-hydroxypyridine, it is clear that the operation of the concerted mechanism is facilitated by the presence of an acidic and a basic group in the same catalyst molecule. Similarly, carboxylic acids are much more effective catalysts than phenols of comparable strength, perhaps because the carboxyl group can act simultaneously as an acid and as a base.[28] This kind of catalysis has been termed *bifunctional catalysis* and has been widely invoked to explain the abnormally high catalytic activity of some species containing both acidic and basic groups, including enzymes.[29]

Nevertheless, it now seems doubtful whether the third-order kinetics observed by Swain and Brown[27] necessarily imply a concerted acid-base catalysis, and whether the abnormally high activity of some bifunctional catalysts can be attributed simply to the presence of acidic and basic centres in the same molecule. It was first pointed out by Pocker[30] that a

[25] T. M. Lowry, *J. Chem. Soc.*, 2554 (1927).
[26] E. M. Richards and T. M. Lowry, *J. Chem. Soc.*, 1385 (1925); T. M. Lowry and I. J. Faulkner, *J. Chem. Soc.*, 2883 (1925).
[27] C. G. Swain and J. F. Brown, *J. Am. Chem. Soc.*, **74**, 2534, 2538 (1952).
[28] See also the work of Eastham and his collaborators[9] on the mutarotation of tetramethylglucose in nitromethane solution.
[29] For references to bifunctional catalysis in non-enzymic reactions, see P. R. Rony, *J. Am. Chem. Soc.*, **91**, 6090 (1969).
[30] Y. Pocker, *Chem. and Ind.*, 968 (1960).

rate proportional to the product of the concentrations of amine and phenol could be attributed to base catalysis by the phenoxide ion in an ion-pair such as $PhO^- \cdot {}^+NH_3R$, and this is supported by the recent demonstration[31] that ion-pairs such as $PhO^- \cdot {}^+NR_4$ are effective catalysts for the mutarotation of tetramethylglucose in benzene, although they contain no acidic group. It also appears[29,32] that bifunctional catalysts are effective only when they can interact with the substrate without the formation of high-energy dipolar intermediates; this implies that they can exist in two tautomeric forms of comparable energy. Thus catalysis by carboxylic acids, 2-hydroxypyridine, pentane-2,4-dione, and pyrazole in the mutarotation reaction can be represented by the following schemes:

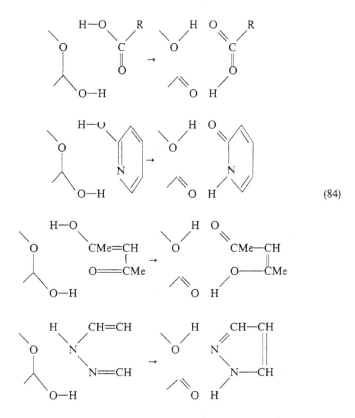

(84)

[31] P. R. Rony, W. E. McCormack, and S. W. Wunderly, *J. Am. Chem. Soc.*, **91**, 4244 (1969).
[32] P. R. Rony, *J. Am. Chem. Soc.*, **90**, 2824 (1968).

The above catalysts are all effective ones, but 2-aminophenol, 8-hydroxyquinoline, and catechol are much less so, since their action would involve the formation of high-energy species, namely,

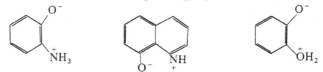

For the first, third, and fourth catalysts in (84) the two tautomers are chemically identical, and the same is true for ions such as $HCO_3^-$, $HPO_4^{2-}$, $H_2PO_4^-$, and $H_2AsO_4^-$, which have been reported to have an abnormally high catalytic activity in some reactions.[33] It is clear that the effectiveness of this kind of catalyst is related to its particular electronic structure rather than to its acid-base properties, and the process is more appropriately described as *tautomeric catalysis* than as bifunctional or concerted acid-base catalysis. It is of interest that a theoretical treatment of some molecules in which acidic and basic groups form part of the same $\pi$-electron system shows some parallelism between catalytic activity and the coupling constants of the molecular orbital theory;[34] moreover, a very general treatment of concerted proton transfers indicates that simple bifunctional acid-base catalysis is likely to be of importance only under very restricted conditions.[35]

There is of course ample evidence that acid-base catalysis in solvents of low dielectric constant does not necessarily involve a concerted process. Such a process cannot operate when catalysis is effected by a single acid or base present in an aprotic solvent, and there are many examples of this, including typical prototropic reactions such as the halogenation of acetone,[36] the racemization and inversion of optically active ketones,[37] and the mutarotation of nitrocamphor.[38] Moreover, in the isomerization of mesityl oxide oxalic ester in chlorobenzene,[39] which depends kinetically on the interconversion of two isomeric enols, the velocity in a solution containing both an amine and an acid is no greater than the sum of the velocities for the two catalysts separately, in contrast to the behaviour found by Swain for the mutarotation reaction.

[33] For references see Ref. 29.
[34] H. J. Gold, *J. Am. Chem. Soc.*, **90**, 3402 (1968).
[35] J. E. Critchlow, *J. Chem. Soc., Faraday Trans.* I, 1774 (1972); W. P. Jencks, *Chem. Reviews*, **72**, 705 (1972).
[36] R. P. Bell and A. D. S. Tantram, *J. Chem. Soc.*, 370 (1948).
[37] R. P. Bell and E. F. Caldin, *J. Chem. Soc.*, 382 (1938); R. P. Bell, O. M. Lidwell, and J. Wright, *J. Chem. Soc.*, 1861 (1938).
[38] R. P. Bell and J. A. Sherred, *J. Chem. Soc.*, 1202 (1940).
[39] R. P. Bell and S. M. Rybicka, *J. Chem. Soc.*, 24 (1947).

The status of the concerted mechanism has been discussed at some length because unduly dogmatic statements have been made from time to time both for and against its validity. As will have been seen, there is some evidence that it operates in non-aqueous solvents, but it appears to be of minor importance for most reactions in water.

# 9 Examples of Reactions Catalysed by Acids and Bases

Any comprehensive account of reactions catalysed by acids and bases would cover a large proportion of organic chemistry and biochemistry, and would demand at least a whole book; in fact, two excellent books giving a general coverage of the subject have recently appeared.[1,2] This chapter will be devoted to a fairly detailed consideration of a small number of reactions, chosen so as to illustrate the variety of kinetic behaviour forecast by the general treatment in the last chapter. They will also show in small measure the variety of chemical change which can result from the removal or addition of a proton. Some of the work described involves a study of series of acidic or basic catalysts, and some mention will be made of the relation of catalytic activity to acid-base strength, though a full treatment of this topic will be deferred until Chapter 10. Similarly, hydrogen isotope effects will be mentioned when they throw light on the mechanism of the reaction, but will not be considered quantitatively until Chapter 12.

In catalysis by proton acids there is rarely any doubt about the stages through which the reaction passes, though it may be more difficult to determine which is the rate-limiting step. In catalysis by bases, on the other hand, there is often a fundamental ambiguity between general base catalysis and *nucleophilic catalysis*; this is because the base will always possess unshared electron pairs, and may act by attaching itself to an electrophilic centre in the substrate rather than by abstracting a proton. This ambiguity is particularly marked in reactions of carbonyl compounds, such as esters, and a decision between the two mechanisms may be difficult, frequently involving arguments about the effect of structural variations in the catalyst or the substrate. In this chapter we shall confine ourselves as far as possible to reactions for which a proton-transfer mechanism is well established.

[1] W. P. Jencks, *Catalysis in Chemistry and Biochemistry*, McGraw-Hill, New York, 1969.
[2] M. L. Bender, *Mechanism of Homogeneous Catalysis from Protons to Proteins*, Wiley-Interscience, New York, 1971.

*The Decomposition of Nitramide*

This reaction occupies a special position in the development of our ideas about acid-base catalysis. It was employed by Brönsted and Pedersen[3] in the first clear demonstration of general base catalysis, and in the same paper they established a quantitative relationship between base strength and catalytic power, as well as laying the foundation for many subsequent developments. Later work has confirmed the wide applicability of this reaction for studying catalysis by bases in both aqueous and non-aqueous solvents. Nitramide is rather unstable, and the method usually employed for its preparation somewhat laborious, though a recent paper[4] describes a more convenient one-stage method. The simple quantitative behaviour of this decomposition depends partly on the fact that the products (nitrous oxide and water) are neutral and inert, while the small size of the nitramide molecule minimizes any steric effects. Moreover, since none of the derivatives of nitramide decomposes in the same manner, it has not been possible for organic chemists to complicate the issue by investigating the effect of substituents. Apart from an early book[5] no general review of this reaction is available.

Nitramide was first prepared by Thiele and Lachmann[6] who found that in alkaline solution it decomposes rapidly according to the equation $H_2N_2O_2 \rightarrow N_2O + H_2O$; measurement of the volume or the pressure of the nitrous oxide evolved has been used universally for following the kinetics of the reaction. They formulated it as $NH_2NO_2$, in agreement with present views, though several other structures were suggested later. Thus Hantzsch[7] first regarded it as $HON{=}NOH$, the *cis*-isomer of hyponitrous acid, though he later[8] favoured $NH{=}NO \cdot OH$, while measurements of dipole moment[9] were interpreted in terms of $\bar{N}{=}\overset{+}{N}(OH)_2$, a formulation unsupported by any other evidence. The structure $NH_2NO_2$ is now supported by a variety of physical evidence. The infrared spectra of both the solid and solutions[10] are consistent with this structure, and the proton magnetic resonance spectrum[11] reveals the presence of only amino hydrogens. The microwave spectrum of nitramide and its deuteriated analogues can be interpreted completely[12] in terms of the

[3] J. N. Brönsted and K. J. Pedersen, *Z. Phys. Chem.*, **108**, 185 (1924).
[4] S. Tellier-Pollon and J. Heubel, *Rev. Chim. Minérale*, **4**, 413 (1967).
[5] A. Avorio, *Decomposizione Catalytica della Nitramide*, F. Centenari, Rome, 1937.
[6] J. Thiele and A. Lachmann, *Annalen*, **288**, 267 (1895).
[7] A. Hantzsch, *Annalen*, **292**, 340 (1896).
[8] A. Hantzsch, *Ber.*, **63**, 1270 (1930).
[9] E. C. E. Hunter and J. R. Partington, *J. Chem. Soc.*, 309 (1933).
[10] M. Davies and N. Jonathan, *Trans. Faraday Soc.*, **54**, 469 (1958).
[11] J. D. Ray and R. A. Ogg, *J. Chem. Phys.*, **26**, 1452 (1957).
[12] J. K. Tyler, *J. Mol. Spectroscopy*, **11**, 39 (1963).

same structure, and yields accurate values for the bond distances and bond angles.

As discussed in Chapter 8, the fact that the decomposition shows general catalysis by bases proves that a proton-transfer reaction is at least partly rate-determining. Nitramide is a weak acid with $pK = 6.48$ at $25°C$,[13] but the formation of the mono-anion cannot be rate-determining, since this ion is present at equilibrium concentrations in aqueous solution, and the rate of isotopic exchange between nitramide and deuterium oxide is much faster than its rate of decomposition.[14] The generally accepted mechanism, first proposed by Pedersen,[15] assumes that solutions of nitramide contain a small proportion of the *aci*-isomer $HN{=}NO \cdot OH$ (in mobile equilibrium with $NH_2NO_2$) which is presence of any acid-base pair A–B undergoes the rate-determining reaction*

$$B + HN{=}\overset{+}{N}\overset{\displaystyle OH}{\underset{\displaystyle O^-}{\Big\langle}} \rightarrow A + N{\equiv}\overset{+}{N}{-}O^- + OH^-. \qquad (85)$$

The observation[16] that the decomposition of isotopically labelled $^{15}NH_2{}^{14}NO_2$ produces only the species $^{15}N^{14}NO$ is in accord with this mechanism, and also excludes the structure $HON{=}NOH$. A recent suggestion[17] that the decomposition involves the reaction $[N{=}NO_2]^{2-} \rightarrow N_2O + O^{2-}$ is inconsistent with the observed dependence of rate upon base concentration.

Turning to the kinetic studies, Brönsted and Pedersen[3] found that in dilute solutions of strong acids the rate of decomposition at $15°C$ was independent of hydrogen ion concentration over a wide range ($10^{-5}$–$0.4M$). They attributed this rate to basic catalysis by water molecules, and concluded that there was no acid catalysis; they also found that added neutral salts had little effect on the rate. Subsequent work in more

[13] L. K. J. Tong and A. R. Olson, *J. Am. Chem. Soc.*, **63**, 3406 (1941); cf. J. N. Brönsted and C. V. King, *J. Am. Chem. Soc.*, **49**, 193 (1927).

[14] V. K. La Mer and S. Hochberg, *J. Am. Chem. Soc.*, **61**, 2552 (1939).

[15] K. J. Pedersen, *J. Phys. Chem.*, **38**, 581 (1934).

*The ion $^-N{=}NO \cdot OH$ could be postulated as a short-lived intermediate without altering the scheme essentially. It is also possible that this ion is in mobile equilibrium with $B + HN{=}NO \cdot OH$, and hence with $B + NH_2NO_2$, followed by the rate-determining reaction $^-N{=}NO \cdot OH + A \rightarrow N{\equiv}\overset{+}{N} \cdot O^- + H_2O + B$. This scheme would also lead to the observation of general base catalysis, and there seems to be no obvious reason for preferring either of these possibilities.

[16] K. Clusius, *Helv. Chim. Acta*, **44**, 1149 (1961).

[17] P. Vast, *Bull. Soc. Chim. France*, 2136 (1970).

concentrated solutions[18,19] has revealed a small effect of added salts and strong acids, but these effects can usually be neglected, and it is doubtful whether acid catalysis has been established in aqueous solution. Most of Brönsted and Pedersen's experiments were in buffer solutions prepared from weak acids. Provided that these solutions are sufficiently acid to suppress catalysis by hydroxide ions, the rate is independent of the concentrations of hydrogen ions and of the acid constituent of the buffer, being given by

$$k = k_0 + k_B[B] \tag{86}$$

where $k_0$ is the velocity constant of the water-catalysed reaction (measured in dilute solutions of strong acids) and $k_B$ is termed the catalytic constant of the base. Brönsted and Pedersen measured the catalytic constants of fifteen carboxylate ions and two phosphate ions, and later work, mostly at 25°C, has added a few more anion bases,[20-25] besides providing information about activation energies[20] and the specific effects of some metallic cations.[21]

Brönsted and Pedersen[3] showed that Equation (86) is also obeyed in aniline buffers, and shortly afterwards[26] the same was found to be the case for catalysis by seven substituted anilines. A little later[27] the same behaviour was observed for seven cation bases, typified by $[Co(NH_3)_5OH]^{2+}$, which can accept a proton to give $[Co(NH_3)_5H_2O]^{3+}$. The parallel catalytic effect exerted by these bases of very different charges and chemical nature provided a firm foundation for the theory of general catalysis by acids and bases.

In all the catalytic solutions so far considered, catalysis by hydroxide ion (or the kinetically equivalent decomposition of the nitramide anion) could be neglected, and Brönsted and Pedersen were unable to give a value for the catalytic constant of hydroxide ion. Measurements in more alkaline solutions were made at 25°C by Tong and Olson.[28] Under these conditions the observed velocity constant is given by

$$k = k_0 + k_N[N^-] + k_{OH}[OH^-] + k_B[B] \tag{87}$$

[18] C. A. Marlies and V. K. LaMer, *J. Am. Chem. Soc.*, **57**, 1812 (1935).
[19] V. Beretta, *Rend. Accad. Sci. Napoli*, **8**, 36 (1938).
[20] R. P. Bell and E. C. Baughan, *Proc. Roy. Soc.*, A, **158**, 464 (1937).
[21] R. P. Bell and G. M. Waind, *J. Chem. Soc.*, 2357 (1951).
[22] R. P. Bell and J. C. McCoubrey, *Proc. Roy. Soc.*, A, **234**, 192 (1956).
[23] B. Perlmutter-Hayman and M. A. Wulff, *Israel J. Chem.*, 153 (1969).
[24] R. P. Bell, B. G. Cox, and B. A. Timimi, *J. Chem. Soc.*, B, 2247 (1971).
[25] R. P. Bell, B. G. Cox, and J. B. Henshall, *J. Chem. Soc., Perk. Trans. 2*, 1232 (1972).
[26] J. N. Brönsted and H. C. Duus, *Z. Phys. Chem.*, **117**, 299 (1925).
[27] J. N. Brönsted and K. Volqvartz, *Z. Phys. Chem.*, **155**, A, 211 (1931).
[28] L. K. J. Tong and A. R. Olson, *J. Am. Chem. Soc.*, **63**, 3406 (1941).

where $N^-$ is the nitramide anion. The second term represents basic catalysis by the nitramide anion, and the last term is absent if no buffer system has been added. Tong and Olson obtained values for $k_N$ and $k_{OH}$, and catalytic constants for the strongly basic anions of four phenols. A similar treatment is needed in interpreting catalysis by amines which are stronger bases than the anilines.[29] Information is also available on the kinetics of the nitramide decomposition in deuterium oxide solution.[30,31]

The decomposition of nitramide has also been investigated extensively in non-aqueous solvents. In isopentanol[32] its behaviour resembles that in water: catalysis by many anion and amine bases was studied, and there is a small but detectable solvent catalysis. The same applies to the more acidic solvent *m*-cresol,[33] in which activation energies have also been measured,[34] though here there is no observable catalysis by the solvent. In both of these solvents there is an appreciable effect of the concentration of the acidic constituent of the buffer. This may indicate some acid catalysis, since the protonated species $HN{=}NO \cdot \overset{+}{O}H_2$ should react with the solvent (or other basic species) more readily than $HN{=}NO \cdot OH$. However, in solvents of such low dielectric constants ($\varepsilon = 14, 12$) there will be complications due to ion-pairing, which was not taken into account in interpreting the results. Information is also available[35] on rates and activation energies for catalysis by solvent, acetate ion, and aniline in mixtures of water with methanol, ethanol, t-butanol, ethylene glycol, glycerol, acetone, and dioxan. The results agree qualitatively with the expected effect of dielectric constant on a reaction with a highly polar transition state.

Catalysis by amines has also been investigated in a number of non-hydroxylic solvents. Thus Bell and Trotman-Dickenson[36] studied catalysis by 21 primary, secondary, and tertiary amines in anisole at 25°C, and later work[37,38] included the measurement of temperature coefficients and

[29] R. P. Bell and G. L. Wilson, *Trans. Faraday Soc.*, **46**, 407 (1950).
[30] V. K. LaMer and J. Greenspan, *Trans. Faraday Soc.*, **33**, 1266 (1937).
[31] S. Liotta and V. K. LaMer, *J. Am. Chem. Soc.*, **60**, 1967 (1938).
[32] J. N. Brönsted and J. E. Vance, *Z. Phys. Chem.*, **163**, A, 240 (1933).
[33] J. N. Brönsted, A. L. Nicholson, and A. Delbanco, *Z. Phys. Chem.*, **169**, A, 379 (1934).
[34] W. W. Carnie, P. M. Duncan, J. A. Kerr, K. Shannon, and A. F. Trotman-Dickenson, *J. Chem. Soc.*, 3231 (1959).
[35] A. Voipio, *Suomen Kem.*, **29**, B, 170 (1956); **30**, A, 72 (1957); *Ann. Acad. Sci. Fennicae*, A, No. 92 (1958).
[36] R. P. Bell and A. F. Trotman-Dickenson, *J. Chem. Soc.*, 1288 (1949).
[37] R. P. Bell and E. F. Caldin, *Trans. Faraday Soc.*, **47**, 50 (1951).
[38] G. C. Fettis, J. A. Kerr, A. McClure, J. S. Slater, C. Steel, and A. F. Trotman-Dickenson, *J. Chem. Soc.*, 2811 (1957).

isotope effects. In these systems the rate is not always a strictly linear function of the catalyst concentration. This is at least partly due to the acidic properties of nitramide, which reacts to an appreciable extent with the amine, $RHN_2 + NH_2NO_2 \rightleftharpoons RNH_3^+[NHNO_2]^-$. This complication can usually be avoided by using a low concentration of nitramide, but with strong bases such as the tertiary aliphatic amines conversion into the nitramide anion is so complete that only the catalytic effect of this latter species could be determined.[36] Similar but less extensive work has been carried out in benzene[39] and in nitrobenzene,[40] and reference 39 compares rates and activation energies for catalysis by $C_6H_5NMe_2$ in seven different solvents. As might be expected for solvents of diverse chemical types, the rate bears no simple relation to the dielectric constant or any other physical property of the solvent, and specific solvation of the initial state and the transition state is no doubt involved.

The contents of this section should serve to show the versatility of the nitramide reaction for the study of catalysis by bases, and it will be quoted widely in the next chapter as a tool for investigating the relations between catalytic power and the structure and acid-base strength of catalysts. However, much of the work in non-aqueous media refers to solvents in which acid-base equilibria have not been adequately investigated, and it would certainly be of interest to extend kinetic measurements to solvents which have been the subject of modern equilibrium studies, for example the dipolar aprotic solvents mentioned in Chapter 4.

*Reactions of Aliphatic Diazo-compounds*

In contrast to the decomposition of nitramide, these reactions are catalysed by acids and not by bases. The general mechanism may be written as follows:

(a) $\bar{N}{=}\overset{+}{N}{=}CRR' + A \rightleftharpoons N{\equiv}\overset{+}{N}{-}CHRR' + B$

(b) $N{\equiv}\overset{+}{N}{-}CHRR' \rightarrow N_2 + RR'CHX$ 
$\qquad\qquad\qquad\qquad\qquad\qquad\qquad\qquad (88)$

where X is a nucleophile which may be derived from the solvent, for example —OH or —OEt, or from other species present in the solution such as halide or acetate. The cation on the right-hand side of (88a) is never detectable by direct means, since even in the absence of further reaction its equilibrium concentration will be very low.

[39] E. F. Caldin and J. Peacock, *Trans. Faraday Soc.*, **51**, 1217 (1955).
[40] D. D. Callander, W. D. Ferguson, G. C. Fettis, and A. F. Trotman-Dickenson, *J. Chem. Soc.*, 3834 (1960).

Reaction (88b) may involve the reaction of this cation with a nucleophilic reagent, or, alternatively, it may take place in two stages:

(c) $N{\equiv}\overset{+}{N}{-}CHRR' \rightarrow N_2 + \overset{+}{C}HRR'$

(d) $\overset{+}{C}HRR' \rightarrow RR'CHX.$

$$(89)$$

In the latter event the free carbonium ion will certainly react very rapidly with nucleophilic reagents, and either (a) or (c) will be the rate-determining step. Much argument has centred round the question of whether the nucleophilic reagent is kinetically involved in the further reaction of $N_2^+CHRR'$, which is particularly difficult to resolve when it reacts mainly with the solvent,[41-43] but we shall be mainly concerned with the prior question of whether the proton-transfer reaction (88a) is partly or wholly rate-determining. Three specific diazo-compounds will be chosen to illustrate different types of behaviour.

The decomposition of *ethyl diazoacetate* in aqueous solution was one of the earliest acid-catalysed reactions to be kinetically investigated. The major process is the production of ethyl glycollate by the overall reaction

$$N_2CHCO_2Et + H_2O \rightarrow N_2 + CH_2OHCO_2Et$$

though in the presence of some anions (e.g., halides and sulphates) there is a small amount of reaction according to

$$N_2CHCO_2Et + H^+ + X^- \rightarrow N_2 + CH_2XCO_2Et$$

This side-reaction is, however, absent for perchlorates and picrates, and takes place to only a small extent with nitrates.[44,45]

Early kinetic work used the evolution of nitrogen for following the reaction, though spectrophotometry has been used in recent years. The reaction follows strictly first-order kinetics, and in dilute solutions of both strong and weak acids the rate is directly proportional to the hydrogen ion concentration.[46-49] The same is true for reaction in buffer solutions, apparent early discrepancies being due to a neglect of secondary salt effects. The decomposition of ethyl diazoacetate was in fact widely used

[41] R. A. More O'Ferrall, *Adv. Phys. Org. Chem.*, **5**, 331 (1967).
[42] W. J. Albery, J. E. C. Hutchins, R. M. Hyde, and R. H. Johnson, *J. Chem. Soc.*, B, 219 (1968).
[43] W. J. Albery and M. H. Davies, *Trans. Faraday Soc.*, **65**, 1066 (1969).
[44] G. Bredig and R. F. Ripley, *Ber.*, **40**, 4015 (1907).
[45] H. Lachs, *Z. Phys. Chem.*, **73**, 291 (1910).
[46] G. Bredig and W. Fraenkel, *Z. Elektrochem.*, **11**, 525 (1905).
[47] W. Fraenkel, *Z. Phys. Chem.*, **60**, 202 (1907).
[48] E. Spitalsky, *Z. Anorg. Chem.*, **54**, 278 (1907).
[49] E. A. Moelwyn-Hughes and P. Johnson, *Trans. Faraday Soc.*, **37**, 282 (1941).

by Brönsted and his collaborators[50] for establishing the laws governing such salt effects.

The position is essentially the same for the alcoholysis of ethyl diazoacetate, where the main product is $EtOCH_2CO_2Et$, though here the salt effects are much larger, and failure to allow for them may give rise to misleading conclusions. Table 21 contains results obtained by Snethlage[51] for ethanolysis at 25° in solutions of picric acid to which varying amounts of $p$-toluidine picrate had been added. These results were at one time widely quoted in support of what was then called the 'dual theory' of catalysis; this supposed that both hydrogen ions and undissociated acid molecules could act as catalysts, in conformity with what was later termed general acid catalysis.

Table 21. ETHANOLYSIS OF ETHYL DIAZOACETATE AT 25°C

$9.09 \times 10^{-3}$M picric acid $+ x$-molar $p$-toluidine picrate
$k$ = first-order velocity constant, $s^{-1}$

| $10^4x$ | $10^6k$ | $10^4K_c$ | $f_{\pm}(HCl)$ | $10^4(K_c f_{\pm}^2)$ |
|---|---|---|---|---|
| 0 | 967 | 3.31 | 0.726 | 1.74 |
| 9.1 | 808 | 3.79 | 0.672 | 1.71 |
| 18.2 | 683 | 4.70 | 0.635 | 1.89 |
| 27.3 | 625 | 4.75 | 0.606 | 1.74 |
| 36.4 | 537 | 4.85 | 0.581 | 1.64 |
| 45.5 | 500 | 5.31 | 0.559 | 1.66 |
| 90.0 | 367 | 6.78 | 0.485 | 1.59 |
| 136.3 | 342 | 9.35 | 0.443 | 1.83 |
| 227.3 | 300 | 13.3 | 0.377 | 1.85 |
| 454.5 | 267 | 23.1 | 0.312 | 2.25 |
| 909.0 | 242 | 42.1 | 0.255 | 2.17 |

It is clear from Table 21 that the addition of salt decreases the reaction velocity much less than would be expected from the classical law of mass action if catalysis were due solely to hydrogen ions. In fact, if the velocity is plotted against the reciprocal of the salt concentration, the curve obtained appears to indicate a finite limiting velocity in presence of an infinite amount of salt; this amounts to about 22% of the velocity in absence of salt, and was attributed to catalysis by undissociated picric

---

[50] J. N. Brönsted and C. E. Teeter, *J. Phys. Chem.*, **28**, 579 (1924); J. N. Brönsted and H. Duus, *Z. Phys. Chem.*, **117**, 299 (1925); J. N. Brönsted and C. V. King, *Z. Phys. Chem.*, **130**, 699 (1927); J. N. Brönsted and K. Volquartz, *Z. Phys. Chem.*, **134**, 97 (1928).
[51] H. C. S. Snethlage, *Z. Elektrochem.*, **18**, 539 (1912).

acid. However, a different picture emerges when activity coefficients are taken into account. Assuming for the moment that the hydrogen ion is the only catalytic species, its catalytic constant can be calculated from the velocity with no picrate added, together with the optically determined degree of dissociation of picric acid in ethanol[52] at the appropriate concentration. This catalytic constant can then be used to deduce the hydrogen ion concentrations in the solutions with added picrate, and hence the values of $K_c = [H^+][Pic^-]/[HPic]$ in these solutions, which are given in the third column of Table 21. These 'constants' increase rapidly with increasing salt concentration, but the increase is no larger than might be anticipated from the omission from the equilibrium constant of the factor $f_\pm^2$, which will be considerably less than unity in this medium of dielectric constant 25. Activity coefficients for hydrogen chloride in ethanol at 25°C have been derived from e.m.f. measurements,[53] and the values for the ionic strengths obtaining in the experiments with ethyl diazoacetate are given in the fourth column. Finally, the last column contains the product $K_c f_\pm^2$; if $f_\pm$ has been correctly estimated, this quantity should remain constant, and the fact that it shows no trend with salt concentration supports the original assumption that the hydrogen ion is the only effective catalyst in these solutions. This analysis is not fully quantitative, since at the ionic concentrations concerned the activity coefficients will depend to some extent on the individual nature of the ions; further, no account has been taken of the activity coefficient of undissociated picric acid, or of any primary kinetic salt effect. However, semi-quantitative confirmation comes from the spectrophotometric determination[54] of $K_c$ at 30°C in ethanol containing 0.1M lithium perchlorate; this gives $K_c = 3.5 \times 10^{-3}$, which is similar to the value in Table 21 for an ionic strength of 0.09M. It is abundantly clear that the original extrapolation was unjustified, and that there is no evidence for catalysis by undissociated acid in this reaction. This example has been treated fully to illustrate the importance of salt effects in catalysed reactions, especially in non-aqueous solvents.

The observation of specific catalysis by hydrogen ions in both aqueous and ethanolic solution suggests that for ethyl diazoacetate the first step in Equation (88) is at equilibrium, and the second step rate-determining. This is confirmed by several other observations. In the first place, the reaction takes place about three times *faster* in $D_2O$ than in $H_2O$.[43,55] The subject

[52] P. Gross and A. Goldstern, *Monatsh.*, **55**, 316 (1930).
[53] J. W. Woolcock and H. B. Hartley, *Phil. Mag.*, (vii) **5**, 1133 (1928).
[54] J. D. Roberts and W. Watanabe, *J. Am. Chem. Soc.*, **72**, 4869 (1950).
[55] P. Gross, H. Steiner, and F. Krauss, *Trans. Faraday Soc.*, **34**, 351 (1938).

of isotope effects will be treated fully in Chapter 12; it is sufficient to note here that in reactions involving a rate-determining proton transfer the replacement of hydrogen by deuterium always causes a considerable decrease of reaction velocity. Secondly, if the reaction is carried out in a deuteriated solvent, for example EtOD, the hydrogen of the CH group in ethyl diazoacetate exchanges faster than the ester decomposes, and both the hydrogens in the $CH_2$ group of the product $EtOCH_2CO_2Et$ exchange equivalently.[56] The last observation shows that the protonated ester contains two equivalent hydrogens, and excludes a formulation such as

$$N_2 {=\!\!=} CHCO_2Et$$
$$\vdots$$
$$\overset{+}{H}$$

Finally, the fact that the addition of nucleophiles ($Cl^-$, $Br^-$, $I^-$) increases the rate of disappearance of the ester[42,44,45,57] shows that the protonation step cannot be wholly rate-determining.

The decomposition of *diphenyldiazomethane*, $Ph_2CN_2$, contrasts in many ways with that of ethyl diazoacetate. In the absence of acids it can decompose either thermally or photochemically, giving a variety of products which can be attributed to the initial formation of the carbene $Ph_2C$. In presence of acids, however, the thermal and photochemical reactions can normally be neglected, and a different pattern of products is observed, attributable to the formation of the cation $Ph_2CHN_2^+$. Much of the systematic kinetic work has been carried out in ethanol solution, particularly by Roberts and his collaborators.[58-63] The reaction is always first-order with respect to diazo-compound, the disappearance of which can be conveniently followed spectrophotometrically. For catalysis by toluene-*p*-sulphonic acid, which is completely dissociated in ethanol, the rate is proportional to the hydrogen ion concentration, while for the much weaker picric acid it is proportional to the concentration of undissociated acid. With both of these catalysts the only product is the ether $Ph_2CHOEt$. These facts suggest general acid catalysis, with the proton

[56] J. D. Roberts, C. M. Regan, and I. Allen, *J. Am. Chem. Soc.*, **74**, 3679 (1952).
[57] W. J. Albery and R. P. Bell, *Trans. Faraday Soc.*, **57**, 1941 (1961).
[58] J. D. Roberts, E. A. McElhill, and R. Armstrong, *J. Am. Chem. Soc.*, **71**, 2923 (1949).
[59] J. D. Roberts and W. Watanabe, *J. Am. Chem. Soc.*, **72**, 4869 (1950).
[60] J. D. Roberts and R. H. Mazur, *J. Am. Chem. Soc.*, **73**, 2509 (1951).
[61] J. D. Roberts, W. Watanabe, and R. E. McMahon, *J. Am. Chem. Soc.*, **73**, 760, 2521 (1951).
[62] J. D. Roberts and C. M. Regan, *Analyt. Chem.*, **24**, 360 (1952); *J. Am. Chem. Soc.*, **74**, 3695 (1952).
[63] J. D. Roberts, C. M. Regan, and I. Allen, *J. Am. Chem. Soc.*, **74**, 3679 (1952).

transfer (88a) rate-determining, and this is confirmed by the observation that the use of EtOD as solvent decreases catalysis by both strong and weak acids by a factor of 2–4. In presence of carboxylic acids the position is complicated by the fact that there are now two products, $Ph_2CHOEt$ and $Ph_2CHOCOR$. However, the proportion of ester (about 65%) is almost independent of temperature and of the nature of carboxylic acid, and it is generally agreed[64,65] that both products arise from a common intermediate which is formed in a rate-determining proton transfer from the carboxylic acid to the diazo-compound. In confirmation of this view, the rates for a series of carboxylic acids ran parallel to their acid strengths. However, the exact nature of the fast steps following the proton transfer presents a more speculative problem. The approximate constancy of the product ratio and the fact that it is not affected by the addition of carboxylate salts speaks against a competition by nucleophilic reagents for either of the free ions $Ph_2CHN_2^+$ or $Ph_2CH^+$, and it is likely that ion pairs such as $RCO_2^-$ $^+N_2CHPh_2$ or $RCO_2^-$ $^+CHPh_2$ are involved. It would be of interest to study this reaction in aqueous solution, where such ion pairs are less likely to be important.

Extensive studies have been made of the reaction of diphenyldiazomethane with carboxylic acids in non-aqueous solvents other than ethanol.[66] As would be expected, in a series of alcohols ROH the ratio of ether to ester formed depends upon the nature of R, but the velocity does not depend in any simple way upon the properties of the solvent, and specific solvation of the reactants or the transition state is probably involved. In non-hydroxylic solvents only one product, the ester, is formed, but there are often complications owing to the self-association of carboxylic acids.

As our third example of reactions of diazo-compounds we shall consider the decomposition of the diazoacetate ion, which is extremely sensitive to catalysis by hydrogen ions and other acidic species. It was first studied kinetically by King and Bolinger,[67] but their results show certain anomalies which remained unexplained for many years. In the solutions studied, the only detectable process is $N_2CHCO_2^- + H_2O \rightarrow CH_2OHCO_2^- + N_2$, and the reaction can be followed conveniently by measuring the nitrogen evolved. Diazoacetates are fairly stable in concentrated alkalis, but

[64] K. Bowden, A. Buckley, and N. B. Chapman, *J. Chem. Soc.*, 3380 (1964).
[65] R. A. More O'Ferrall, Wo Kong Kwok, and S. I. Miller, *J. Am. Chem. Soc.*, **86**, 5553 (1964).
[66] For the reaction with benzoic acid in 42 solvents, see N. B. Chapman, M. R. J. Dack, and J. Shorter, *J. Chem. Soc.*, B, 834 (1971), and earlier papers.
[67] C. V. King and E. D. Bolinger, *J. Am. Chem. Soc.*, **58**, 1533 (1936).

catalysis by hydrogen ions is conveniently measurable in the range $[H^+] = 10^{-13}–10^{-10}$, i.e., in alkaline solution throughout. However, the plot of reaction velocity against $[H^+]$ is far from linear, being strongly convex towards the concentration axis. Further, although measurements in buffer solutions reveal qualitatively general acid catalysis by the species $MeCO_2H$, $C_6H_5OH$, $NH_4^+$, piperidinium ion, and $HPO_4^{2-}$, the observed velocity constants cannot be represented by the usual type of expression $k = k_0 + k_H[H^+] + k_A[A]$. If $k$ is plotted against $[A]$ at constant $[H^+]$ (i.e., constant buffer ratio $[A]/[B]$), curves concave to the concentration axis are obtained instead of the usual straight lines, and the catalytic effect of an acid A appears to depend on the concentration of the corresponding base B; thus a plot of $k$ against $[A]$ and constant $[B]$ does give a straight line, whose slope depends upon the value of $[B]$.

This behaviour was left unexplained by the authors, but it can be accounted for by the reaction scheme

$$N_2CHCO_2^- + \sum A_i \underset{k_{-1}}{\overset{k_1}{\rightleftharpoons}} N\equiv\overset{+}{N}CH_2CO_2^- + \sum B_i,$$

$$N\equiv\overset{+}{N}CH_2CO_2^- + H_2O \xrightarrow{k_2} N_2 + CH_2OHCO_2H$$

in which $k_1 \ll k_{-1} \sim k_2$. This is the exact analogue of case (a) (Equation 55) for basic catalysis, and if the solution contains only one acid-base pair A–B in addition to the solvent, we obtain for the observed velocity constant

$$\frac{1}{k} = \frac{1}{k_0 + k_H[H^+] + k_A[A]} + \frac{K}{k_2[H^+]}$$

where $K$ is the equilibrium constant $[H^+][N_2CHCO_2^-]/[N_2^+CH_2CO_2^-]$. Remembering that $[H^+] = K_A[A]/[B]$, where $K_A$ is the acidity constant of A, it will be seen that (90) predicts qualitatively all the kinetic features* mentioned, and in particular that the rate will go to zero in strongly alkaline solutions rather than to the 'spontaneous' rate $k_0$. In fact it is found[68] that all the velocity constants reported by King and Bolinger[67] (about a hundred in all) can be represented quantitatively by Equation (90), which involves only one constant $(k_2/K)$ in addition to those required in any conventional example of general acid catalysis. Essentially the same picture is presented by recent results[69] of Kreevoy and Konasewich, who

---

*The linear relation between $k$ and $[A]$ at constant $[B]$ follows only if $k_0 \ll k_H[H^+] + k_A[A]$, which is the case for the buffers used.
[68] R. P. Bell and P. T. McTigue, *J. Chem. Soc.*, 2983 (1960).
[69] M. M. Kreevoy and D. E. Konasewich, *J. Phys. Chem.*, **74**, 4464 (1970); *Adv. Chem. Phys.*, **21**, 243 (1971).

followed the reaction spectrophotometrically and used automatic pH control, thus obtaining more reliable values for the velocity constants.

In its kinetic behaviour the diazoacetate ion thus occupies a position intermediate between ethyl diazoacetate and diphenyldiazomethane. Equation (90) predicts specific hydrogen ion catalysis ($k = k_2[H^+]/K$) at very low acidities, and general acid catalysis ($k = k_0 + k_H[H^+] + k_A[A]$) at sufficiently high acidities. In practice, both terms of (90) contribute significantly in the range of acidities corresponding to convenient reaction rates, leading to the more complex behaviour described above.

The acid-catalysed reactions of some other classes of diazo-compound have also been investigated. Thus pre-equilibrium protonation, and hence specific hydrogen ion catalysis, has been observed for α-diazo-sulphones,[70] for a number of diazo-ketones of the general formula $RCOCHN_2$,[71] and for $CF_3CHN_2$.[72] On the other hand, diazo-ketones $CH_3COCRN_2$,[73,74] $N_2CMeCO_2Et$,[74] ring-substituted phenyldiazomethanes,[75] and benzoyl-phenyldiazomethane[76] all react through a rate-determining proton transfer, and therefore exhibit general acid catalysis. Finally, the decomposition reactions of the ions $N_2C(CO_2^-)_2$[77] and $N_2(CO_2^-)_2$[78] exhibit many of the complex features already described for the diazoacetate ion, though no satisfactory interpretation of their behaviour has yet been given.

*The ionization and enolization of carbonyl compounds*
This class of reaction has been widely quoted in the last chapter as an example of proton transfer leading to subsequent chemical change, and of prototropic isomerization. It was also used to illustrate the problems involved in deciding for or against a ternary catalytic mechanism in aqueous solution. Like the decomposition of nitramide, it occupies an

[70] B. Zwanenburg and J. B. F. N. Engberts, *Rec. Trav. Chim.*, **84**, 165 (1965); J. B. F. N. Engberts, G. Zuidema, B. Zwanenburg, and J. Strating, *Rec. Trav. Chim.*, **88**, 641 (1969).

[71] C. E. McCauley and C. V. King, *J. Am. Chem. Soc.*, **74**, 6221 (1952); H. Dahn, and H. Gold, *Helv. Chim. Acta*, **56**, 983 (1963; H. Dahn, A. Donzel, A. Merbach, and H. Gold, *Helv. Chim. Acta*, **56**, 994 (1963).

[72] H. Dahn, H. Gold, M. Ballenegger, J. Lenoir, G. Diderich, and R. Malherbe, *Helv. Chim. Acta*, **51**, 2065 (1968).

[73] H. Dahn and M. Ballenegger, *Helv. Chim. Acta*, **52**, 2417 (1969).

[74] W. J. Albery, A. N. Campbell-Crawford, and K. S. Hobbs, *J. Chem. Soc. Perkin Trans.* II, 2180 (1972).

[75] H. Dahn and G. Diderich, *Helv. Chim. Acta*, **54**, 1950 (1971); G. Diderich and H. Dahn, *Helv. Chim. Acta*, **55**, 1 (1972).

[76] J. B. F. N. Engberts, N. F. Bosch, and B. Zwanenburg, *Rec. Trav. Chim.*, **85**, 1068 (1966).

[77] C. V. King and P. Kulka, *J. Am. Chem. Soc.*, **72**, 1906 (1950); C. V. King, P. Kulka, and A. Mebane, *J. Am. Chem. Soc.*, **74**, 3128 (1952).

[78] C. V. King, *J. Am. Chem. Soc.*, **62**, 379 (1940); C. V. King and J. J. Josephs, *J. Am. Chem. Soc.*, **66**, 767 (1944).

important place in the history of acid-base catalysis, since the work of Dawson and Powis[79] on the acetone–iodine reaction provided the first definite proof that undissociated acid molecules can act as catalysts in aqueous solution.*

It is only rarely that the kinetics of keto–enol isomerization have been studied directly, largely because the equilibrium ion solution is nearly always one-sided, usually favouring the keto form. Instead, use has commonly been made of rates of racemization, hydrogen isotope exchange, or rapid reaction of the enol or enolate ion with electrophilic reagents such as the halogens. There is good evidence[80] that these rates are all equal in hydroxylic solvents for reactions of the group $\gtrdot$CHCO—, though the same is not always true in aprotic solvents or for other systems. Extensive use has been made of *rates of halogenation*, and we shall now summarize some of the experimental results, with special reference to acetone.

It was first shown by Lapworth[81] that in aqueous solutions of strong acids acetone reacts with bromine and chlorine at a rate which is inde-

Table 22. HALOGENATION OF ACETONE IN AQUEOUS SOLUTION AT 25°C CATALYSED BY HYDROGEN IONS

$v = k_H[H^+][Me_2CO]$,     $k_H$ in $dm^3\ mol^{-1}\ s^{-1}$

| Halogen | Method | $10^7 k_H$ | Reference |
|---------|--------|-----------|-----------|
| Iodine | Chemical analysis | 285 | (a) |
| Iodine | Chemical analysis | 283 | (b) |
| Iodine | Spectrophotometric | 288 | (c) |
| Bromine | Chemical analysis | 283 | (d) |
| Bromine | Chemical analysis | 287 | (e) |
| Bromine | Spectrophotometric | 283 | (f) |
| Chlorine | Chemical analysis | 285 | (g) |
| Chlorine | Chemical analysis | 292 | (h) |
| Chlorine | Spectrophotometric | 285 | (h) |

For references, see p. 193.

[79] H. M. Dawson and F. Powis, *J. Chem. Soc.*, 2135 (1913).

* Dawson's quantitative conclusions need some modification in the light of modern views, since he assumed that the degree of dissociation of an acid was given directly by the conductivity ratio $\Lambda/\Lambda_0$, and he took no account of primary or secondary salt effects. However, recalculation of his results for solutions of carboxylic acids (R. P. Bell, *Acid-Base Catalysis*, Oxford, 1941, pp. 52–56) confirms the conclusion that a large part of the catalysis is due to acid molecules, though some of the numerical values are considerably changed. On the other hand, the original suggestion of catalysis by molecules of strong acids such as HCl could not be maintained, as subsequently confirmed by Dawson himself [H. M. Dawson and J. S. Carter, *J. Chem. Soc.*, 2782 (1926)].

[80] C. K. Ingold, *Structure and Mechanism in Organic Chemistry*, 2nd edn., Bell, London, 1969, p. 835.

[81] A. Lapworth, *J. Chem. Soc.*, 30 (1904).

pendent of the nature or concentration of the halogen, but proportional
to the hydrogen ion concentration, and he correctly interpreted this in
terms of a rate-determining acid-catalysed enolization, followed by very
rapid reaction of the enol with halogen. Later work has fully confirmed
this, and Table 22 shows the excellent agreement between the results
obtained by different methods.

Extensive work, especially by Dawson and his collaborators,[82] has
shown that in buffer solutions of carboxylic acids the halogenation of
acetone is catalysed by both carboxylic acids and carboxylate ions, as well
as by hydrogen and hydroxide ions. In this situation of general catalysis
by both acids and bases the observed velocity constant will be given by an
expression of the form

$$k = k_0 + k_H[H^+] + k_{OH}[OH^-] + k_{HA}[HA] + k_A[A^-] \qquad (91)$$

where $k_0$ refers to the 'spontaneous' or solvent-catalysed reaction, and
HA and A$^-$ are the acidic and basic constituents of the buffer. The problem
of detecting and measuring the catalytic effects of the different species is
thus a complicated one, though it can generally be much simplified by a
judicious choice of experiments.

In the first place it may be possible to choose a pH range such that one
or more of the first three terms of (91) is negligible. In any case, if we write
$[H^+][OH^-] = K_w$, $[H^+][A^-]/[HA] = K_A$, $[HA]/[A^-] = r$, (91) becomes

$$k = k_0 + k_H K_A r + k_{OH} K_w/rK_A + [HA](k_{HA} + k_A/r) \qquad (92)$$

If it can be assumed that $K_A$ and $K_w$ are true constants, the first three
terms of this expression can be kept constant by maintaining a constant
buffer ratio $r$, and the observed rate is given by

$$k = k'(r) + [HA](k_{HA} + k_A/r) \qquad (93)$$

where $k'(r)$ depends only upon $r$. A series of experiments with constant $r$
and variable [HA] thus gives $k_{HA} + k_A/r$, and from several such series with
different values of $r$, $k_{HA}$ and $k_A$ can be obtained separately.

The validity of this method depends upon the assumption that $K_w$ and
$K_A$ remain constant throughout each series. Since, as defined, they do not
contain activity coefficients, this assumption will be valid if the ionic

[82] H. M. Dawson and F. Powis, *J. Chem. Soc.*, 2135 (1913); H. M. Dawson and
C. K. Reiman, *J. Chem. Soc.*, 1426 (1915); H. M. Dawson and N. C. Dean, *J. Chem.
Soc.*, 2872 (1926); H. M. Dawson and C. R. Hoskins, *J. Chem. Soc.*, 3166 (1926);
H. M. Dawson, *J. Chem. Soc.*, 213, 756, 1146 (1927); H. M. Dawson and A. Key,
*J. Chem. Soc.*, 543, 1239, 1248 (1928); H. M. Dawson, G. V. Hall, and A. Key,
*J. Chem. Soc.*, 2844 (1928); H. M. Dawson, C. R. Hoskins, and J. E. Smith, *J. Chem.
Soc.*, 1884 (1929); F. O. Rice and H. C. Urey, *J. Am. Chem. Soc.*, **52**, 95 (1930);
G. F. Smith, *J. Chem. Soc.*, 1744 (1934); R. P. Bell and O. M. Lidwell, *Proc. Roy. Soc.*,
A, **176**, 88 (1940).

strength is maintained at a constant and not too high value in each series by the addition of a neutral salt. This procedure also minimizes variations caused by primary salt effects upon the velocity constants, and does not demand a knowledge of the actual values of $K_w$ or $K_A$ under experimental conditions. Much of the earlier work on the halogenation of acetone was not planned in this way, and is therefore difficult to interpret quantitatively.

In interpreting the halogenation kinetics of acetone (or of $CH_3$ or $CH_2$ groups in analogous compounds) it is often necessary to consider the consecutive introduction of more than one halogen atom. This is particularly true for basic catalysis, since under these conditions $CH_3COCH_2Cl$ and $CH_3COCHCl_2$ halogenate respectively about 3000 and 60,000 times faster than acetone itself. Except in the very early stages of the reaction it may be legitimate to assume that the introduction of the second and third halogen atoms is very much faster than the first stage, and under these conditions the rate of consumption of halogen will be three times the rate of the initial ionization or enolization. Alternatively (and especially when the reaction is catalysed mainly by acids) it may be more convenient to use halogen concentrations very much smaller than the ketone concentration, when the initial rate of halogenation can be equated to the rate of enolization or ionization. If neither of these conditions is fulfilled the observed halogenation kinetics will be complex, though in principle the curves can be analysed to yield the velocity constants of the successive reactions.

The halogenation of acetone is also catalysed by a number of alkyl-pyridines,[83] but apparent acid catalysis by ammonium ions[84] was traced to a specific mechanism,

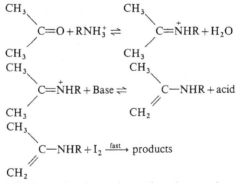

since no catalysis is shown by the cations of tertiary amines.

[83] J. A. Feather and V. Gold, *J. Chem. Soc.*, 1752 (1965).
[84] M. L. Bender and A. Williams, *J. Am. Chem. Soc.*, **88**, 2504 (1966).

The measurement of catalysis by hydroxide ions offers some special problems. As shown by Equations (91)–(93), if the rate at a constant buffer ratio is plotted against $[HA]$ (or $[A^-]$) the intercept will be $k_0 + k_H K_A r + k_{OH} K_w / r K_A$, and a series of intercepts for different buffer ratios should yield $k_{OH}$, since $k_H$ can be determined in separate experiments in solutions of strong acids. However, these intercepts represent only a small proportion of the observed rate, and their interpretation demands a knowledge of $K_A$ and $K_w$ under the experimental conditions, so that the resulting values of $k_{OH}$ are often uncertain. The position can be improved by using a buffer containing sterically hindered acids and bases (for example 2,6-disubstituted pyridines), thus reducing the contribution from $k_{HA}$ and $k_A$. An extension of this idea is to use a heterogeneous buffer system such as $Zn(OH)_2(s) \rightleftharpoons Zn^{2+} + 2OH^-$, in which the concentration of hydroxide ion is controlled by varying $[Zn^{2+}]$, but even here catalysis by dissolved species makes a considerable contribution.[85] If halogenation is carried out in solutions of sodium or potassium hydroxide, the substance $CX_3COCH_3$ which is first formed is rapidly hydrolysed to the haloform $CHX_3$. Since the halogens in these solutions exist predominantly as hypohalite ions $XO^-$, the overall reaction becomes $CH_3COCH_3 + 2XO^- \rightarrow CHX_3 + CH_3CO_2^- + 2OH^-$, though the rate is still proportional to $[OH^-]$ and independent of $[XO^-]$. In these solutions the effective halogenating agent is probably the halogen molecule, present in very low concentrations by virtue of the equilibrium $XO^- + X^- + H_2O \rightleftharpoons X_2 + 2OH^-$, and there is a danger that the enolate ion may not be removed fast enough for its formation to remain the rate-determining step. This is particularly the case for halogenation of ketones by hypochlorite, in which the reaction is usually first-order in $[ClO^-]$, rather than zero-order;[86,87] under these conditions the observed velocity will of course give no information about the rate of the process $CH_3COCH_3 + OH^- \rightarrow [CH_3COCH_2]^- + H_2O$. Nevertheless, the most reliable information about this rate probably derives from halogenation in alkaline hypobromite and hypoiodite solutions, the rates of which are independent of hypohalite concentration.[88,89]

There is a wealth of information about the behaviour of keto-compounds other than acetone, and we shall list here only investigations in which a

[85] R. P. Bell and J. E. Prue, *Trans. Faraday Soc.*, **46**, 5 (1950).

[86] P. D. Bartlett and J. R. Vincent, *J. Am. Chem. Soc.*, **57**, 1596 (1935).

[87] R.-R. Li and S. I. Miller, *J. Chem. Soc.*, B, 2269 (1971).

[88] R. P. Bell and H. C. Longuet-Higgins, *J. Chem. Soc.*, 636 (1946).

[89] J. R. Jones, *Trans. Faraday Soc.*, **61**, 65 (1965); **65**, 2138 (1969).

systematic study has been made of catalysis by several acids or bases in aqueous solution. All the compounds concerned are more acidic than acetone, either because of the presence of electronegative substituents such as halogens, or (particularly in $\beta$-diketones and $\beta$-keto-esters) because of the sharing of the negative charge in the anion between two oxygen atoms (cf. p. 105 ). Their halogenation is therefore more susceptible to basic catalysis, and for many of them the base-catalysed water reaction is fast enough to obscure any catalysis by acids, including the hydrogen ion. The substances investigated may be classified as aliphatic and alicyclic mono-ketones,[91,97,102,104] diketones,[91,94,100] keto-esters,[90,93,94,95 101,103,105] nitroacetone,[99] and propan-2-one-1-sulphonate,[98,106]. Similar kinetic behaviour is shown by a number of esters such as ethyl malonate,[92,94] methyl methanetricarboxylate,[106] and ethyl nitroacetate.[96,106]

We have seen (p. 152 ) that measurements on the acetone–iodine reaction provide no evidence that an appreciable fraction of the reaction takes place by a ternary mechanism, and it is also of interest that no abnormal catalytic activity is exhibited by the mono-anions of dicarboxylic acids, $CO_2^- \cdots CO_2H$, which might have been expected to act as bifunctional catalysts.[107] There is thus little doubt that the rate-determining step in the base-catalysed halogenation of keto-compounds (and in the analogous racemization and isotope exchange reactions) is the transfer of a proton from the substrate to the catalyst, producing the enolate ion. In the acid-catalysed reaction it is necessary to produce the enol before halogenation can take place, and the observation of general acid catalysis is consistent with two mechanisms: (a) a rate-determining addition of a

[90] K. J. Pedersen, *J. Phys. Chem.*, **37**, 751 (1933); **38**, 601 (1934).
[91] R. P. Bell and O. M. Lidwell, *Proc. Roy. Soc.*, A, **176**, 88 (1940).
[92] R. P. Bell, D. H. Everett, and H. C. Longuet-Higgins, *Proc. Roy. Soc.*, A, **186**, 433 (1946).
[93] R. P. Bell, R. D. Smith, and L. A. Woodward, *Proc. Roy. Soc.*, A, **192**, 479 (1948).
[94] R. P. Bell, E. Gelles, and E. Möller, *Proc. Roy. Soc.*, A, **198**, 310 (1949).
[95] R. P. Bell and H. L. Goldsmith, *Proc. Roy. Soc.*, A, **216**, 322 (1952).
[96] R. P. Bell and T. Spencer, *Proc. Roy. Soc.*, A, **251**, 41 (1959).
[97] R. P. Bell and J. Hansson, *Proc. Roy. Soc.*, A, **255**, 214 (1960).
[98] R. P. Bell and G. A. Wright. *Trans. Faraday Soc.*, **57**, 1386 (1961).
[99] R. P. Bell and R. R. Robinson, *Proc. Roy. Soc.*, A, **270**, 411 (1962).
[100] T. Riley and F. A. Long, *J. Am. Chem. Soc.*, **84**, 522 (1962).
[101] R. P. Bell and J. E. Crooks, *Proc. Roy. Soc.*, A, **286**, 285 (1965).
[102] J. A. Feather and V. Gold, *J. Chem. Soc.*, 1752 (1965).
[103] R. P. Bell and H. F. F. Ridgewell, *Proc. Roy. Soc.*, A, **298**, 178 (1967).
[104] G. E. Lienhard and Tung-Chia Wang, *J. Am. Chem. Soc.*, **91**, 1146 (1969).
[105] R. P. Bell and P. de Maria, *Trans. Faraday Soc.*, **66**, 930 (1970).
[106] R. P. Bell and D. J. Barnes, *Proc. Roy. Soc.*, A, **318**, 421 (1970).
[107] G. E. Lienhard and F. H. Anderson, *J. Org. Chem.*, **32**, 2229 (1967).

proton to the carboxyl group, followed by rapid proton loss from the carbon atom; (b) equilibrium proton addition to the carbonyl group, followed by rate-determining proton loss from carbon to the conjugate base of the acid catalyst. As already shown (p.143 ), these two mechanisms are kinetically equivalent, although they correspond to different structures for the transition state. Since proton transfers to and from oxygen are normally very fast, (b) appears to be more likely, and is in fact supported by several pieces of experimental evidence. If proton addition to the carbonyl group were rate-determining we should expect a parallelism between the basic strengths of a series of ketones and their rates of acid-catalysed halogenation. Zucker and Hammett[108] found that no such parallelism existed for a series of six substituted acetophenones; this is consistent with (b) according to which the rate is given (cf. Equation 63) as the ratio of a velocity constant and an equilibrium constant which will be affected in the same direction by substitution. Further, ketonization, the reverse of enolization, is closely analogous to the first step in the hydrolysis of vinyl ethers, i.e.,

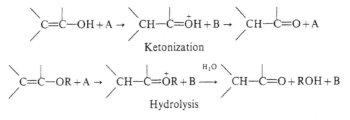

Ketonization

Hydrolysis

The latter process is known to involve rate-determining proton transfer to carbon,[109] and since the two reactions are kinetically very similar[104,110] they probably have similar rate-determining steps.

In non-aqueous solvents the addition of acids to optically active ketones causes an instantaneous change of rotation followed by a slow racemization,[111] again suggesting a pre-equilibrium preceding the rate-determining step. Finally, and most convincingly, the acid-catalysed bromination of acetone is *faster* by a factor of 2 in $D_2O$ than in $H_2O$,[112] which is quite inconsistent with a single rate-determining proton transfer.

[108] L. Zucker and L. P. Hammett, *J. Am. Chem. Soc.*, **61**, 2785 (1939).
[109] P. Salomaa, A. Kankaaperä, and M. Lajunen, *Acta Chem. Scand.*, **20**, 1790 (1966); A. J. Kresge and Y. Chiang, *J. Chem. Soc.* B, 53 (1967); M. M. Kreevoy and R. Eliason, *J. Phys. Chem.*, **72**, 1313 (1968).
[110] J. E. Dubois and J. Toullec, *Chem. Comm.*, 478 (1969).
[111] R. P. Bell and E. F. Caldin, *J. Chem. Soc.*, 382 (1938); R. P. Bell, O. M. Lidwell, and J. Wright, *J. Chem. Soc.*, 1861 (1938).
[112] O. Reitz, *Z. Phys. Chem.*, A, **179**, 119 (1936).

If a molecule containing the group $>$CHCO— also contains an acidic
or basic group such as —$CO_2H$, —$CO_2^-$, or —$NR_2$ there arises the
possibility of *intramolecular catalysis* of enolization or ionization. Such
catalysis will be sensitive to the geometry and rigidity of the molecule, and
is of interest as a model for enzyme action.[113] Intramolecular catalysis is
well established for the iodination of keto-carboxylic acids[114] of the
general formula $CH_3CO(CH_2)_nCO_2H$. Iodination takes place at the
carbon atom adjacent to the carbonyl group, and is mainly due to intramolecular base catalysis by —$CO_2^-$, or intramolecular acid catalysis by
—$CO_2H$. The rates are much too large to be attributed to water catalysis,
or to any other intermolecular process, and for both acid and base catalysis a maximum rate is observed when $n = 3$; this corresponds to a cyclic
transition state with a five-membered ring (in addition to the proton which
is being transferred). Similarly, intramolecular base catalysis is observed
in the iodination of $CHR_2COCR_2CH_2NR_2$,[115] $CH_3CO(CH_2)_2NEt_2$ and
$CH_3CO(CH_2)_3NEt_2$.[116] In analogous rigid systems, intramolecular base
catalysis is mainly responsible for the iodination of the anions of
*o*-carboxyisobutyrophenone[117] and *o*-carboxyacetophenone.[118] In the
latter case the transition state can be written as

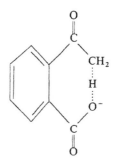

[113] See T. C. Bruice and S. J. Benkovic, *Bioorganic Mechanisms*, Vol. 1, pp. 119ff.,
Benjamin, New York, 1966; W. P. Jencks, *Catalysis in Chemistry and Enzymology*,
McGraw-Hill, New York, 1969, Ch. 1.
[114] R. P. Bell and M. A. D. Fluendy, *Trans. Faraday Soc.*, **59**, 1623 (1963).
[115] J. K. Coward and T. C. Bruice, *J. Am. Chem. Soc.*, **91**, 5339 (1969).
[116] R. P. Bell and B. A. Timimi, to be published.
[117] E. T. Harper and M. L. Bender, *J. Am. Chem. Soc.*, **87**, 5625 (1965).
[118] R. P. Bell, B. G. Cox, and J. B. Henshall, *J. Chem. Soc., Perk. Trans. 2*, 1232
(1972).

The ring now contains six atoms in addition to the proton, but two of its angles will be held close to 120° by the benzene ring. These processes owe some of their efficiency, compared with intermolecular catalysis, to the fact that they avoid the loss of translational entropy which occurs when two solute species are brought together.[119] However, it is not yet known whether the activation energies are the same for analogous intermolecular and intramolecular processes, and energy effects may also be involved.

It has so far been assumed that the reaction of halogen with enol or enolate is so fast that it plays no part in determining the observed rate, though we have seen that this is not the case for halogenation of ketones by hypochlorite. The assumption is certainly valid for solutions containing an appreciable concentration of molecular halogen (say $10^{-3}$M or greater), but if this concentration is progressively reduced there must come a point at which the halogenation reaction becomes at least partly rate-determining. It is a simple matter to derive the general kinetic expression for a particular situation, and we shall give the result for the common case in which a keto-compound HS undergoes base-catalysed halogenation in a buffer solution A–B in which catalysis by hydroxide ion can be neglected, and both the enol SH and the enolate ion $S^-$ can react with halogen. We shall also assume that the equilibrium $SH \rightleftharpoons S^- + H^+$ between the enol and its ion can be regarded as instantaneous on the relevant time-scale; this is of course not true for the equilibrium $HS \rightleftharpoons S^- + H^+$ involving the keto-form. The kinetic scheme is then

$$HS \underset{k_H}{\overset{k_0}{\rightleftharpoons}} S^- + H^+, \qquad HS + B \underset{k_A}{\overset{k_B}{\rightleftharpoons}} S^- + A$$

$$SH \underset{k_{SH}}{\rightleftharpoons} S^- + H^+ \text{ (fast)} \qquad\qquad (94)$$

$$S^- + X_2 \xrightarrow{k_2} SX + X^-, \qquad SH + X_2 \xrightarrow{k_3} SX + X^- + H^+$$

On the assumption that the concentrations of enolate $S^-$ and enol SH are throughout very small, a steady-state treatment gives for the observed reaction velocity $v = (d[SX]/dt)/[HS]$,

$$\frac{1}{v} = \frac{1}{k_0 + k_B[B]} + \frac{1}{[X_2](k_3 K_E + k_2 K_{HS}/[H^+])} \qquad (95)$$

where $K_{HS}$ is the acidity constant of the keto-form, and $K_E = K_{HS}/K_{SH}$

[119] M. I. Page and W. P. Jencks, *Proc. Nat. Acad. Sci.*, **68**, 1671 (1971).

is the equilibrium enol content of the system.* Analogous equations are easily derived for acid catalysis.

Equation (95) shows that for sufficiently high halogen concentrations $v = k_0 + k_B[B]$, corresponding to zero-order with respect to halogen, as usually assumed; this assumption is usually valid when $[X_2] > 10^{-3}$M. At the other extreme, sufficiently low values of $[X_2]$ should give reactions which are first-order in $X_2$, and experiments at different acidities will yield absolute values of $k_2$ and $k_3$ provided that the equilibrium constants $K_E$ and $K_{HS}$ are known. This second possibility has been realized in practice by using measurements of redox potential to follow halogen concentrations in the range $10^{-5}$–$10^{-8}$M, for example, in the acid-catalysed bromination[120] and chlorination[121] of acetone. The intermediate situation, in which both terms of (95) must be retained, has been observed in the base-catalysed bromination of ethyl malonate[122,123] and methyl methanetricarboxylate.[123] It is not always practicable to use halogen concentrations high enough to ensure zero-order kinetics, and in this case Equation (95) shows that the rate of ionization or enolization can be obtained by extrapolating a linear plot of $1/v$ against $1/[X_2]$.[124] Since the iodination of ketonic substances is frequently appreciably reversible,[125] very low concentrations of iodine cannot be used, but equivalent information can be obtained by studying the deiodination of the iodo-derivatives, a scavenger such as ascorbic acid being used to remove the iodine as quickly as it is formed. This procedure has been applied to the iodination of ethyl malonate[126] and of 2-ethoxycarbonylcyclohexanone.[127] In all these

---

* In arriving at equation (95) the velocity constants $k_H$ and $k_A$, and the concentration [A] are eliminated by making use of the equilibrium conditions $[S^-]_e/[HS]_e = k_0/k_H[H^+] = k_B[B]/k_A[A]$. It should be noted that the halogenation velocity constants $k_2$ and $k_3$ are frequently composite quantities, since the solution may contain more than one halogenating species. In solutions of bromine and iodine these comprise the halogen molecule $X_2$ and the trihalide ion $X_3^-$. If we replace $[X_2]$ in Equation (95) by the stoichiometric halogen concentration $[X_2]^* = [X_2] + [X_3^-]$, then it is easily shown that the effective value of $k_2$ is given by $k_2(1 + K[X^-]) = k_2' + k_2''K[X^-]$, where $K = [X_3^-]/[X_2][X^-]$ and $k_2'$ and $k_2''$ refer to the reactions $S^- + X_2$ and $S^- + X_3^-$ respectively; a similar equation holds for $k_3$. It is then possible to obtain the separate values for $k'$ and $k''$ by making measurements at different values of $[X^-]$. For the sake of simplicity we shall ignore this complication, and deal only with the effective values of $k_2$ and $k_3$.

[120] R. P. Bell and G. G. Davis, *J. Chem. Soc.*, 902 (1962).
[121] R. P. Bell and K. Yates, *J. Chem. Soc.*, 1927 (1962).
[122] R. P. Bell and M. Spiro, *J. Chem. Soc.*, 429 (1953).
[123] R. P. Bell and D. J. Rawlinson, *J. Chem. Soc.*, 726 (1961).
[124] R. P. Bell and B. G. Cox, *J. Chem. Soc.*, B, 654 (1971).
[125] R. P. Bell and E. Gelles, *Proc. Roy. Soc.*, A, **210**, 310 (1952).
[126] R. P. Bell and P. Engel, *J. Chem. Soc.*, 247 (1957).
[127] R. P. Bell and D. C. Vogelsong, *J. Chem. Soc.*, 243 (1958).

experiments $k_3$ is found to lie in the range $10^4$–$10^6$ $dm^3$ $mol^{-1}$ $s^{-1}$, while $k_2$ is close to the diffusion-controlled value of $10^9$–$10^{10}$ $dm^3$ $mol^{-1}$ $s^{-1}$. Rates of the same order of magnitude have been observed directly for the analogous reactions of bromine with enol ethers,[128] phenols and phenoxide ions.[129]

Enols and enolate ions can of course react with electrophilic reagents other than halogens, and it has been found[130] that the rate of oxidation of cyclohexanone or acetone by mercuric salts in acid solution is independent of the concentration of the oxidizing agent and equal to the rate of bromination under the same conditions; clearly the rate of enolization is again rate-determining. Enolate ions can also react with the electrophilic carbonyl group of the same or a different carbonyl compound, giving rise to a base-catalysed *aldol condensation*. The simplest example is the condensation of two molecules of acetaldehyde to give acetaldol, for which the mechanism is

(a) $CH_3CHO + \sum B_i \overset{k_1}{\underset{k_{-1}}{\rightleftharpoons}} CH_2CHO^- + \sum A_i$

(b) $CH_3CHO + CH_2CHO^- \overset{k_2}{\longrightarrow} CH_3\underset{\underset{O^-}{|}}{C}HCH_2CHO$ \hfill (96)

(c) $CH_3\underset{\underset{O^-}{|}}{C}HCH_2CHO + \sum A_i \rightleftharpoons CH_3CHOHCH_2CHO + \sum B_i$

where the third step is much faster than the other two. This differs from the reaction scheme (52) in that step (b) now involves a second molecule of substrate, its rate being given by $k_2[S^-][SH]$. Assuming as before that $k_1 \ll k_{-1} \sim k_2[SH]$, we find for the velocity in a simple buffer solution,

$$-\frac{2}{d[SH]/dt} = \frac{1}{[SH](k_0 + k_{OH}[OH^-] + k_B[B])}$$

$$+ \frac{K_w}{[SH]^2 k_2 K_{SH}[OH^-]} \hfill (97)$$

where the factor 2 on the left arises from the fact that reaction (96b) consumes two molecules of acetaldehyde. The form of (97) suggests that

[128] D. R. Marshall and T. R. Roberts, *J. Chem. Soc.*, B, 797 (1971).
[129] R. P. Bell and D. J. Rawlinson, *J. Chem. Soc.*, 726 (1961).
[130] A. J. Green, T. J. Kemp, J. S. Littler, and W. A. Waters, *J. Chem. Soc.*, 2722 (1964).

the course of a single reaction will not be expressible by a simple kinetic order, and since further the later stages of the reaction show chemical complications the most useful information can be obtained from initial rates.

Although Equation (97) has not been confirmed in full, kinetic studies[131] reveal the following points which are consistent with it.

(1) In sodium hydroxide solutions the apparent order with respect to aldehyde is two at low aldehyde concentrations, falling somewhat with increasing concentrations.

(2) At constant aldehyde concentration the rate is not quite linear in the hydroxide ion concentration but falls off more rapidly at low concentrations and becomes zero in slightly acid solution.

(3) Measurements in buffer solutions give evidence for general base catalysis by the species $CO_3^{2-}$ and $B(OH)_4^-$.

Further support comes from a study of deuterium exchange. If the condensation takes place in deuterium oxide, it is clear that every reversal of reaction (96a) will lead to the introduction of deuterium into acetaldehyde, and hence ultimately into the aldol. At high aldehyde concentrations $k_2[SH] \gg k_{-1}$, so that (96a) will rarely be reversed, but with decreasing aldehyde concentration the extent of deuteriation should increase. Bonhoeffer and Walters[132] showed that in 10M acetaldehyde no detectable amount of deuterium is attached to the carbon of the aldol produced, and more recent work[133] shows that in the range 0.05–1.4M the extent to which the unchanged aldehyde is deuteriated increases with decreasing concentration and is quantitatively consistent with the above reaction scheme.

The correspondence between isotope exchange and kinetic behaviour extends to the analogous aldol condensation of acetone.* This is well established as a second-order reaction catalysed specifically by hydroxide ions, corresponding to $k_2[SH] \ll k_{-1}$ at all values of $[SH]$ studied. This should lead to facile isotope interchange, and in fact it was found[132] that a molar solution of acetone in alkaline $D_2O$ takes up deuterium about 1000 times as quickly as it undergoes condensation.

[131] R. P. Bell, *J. Chem. Soc.*, 1637 (1937); A. Broche, *Colloque Nationale de Cinétique*, Strasbourg, 1953; A. Broche and R. Gibert, *Bull. Soc. Chim. France*, 131 (1955); R. P. Bell and P. T. McTigue, *J. Chem. Soc.*, 2983 (1960); L. C. Gruen and P. T. McTigue, *Austral. J. Chem.*, **17**, 953 (1964).
[132] K. F. Bonhoeffer and W. D. Walters, *Z. Phys. Chem.*, **181**, 441 (1938).
[133] R. P. Bell and M. J. Smith, *J. Chem. Soc.*, 1691 (1958).
*It is more convenient here to study the reverse reaction, the depolymerization of diacetone alcohol, but the known equilibrium constant can be used to deduce the velocity of the forward reaction.

When acetaldehyde and formaldehyde are present together in alkaline solution, addition takes place to give $\beta$-hydroxypropionaldehyde, the first stage in the production of pentaerythritol. The rate of this reaction is proportional to the acetaldehyde concentration and to $[OH^-]$, while its order with respect to formaldehyde changes from unity at low $[CH_2O]$ to zero at high $[CH_2O]$.[134] These observations are consistent with a mechanism analogous to (96), in which the anion of acetaldehyde adds on to a molecule of formaldehyde.

*The reversible addition of hydroxy-compounds to the carbonyl group*
In the reactions considered so far, the transfer of protons between the catalyst and the substrate can be regarded as a separate process, which may constitute a pre-equilibrium, or may be wholly or partly rate-determining. There also exist many reactions in which the proton-transfers are concerted with the making or breaking of other bonds in the system, and the reactions considered in this section represent one example of such a situation.

The addition of nucleophilic reagents to the carbonyl group is of very common occurrence in organic chemistry. In many instances it is followed by loss of water, for example in the formation of oximes, hydrazones, and semicarbazones, frequently leading to rather complex kinetics. Similarly, if the carbon atom of the carbonyl group bears a group which can be readily lost, the formation of a tetrahedral intermediate often represents an intermediate stage in a replacement reaction of the type

$$\underset{\displaystyle R\overset{\textstyle O}{\overset{\|}{C}}X + Y^-}{} \rightleftharpoons \underset{\displaystyle R\overset{\textstyle O^-}{\overset{|}{\underset{|}{\underset{Y}{C}}}}X}{} \rightleftharpoons \underset{\displaystyle R\overset{\textstyle O}{\overset{\|}{C}}Y + X^-}{}$$

as in the hydrolysis and aminolysis reactions of esters and other carboxylic derivatives. The general aspects of carbonyl group reactions have been reviewed recently by several authors,[135-137] and we shall confine ourselves to reactions of the type $R_2R_3CO + R_1OH \rightleftharpoons R_2R_3C(OR_1)OH$, in which $R_1$, $R_2$, and $R_3$ are hydrogen atoms or groups which do not lead to any further reaction. Special attention will be paid to the *reversible*

[134] R. P. Bell and P. T. McTigue (131); Y. Ogata, A. Kawasaki, and K. Yokoi, *J. Chem. Soc.*, B, 1013 (1967); P. T. McTigue, J. Kirsanova, and H. Tankey, *J. Austral. Chem.*, **17**, 499 (1964).
[135] M. L. Bender, *Chem. Rev.*, **60**, 53 (1960).
[136] W. P. Jencks, *Progr. Phys. Org. Chem.*, **2**, 63 (1964).
[137] S. Johnson, *Adv. Phys. Org. Chem.*, **5**, 237 (1967).

*hydration of aldehydes and ketones*, representing the simplest reaction of this class.[138]

Aliphatic aldehydes are hydrated to a considerable extent in aqueous solution, and although the reaction is fast at ordinary temperatures, early observations of the heat evolution and density changes accompanying the dissolution of acetaldehyde in water[139] showed that several minutes were required to reach equilibrium. This reaction is inconveniently fast in aqueous solution at room temperature, especially in the presence of catalysts, and the first systematic kinetic study was made by Bell and Higginson[140] in a 92.5% acetone–water mixture. The reaction was initiated by adding a concentrated aqueous solution of acetaldehyde to a large excess of acetone, and the dehydration process was followed dilatometrically. Catalysis by 52 acids was investigated at 25°C, and qualitative evidence obtained for catalysis by bases. Dilatometric measurements in aqueous buffer solutions[141] provided definite evidence for general acid-base catalysis, and more extensive studies were made[142,143] using a thermal method suitable for studying reactions with half-time down to 1 second or less. Similar investigations have been made for other aliphatic aldehydes using either thermal methods or the u.v. absorption of the carbonyl group.[144,145] Supporting evidence for some of these kinetic results comes from measurements of n.m.r. line broadening (for example of the CH or $CH_3$ protons in $CH_3CHO$) in the presence of catalysts,[146-148] and this technique has recently been extended to use the broadening of resonance lines for $^{17}O$ in the carbonyl group.[149]

Aqueous solutions of formaldehyde contain only about 0.05% of un-hydrated carbonyl compound, and the rate of hydration is very high even in the absence of added catalyst. However, it is possible to measure the rate of dehydration of $CH_2(OH)_2$ if steps are taken to remove unhydrated $CH_2O$ as quickly as it is formed. One method of doing this is by polaro-

[138] For a review, see R. P. Bell, *Adv. Phys. Org. Chem.*, **4**, 1 (1966).

[139] W. H. Perkin, *J. Chem. Soc.*, 808 (1887); H. T. Brown and P. S. U. Pickering, *J. Chem. Soc.*, 774 (1897).

[140] R. P. Bell and W. C. E. Higginson, *Proc. Roy. Soc.*, A, **197**, 141 (1949).

[141] R. P. Bell and B. de B. Darwent, *Trans. Faraday Soc.*, **46**, 34 (1950).

[142] R. P. Bell and J. C. Clunie, *Proc. Roy. Soc.*, A, **212**, 33 (1952).

[143] R. P. Bell, M. H. Rand, and K. M. A. Wynne-Jones, *Trans. Faraday Soc.*, **52**, 1093 (1956).

[144] L. C. Gruen and P. T. McTigue, *J. Chem. Soc.*, 5224 (1963).

[145] Y. Pocker and D. G. Dickerson, *J. Phys. Chem.*, **73**, 405 (1969).

[146] P. G. Evans, M. M. Kreevoy, and G. R. Miller, *J. Phys. Chem.*, **69**, 4325 (1965).

[147] M.-L. Ahrens and H. Strehlow, *Disc. Faraday Soc.*, **39**, 112 (1965).

[148] J. Hine and J. G. Houston, *J. Org. Chem.*, **30**, 1328 (1965).

[149] P. Greenzaid, Z. Luz, and D. Samuel, *J. Am. Chem. Soc.*, **89**, 756 (1967).

graphic reduction,[150] but it is difficult to obtain accurate kinetic information by this method, and it is preferable to use a chemical scavenger to
react with the $CH_2O$ as soon as it is formed. It was shown by Hénaff[151]
that the rate of reaction of several carbonyl reagents (bisulphite, hydra-
zine, phenylhydrazine, semicarbazide, and hydroxylamine) with aqueous
formaldehyde solutions is independent of the nature and concentration
of the reagent, and is therefore determined by the rate of dehydration of
methylene glycol. This principle has been used to make a detailed study of
this reaction,[152] and of the analogous reaction of glycollaldehyde.[153]

The above investigations give ample evidence that the reversible hydration of aldehydes exhibits general catalysis by acids and bases, and we
shall now consider the reaction mechanism. A reasonable sequence is as
follows:

*Acid catalysis*

$$\begin{array}{c} \diagdown \\ CO + H_2O + A \\ \diagup \end{array} \rightleftharpoons \begin{array}{c} \diagdown \\ C(OH)\overset{+}{O}H_2 + B \\ \diagup \end{array} \qquad\qquad (i)$$

$$\begin{array}{c} \diagdown \\ C(OH)\overset{+}{O}H_2 + B \\ \diagup \end{array} \rightleftharpoons \begin{array}{c} \diagdown \\ C(OH)_2 + A \\ \diagup \end{array} \qquad\qquad (ii)$$

(98)

*Base catalysis*

$$\begin{array}{c} \diagdown \\ CO + H_2O + B \\ \diagup \end{array} \rightleftharpoons \begin{array}{c} \diagdown \\ C(OH)O^- + A \\ \diagup \end{array} \qquad\qquad (iii)$$

$$\begin{array}{c} \diagdown \\ C(OH)O^- + A \\ \diagup \end{array} \rightleftharpoons \begin{array}{c} \diagdown \\ C(OH)_2 + B. \\ \diagup \end{array} \qquad\qquad (iv)$$

Of these reactions, (ii) and (iv) are simple proton-transfers to and from
oxygen, and experience shows that such equilibria are set up very rapidly.
On the other hand, (i) and (iii) involve both proton-transfer and the making
of covalent bonds between carbon and oxygen, so that they represent
reasonable rate-limiting steps. They are both formally termolecular processes* in one direction, and it is natural to enquire whether either of them

[150] R. Brdicka, *Coll. Czech. Chem. Comm.*, **20**, 387 (1955).
[151] P. L. Hénaff, *Compt. Rend.*, **256**, 1752 (1963).
[152] R. P. Bell and P. G. Evans, *Proc. Roy. Soc.*, A, **291**, 297 (1966).
[153] P. E. Sörensen, *Acta Chem. Scand.*, **26**, 3357 (1972).
*The use of the term termolecular does not imply a simultaneous ternary collision
between the three species, but only that all three are present in the transition state.
Since one of the species is in each case water, it is easy to envisage a preliminary step
in which it is hydrogen-bonded to one of the solute species before the arrival of the
other.

can be split up into two consecutive bimolecular processes, one of which is rate-limiting. On paper this can be done in six different ways for each of the reactions (i) and (iii), but a detailed examination shows[138] that not one of these dissections is consistent with the observed fact of general catalysis. We must therefore conclude that the rate-determining proton transfer is in fact concerted with the making and breaking of carbon–oxygen bonds.

It is in fact doubtful whether the separation of the steps (i) and (ii) or (iii) and (iv), implied by (98), corresponds to reality. It was first pointed out by Eigen[154] that, if reasonable estimates are made of the equilibrium constants of reactions (ii) and (iv), the observed rates for some catalysts would demand velocity constants greater than the diffusion-controlled value for the reverse of reactions (i) and (iii); in other words, there would be insufficient time for the species B or A to become free and move to the other oxygen atom of the hydrate. Eigen therefore proposed that the whole process (i)+(ii) or (iii)+(iv) takes place during a single encounter; this is often referred to as a *one-encounter* or *intimate* mechanism. In aqueous solution this is most readily envisaged by incorporating one or more extra water molecules into the transition state; thus with two extra molecules the solvent-catalysed reaction can be pictured as

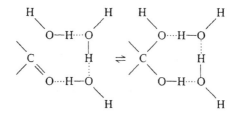

(99)

where the broken lines represent hydrogen bonds. Similar cyclic transition states have been proposed for the exchange of protons between hydrogen peroxide and water[155] and between carboxylic acids and water or methanol.[156]

When water is the solvent it is difficult to obtain evidence about the number of water molecules in the transition state, but interesting information has been obtained from a study of the hydration of 1,3-dichloroacetone by

[154] M. Eigen, *Disc. Faraday Soc.*, **39**, 7 (1965).
[155] M. Anbar, A. Loewenstein, and S. Meiboom, *J. Am. Chem. Soc.*, **80**, 2630 (1958).
[156] E. Grunwald, C. F. Jumper, and S. Meiboom, *J. Am. Chem. Soc.*, **85**, 522 (1963); E. Grunwald and S. Meiboom, *J. Am. Chem. Soc.*, **85**, 2047 (1963); Z. Luz and S. Meiboom, *J. Am. Chem. Soc.*, **85**, 3923 (1963).

water dissolved in organic solvents. This reaction can conveniently be followed spectrophotometrically in dioxan containing a few per cent of water, and catalysis by a large number of acids and bases was studied.[157] It was then shown[158] that in dioxan and acetonitrile solution the kinetic orders with respect to water for the hydration and dehydration reactions are very close to 3 and 2 respectively when no acid or base catalyst is added. This accords well with the reaction scheme (99), and further confirmation is obtained from studies of activation entropies[159] and of hydrogen isotope effects.[160]

The presence of an acid catalyst lowers the order with respect to water, which suggests that one or more water molecules in the transition state of (99) is replaced by a molecule of catalyst.* There has been a good deal of argument as to whether the three proton-transfers in (99) take place synchronously or successively, but there is little firm evidence on this point and we shall not discuss it here.

The addition of water to carbon dioxide, $CO_2 + H_2O \rightleftharpoons OC(OH)_2 \rightleftharpoons H^+ + HCO_3^-$, is formally very similar to its addition to aldehydes and ketones, although here only 0.2% of the carbon dioxide is hydrated at equilibrium, and observations make use of the further equilibrium with $H^+$ and $HCO_3^-$. A summary of work up to 1958 has been given by Edsall and Wyman,[161] and several later kinetic studies have been made:[162-165] the hydration process shows general catalysis by basic anions. It is particularly interesting that the enzyme carbonic anhydrase, which is active in maintaining the carbon dioxide–bicarbonate equilibrium in the body, is also an effective catalyst for the hydration of acetaldehyde and other carbonyl compounds.[166]

Similar considerations apply to the *addition of alcohols to aldehydes* to give semi-acetals, $RCHO + R'OH \rightleftharpoons RCH(OH)OR'$, though simple re-

[157] R. P. Bell and M. B. Jensen, *Proc. Roy. Soc.*, A, **261**, 38 (1961).
[158] R. P. Bell, J. F. Millington, and J. M. Pink, *Proc. Roy. Soc.*, A, **303**, 1 (1968).
[159] R. P. Bell and P. E. Sörensen, *J. Chem. Soc., Perk. Trans. 2*, 1740 (1972).
[160] R. P. Bell and J. E. Critchlow, *Proc. Roy. Soc.*, A, **325**, 35 (1971).
*It was originally supposed[158] that a carboxylic acid molecule could replace two water molecules in (99), but this argument ignored the hydration of the catalyst in the initial state, and it is more likely that only one molecule is replaced.[159,160]
[161] J. T. Edsall and J. Wyman, *Biophysical Chemistry*, Academic Press, New York, 1958, Vol. I, Ch. 10.
[162] A. Sharma and P. V. Danckwerts, *Trans. Faraday Soc.*, **59**, 386 (1963).
[163] C. Ho and J. M. Sturtevant, *J. Biol. Chem.*, **238**, 3499 (1963).
[164] B. H. Gibbons and J. T. Edsall, *J. Biol. Chem.*, **238**, 3502 (1963).
[165] J. C. Kernohan, *Biochim. Biophys. Acta*, **81**, 346 (1964).
[166] J. Pocker and J. E. Meany, *J. Am. Chem. Soc.*, **87**, 1809 (1965); *J. Phys. Chem.*, **71**, 3113 (1967); **72**, 655 (1968); *Biochemistry*, **4**, 2535 (1965); **6**, 239 (1967).

actions of this kind have been little investigated kinetically. The addition of methanol to acetaldehyde has been studied dilatometrically in methanol solution;[167] general catalysis by both acids and bases is observed, and reaction with a second molecule of alcohol to form acetal is negligibly slow. When the addition of methanol to chloral is studied in dioxan solution[168] the order of the uncatalysed reaction with respect to methanol is close to three, falling to lower values for the acid-catalysed reaction. This suggests a one-encounter transition state containing several alcohol molecules, as already discussed for the hydration reactions. The making and breaking of semi-acetal links is also involved in the reversible dimerization of α-hydroxy-aldehydes and ketones according to the scheme

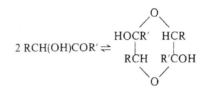

The depolymerization of dimeric dihydroxyacetone[169] and glycollaldehyde[170] has been studied in aqueous solution, and again shows general acid-base catalysis. Qualitatively similar catalytic behaviour has been observed in the mutarotation of optically active α-keto-esters in alcoholic solution,[171] and in exchange reactions between alcohols and esters,[172,173] both of which presumably take place by the reversible addition of alcohol to the carbonyl group.

Much more attention has been paid to the mechanistically similar *mutarotation of glucose and tetramethylglucose*, which have already been mentioned in connection with ternary and bifunctional catalytic mechanisms (p. 155). The mutarotation reactions of sugars have been recently reviewed,[174], and only a brief account will be given here. The interconversion of the cyclic epimers α- and β-glucose takes place through the open-chain aldehydic form, which is shown by polarographic investiga-

[167] G. W. Meadows and B. de B. Darwent, *Trans. Faraday Soc.*, **48**, 1015 (1952).
[168] R. P. Bell and D. G. Horne, *J. Chem. Soc., Perk. Trans.* 2, 1371 (1972).
[169] R. P. Bell and E. C. Baughan, *J. Chem. Soc.*, 1947 (1937).
[170] R. P. Bell and J. P. H. Hirst, *J. Chem. Soc.*, 1777 (1939).
[171] A. McKenzie and A. G. Mitchell, *Biochem. Z.*, **208**, 456 (1929); A. McKenzie and P. D. Ritchie, *Biochem. Z.*, **231**, 412 (1931); **237**, 1 (1931); **250**, 376 (1932).
[172] R. Alquier, *Bull. Soc. Chim. France*, (5)**10**, 197 (1943).
[173] J. R. Schaefgen, F. H. Verhoek, and M. S. Newman, *J. Am. Chem. Soc.*, **67**, 253 (1945).
[174] W. Pigman and H. S. Isbell, *Adv. Carbohydrate Chem.*, **23**, 11 (1968); H. S. Isbell and W. Pigman, *Adv. Carbohydrate Chem.*, **24**, 14 (1969).

tions[175] to be present in very small amount (0.003%). The mutarotation process thus involves the breaking and re-forming of a semi-acetal link,

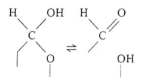

and would therefore be expected to show the type of catalytic behaviour already discussed. It was in fact one of the earliest reactions for which general acid-base catalysis was established; simultaneous publications by Lowry and Smith[176] and by Brönsted and Guggenheim[177] demonstrated catalysis by uncharged acids (e.g., $CH_3CO_2H$), cation acids (e.g., $NH_4^+$), anion bases (e.g., $CH_3CO_2^-$, $SO_4^{2-}$), and cation bases (e.g., $Co(NH_3)_5OH^{2+}$). There is an appreciable water-catalysed reaction, and in the range pH 4–6 catalysis by both hydrogen ions and hydroxide ions is negligible. Hydrogen ion catalysis can be readily studied in dilute solutions of strong acids, but it is more difficult to obtain a reliable value for $k_{OH}$, since $\alpha$- and $\beta$-glucose are acids of appreciable strength ($K = 3.4 \times 10^{-13}$ and $6.7 \times 10^{-13}$ respectively) and in alkaline solutions there is a considerable contribution from base catalysis by glucosate anions. The most reliable study is probably that by Los and Simpson,[178] who used the change in pH in unbuffered alkaline solutions to follow the mutarotation process. Other work on this reaction has concerned catalysis by pyridine bases,[179] for which modest effects of steric hindrance are observed, and by amino-acids[180] and the mono-anions of dicarboxylic acids,[181] neither of which show the abnormal activity which might have been expected if bifunctional catalysis were important. There is also a good deal of information available about activation energies for catalysis by various species.[178,182]

[175] J. M. Los, L. B. Simpson, and K. Wiesner, *J. Am. Chem. Soc.*, **78**, 1564 (1956).
[176] T. M. Lowry and G. F. Smith, *J. Chem. Soc.*, 2539 (1927).
[177] J. N. Brönsted and E. A. Guggenheim, *J. Am. Chem. Soc.*, **49**, 2554 (1927).
[178] J. M. Los and L. B. Simpson, *Rec. Trav. Chim.*, **76**, 267 (1957).
[179] F. Covitz and F. H. Westheimer, *J. Am. Chem. Soc.*, **85**, 1773 (1963).
[180] F. H. Westheimer, *J. Org. Chem.*, **2**, 431 (1938).
[181] G. E. Lienhard and F. H. Anderson, *J. Org. Chem.*, **32**, 2229 (1967).
[182] G. F. Smith and M. C. Smith, *J. Chem. Soc.*, 1413 (1937); J. C. Kendrew and E. A. Moelwyn-Hughes, *Proc. Roy. Soc.*, A, **176**, 352 (1940); P. Johnson and E. A. Moelwyn-Hughes, *Trans. Faraday Soc.*, **37**, 289 (1941); D. G. Hill and B. A. Thumm, *J. Am. Chem. Soc.*, **74**, 1380 (1952); G. Kilde and W. F. K. Wynne-Jones, *Trans. Faraday Soc.*, **49**, 243 (1953).

The evidence quoted for the addition of water and alcohols to carbonyl compounds suggests that the transition state for the mutarotation reaction might also contain one or more solvent molecules, and this idea receives support from kinetic measurements in $H_2O$–$D_2O$ mixtures.[183]

*Electrophilic aromatic substitution*
This class of reaction will be dealt with briefly here since it provides an example of a two-stage process in which only the second stage involves proton transfers; this contrasts with the reactions considered so far, in which proton transfers occur either in the first or in both stages. A general review of this class of reaction has been given by Berliner.[184] In this field particular use has been made of hydrogen isotope effects; these will be mentioned briefly here, and referred to again in Chapter 12.[185] We shall pay special attention to *diazo coupling reactions*, for which the part played by proton transfer can be shown particularly clearly.

It is generally agreed that the substitution of an aromatic species ArH by an electrophilic reagent $X^+$ can be represented as a two-stage process,

(a) $\text{ArH} + X^+ \underset{k_{-1}}{\overset{k_1}{\rightleftharpoons}} X\overset{+}{A}rH$

(b) $X\overset{+}{A}rH + B \xrightarrow{k_2} ArX + A$                      (100)

where the intermediate $X\overset{+}{A}rH$ has a structure such as

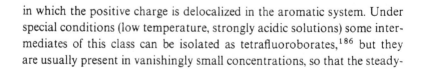

in which the positive charge is delocalized in the aromatic system. Under special conditions (low temperature, strongly acidic solutions) some intermediates of this class can be isolated as tetrafluoroborates,[186] but they are usually present in vanishingly small concentrations, so that the steady-

[183] H. H. Huang, R. R. Robinson, and F. A. Long, *J. Am. Chem. Soc.*, **88**, 1866 (1966).
[184] E. Berliner, *Progr. Phys. Org. Chem.*, **2**, 253 (1964).
[185] For a review of hydrogen isotope effects in aromatic substitution reactions, see H. Zollinger, *Adv. Phys. Org. Chem.*, **2**, 163 (1964).
[186] G. A. Olah and S. J. Kuhn, *J. Am. Chem. Soc.*, **80**, 6535, 6540 (1958).

state approximation can be applied to the above reaction scheme, giving
for the observed velocity constant,

$$k = \frac{d[ArX]/dt}{[ArH][X^+]} = \frac{k_1 k_2 [B]}{k_{-1} + k_2 [B]}. \tag{101}$$

(If more than one base is present, for example, the solvent and a buffer
component, then $k_2[B]$ must be replaced by $\sum k_i[B_i]$.) There are two
limiting cases of (101), as follows

(a) $k_{-1} \ll k_2[B]$, $\quad k = k_1$

(b) $k_{-1} \gg k_2[B]$, $\quad k = (k_1/k_{-1})k_2[B]$. $\qquad$ (102)

In (102a) the formation of the intermediate is rate-determining. No catalysis by bases will be observed, and no considerable hydrogen isotope effect,
since in reaction (100a) there is little change in the binding of the hydrogen.
In (102b) the intermediate is in equilibrium with $ArH + X^+$, and reaction
(100b) is rate-determining; the observed rate should therefore exhibit
general base catalysis and a considerable hydrogen isotope effect. In the
intermediate case represented by (101) general base catalysis should be
observed, but the rate will not be a linear function of $[B]$ and the magnitude of the hydrogen isotope effect will depend upon the nature and
concentration of B.

In nitration and bromination it is often difficult to determine whether
base catalysis is operating, since these reactions may involve pre-
equilibria in which the active forms of the reagents are generated (e.g.,
$NO_2$, $H_2\overset{+}{O}Br$, etc.) and they are frequently carried out in solvents such as
concentrated acids in which it is difficult to make systematic changes of
base concentration. The first distinction between (102a) and (102b) therefore came from the observation by Melander[187] that there was no detectable discrimination between hydrogen and tritium in the bromination and
nitration of several benzene derivatives. This indicates that the formation
of the intermediate is rate-determining, but a similar investigation of
sulphonation[187,188] reveals a small but definite isotope effect, showing
that the loss of the proton must be partly rate-determining.

Systematic investigation is much simpler for another type of electrophilic substitution, namely, diazo coupling, which has been investigated

[187] L. Melander, *Arkiv Kemi*, **2**, 213 (1950).
[188] U. Berglund-Larsson and L. Melander, *Arkiv Kemi*, **6**, 219 (1953); U. Berglund-
Larsson, *Arkiv Kemi*, **10**, 549 (1957).

particularly by Zollinger.[189] These reactions can be studied in dilute aqueous buffer systems of low ionic strength, the electrophilic agent is well defined as the aryldiazonium ion, and any pre-equilibria involving either reagent are well understood. By choosing suitable systems, diazo coupling reactions can be found which illustrate all the kinetic possibilities described above. For example, in the reaction of 1-naphthol-4-sulphonate with *o*-methoxydiazobenzene or *p*-chlorodiazobenzene there is no general base catalysis and no hydrogen isotope effect, corresponding to (102a). On the other hand, the coupling of 1-naphthol-2-sulphonate with 2-diazophenol-4-sulphonate shows general base catalysis by a number of substituted pyridines with rates which are a linear function of pyridine concentration. This reaction also exhibits a large hydrogen isotope effect, which varies with the nature of the catalysing base, and its kinetics thus correspond to (102b). There are also reactions, for example the coupling of 1-naphthol-3-sulphonate with *p*-chlorodiazobenzene, which show general base catalysis, but with non-linear dependence of rate on base concentration, and an isotope effect whose magnitude varies with base concentration. This corresponds to the intermediate case (101), and is particularly interesting because it excludes a ternary mechanism in which the transition state contains both reactants together with the catalysing base, since this would imply a linear dependence of rate upon base concentration. Other examples of the three types of behaviour could be given, and it is possible to rationalize the effect of substituents upon the kinetic behaviour by considering the effect of polar or steric factors upon the constants $k_2$ and $k_{-1}$ in (100).

There are of course many other kinetic studies of electrophilic aromatic substitution, particularly of the halogenation of activated aromatic species such as phenols and anilines. These are frequently complicated by the effect of pH and halide ion concentration upon equilibria between different halogenating species and between different forms of the aromatic substance. The role of bases in removing a proton has been rarely investigated directly, and much of the mechanistic information comes from a study of isotope effects, which will be considered in Chapter 12.

*Hydrogen isotope exchange* in aromatic species constitutes a special case of electrophilic aromatic substitution, in which the electrophile is the

[189] H. Zollinger, *Helv. Chim. Acta*, **38**, 1597, 1617, 1633 (1955); *Diazo and Azo Chemistry*, Interscience, New York, 1961; R. Ernst, O. A. Stamm, and H. Zollinger, *Helv. Chim Acta*, **41**, 2274 (1958); C. Jermini, S. Koller, and H. Zollinger, *Helv. Chim. Acta*, **53**, 72 (1970).

hydrogen ion, or some other acid. Taking benzene as an example, the accepted mechanism is

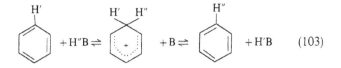

$$+ H''B \rightleftharpoons \quad + B \rightleftharpoons \quad + H'B \qquad (103)$$

where H' and H" are any two isotopes of hydrogen, and HB any acid. Under most conditions only a very small proportion of the protonated form will be present, and a steady-state treatment can be applied, but because of the symmetry of (103) it is not possible to pick out a single rate-determining step. However, because there are considerable kinetic isotope effects for proton-transfers the system is not really symmetrical kinetically, and this has to be taken into account in interpreting the results of exchange experiments.

The reaction scheme (103) implies that isotope exchange should exhibit general acid catalysis. Simple aromatic hydrocarbons require strongly acidic media to produce measurable exchange rates, and it is difficult to distinguish between specific and general acid catalysis because of uncertainties about the acidic species present and the presence of considerable medium effects. However, by choosing suitable substances it is possible to work in dilute aqueous solution and to demonstrate catalysis by a wide range of chemical and charge types. This has been done particularly for alkoxybenzenes by Kresge,[190] and for azulenes by Long.[191]

[190] A. J. Kresge and Y. Chiang, *J. Am. Chem. Soc.*, **81**, 5509 (1959); **83**, 2877 (1961); A. J. Kresge, S. Slae, and D. W. Taylor, *J. Am. Chem. Soc.*, **92**, 6309 (1970).

[191] J. Colapietro and F. A. Long, *Chem. and Ind.*, 1056 (1960); B. C. Challis and F. A. Long, *J. Am. Chem. Soc.*, **85**, 2524 (1963); J. Schulze and F. A. Long, *J. Am. Chem. Soc.*, **86**, 331 (1964); R. J. Thomas and F. A. Long, *J. Am. Chem. Soc.*, **86**, 4770 (1964).

---

References for Table 22, p. 172.

(a) F. O. Rice and M. Kilpatrick, *J. Am. Chem. Soc.*, **45**, 1401 (1923); (b) H. M. Dawson, *Trans. Faraday Soc.*, **24**, 640 (1928); (c) M. Bamford and R. P. Bell, unpublished measurements; (d) H. M. Dawson and A. Key, *J. Chem. Soc.*, 2154 (1928); (e) G. F. Smith, *J. Chem. Soc.*, 1744 (1934); (f) O. Reitz, *Z. Phys. Chem.*, A, **179**, 119 (1937); (g) F. O. Rice and C. F. Fryling, *J. Am. Chem. Soc.*, **47**, 382 (1925); (h) R. P. Bell and K. Yates, *J. Chem. Soc.*, 1927 (1962).

# 10 Rates, Equilibria, and Structures in Proton-Transfers

This chapter will be concerned mainly with the relation between the equilibrium constants of acid-base reactions and their forward and reverse rates. Relations between equilibrium constants and structure have already been considered in Chapter 6, so that the present discussion also implies relations between rates and structure. Moreover, there are many cases in which rates are easier to measure (though more difficult to interpret) than equilibria and can be compared directly with structures. We shall first consider the general basis and experimental evidence for this type of relation, followed by its molecular interpretation, with special reference to exceptional cases. We have seen in the two preceding chapters that the rates of proton-transfer reactions can be measured either directly, or indirectly through the study of acid-base catalysis, and in the following discussion information from both sources will be used indifferently.

One generalization has already emerged (p. 130 ) about the relation between the rate and equilibrium constants for a reaction of the type

$$A_1 + B_2 \underset{k_r}{\overset{k_f}{\rightleftharpoons}} B_1 + A_2, \qquad K = k_f/k_r \tag{104}$$

namely, that for a thermodynamically favourable reaction $(K \gg 1)$ in water at room temperature, $k_f \sim 10^{10}$ dm$^3$ mol$^{-1}$ s$^{-1}$, and hence for the reverse reaction $k_r \sim 10^{10}/K$ dm$^3$ mol$^{-1}$ s$^{-1}$. Systems which behave in this way have been termed *normal acids and bases* by Eigen.[1] However, this generalization is of limited applicability, being mainly valid for the reaction of oxygen or nitrogen acids with one of the solvent species $H_2O$, $H_3O^+$, or $OH^-$ (Tables 17 and 18, p. 127 ). Appreciably lower rates are observed when $A_1$–$B_1$ and $A_2$–$B_2$ are all solute species, even when they are all oxygen or nitrogen acids. This has been illustrated in Table 19 and Figure 8 (p. 130 ); although rates close to the diffusion-controlled limit are

---

[1] M. Eigen, *Angew. Chem. Internat. Edn.*, 1, 3 (1964).

observed for large values of $\Delta pK$,* they are considerably lower over most of the range. Although the theory for diffusion-controlled reactions (Equation 40) predicts some decrease in rate in the immediate neighbourhood of $\Delta pK = 0$, as shown by the broken lines in Figure 8, this is insufficient to account for the observed values.

Even lower rates are encountered for many classes of *carbon acids*, as already illustrated in Table 20 for reactions with solvent species. The same result follows from the very extensive information derived from the base-catalysed reactions of ketones and similar substances, as described in Chapter 9.

Catalytic studies do not normally cover a wide range of base strengths, and a more complete picture is given by the temperature-jump studies of Eigen and his collaborators[2] on the reaction of the keto form of acetylacetone with bases, which cover a range of 16 pK units, and include reaction with 13 oxygen bases, 9 amines, and 8 sulphur bases. These results are plotted in Figure 9, which shows that the rates do not approach the diffusion-controlled limits even for reactions which are thermodynamically very favourable. This behaviour is typical of proton transfers involving ketones, esters, nitro-compounds, and hydrocarbons. There are, however, some carbon acids (e.g., cyano-compounds and sulphones) which behave as 'normal' acids in the sense defined at the beginning of this chapter. The interpretation of these differences will be considered in a later section.

It can be seen from Figures 8 and 9 that a kinetic investigation over a limited range of $\Delta pK$ would yield a linear relation between $\lg k$ and $\Delta pK$, but with a slope less than unity. This type of relation has been known for many years for catalysed reactions in the form of the *Brönsted relation*,* first proposed by Brönsted and Pedersen[3] in 1924 on the basis of their experimental work on the decomposition of nitramide. It relates the effectiveness of a catalyst to its acid-base strength, having the form

$$k_A = G_A K^\alpha, \qquad k_B = G_B(1/K)^\beta \qquad (105)$$

*For a reaction $A_1 + B_2 \rightleftharpoons B_1 + A_2$, $\Delta pK = pK_2 - pK_1 = \log K$, where $K = [B_1][A_2]/[A_1][B_2]$.
[2] M. L. Ahrens, M. Eigen, W. Kruse, and G. Maass, *Ber. Bunsengesell. Phys. Chem.*, **74**, 380 (1970). A similar picture is presented by results for the reaction of the anion of 2,6-dinitrotoluene with a wide range of acids [M. E. Langmuir, L. Dogliotti, E. D. Black, and G. Wettermark, *J. Am. Chem. Soc.*, **91**, 2204 (1969)].
*It is interesting to note that Brönsted and Pedersen[3] realized that their linear relation could not hold over an unlimited range of acid or base strengths, but must have a slope which decreases with increasing reaction velocity. Their argument is essentially the same as that advanced by Eigen[1] some forty years later.
[3] J. N. Brönsted and K. J. Pedersen, *Z. Phys. Chem.*, **108**, 185 (1924).

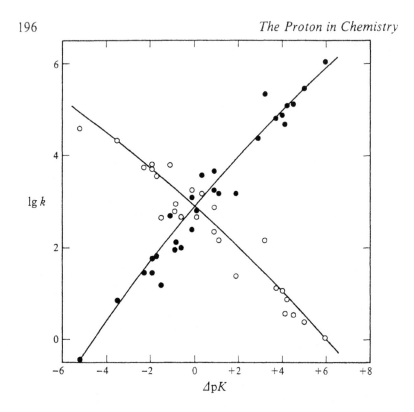

Fig. 9. Velocity constants for the reactions of acetylacetone and its anion with oxygen, nitrogen, and sulphur bases and acids. (Values from M. L. Ahrens, M. Eigen, W. Kruse, and G. Maas.[2])

where $k_A$ and $k_B$ are the catalytic constants for acid and base catalysis and $K$ the conventional strength of the acid A, or of the acid corresponding to the base B. $G_A$ and $\alpha$ are constants for a series of similar catalysts but depend on the nature of the reaction and also on the solvent and the temperature. Analogous statements hold for $G_B$ and $\beta$. The exponents $\alpha$ and $\beta$ are usually positive and less than unity.

We have seen in the last chapter that many catalysed reactions are determined kinetically by a single acid-base reaction between the catalyst and the substrate, and when this is so $k_A$ or $k_B$ in Equation (105) is just the second-order velocity constant for this reaction. The interpretation is not quite so simple when the reaction involves two consecutive proton transfers, of which the first is effectively at equilibrium. Thus for acid catalysis we may have

$$HS + \sum A_i \rightleftharpoons HSH^+ + \sum B_i \text{ (equilibrium)}$$

$$HSH^+ + \sum B_i \xrightarrow{\overrightarrow{k_2}} SH + \sum A_i$$

where $k_2 = \sum \pi_i[B_i]$. The observed reaction velocity is then given by

$$v = [HSH^+]\sum \pi_i[B_i] = [HS]\sum \pi_i[A_i]K'_i \qquad (106)$$

where $K'_i$ is the equilibrium constant $[B_i][HSH^+]/[A_i][HS]$, related to the conventional strength $K_i$ of the acid $A_i$ by $K'_i = K_i/K_{HSH}$. The observed catalytic constant thus becomes

$$k^i_A = \pi_i K_i/K_{HSH}. \qquad (107)$$

Now the second stage of the reaction is equivalent to a one-stage basic catalysis of the substrate $HSH^+$, for which we can write the usual type of Brönsted relation

$$\pi_i = G(1/K_i)^\beta \qquad (108)$$

giving on substitution in (107)

$$k^i_A = \frac{G}{K_{HSH}} K_i^{1-\beta}. \qquad (109)$$

Since $\beta < 1$, this is equivalent to (105), and it is thus understandable that the same type of relation applies to prototropic reactions whether or not there is a pre-equilibrium with the catalyst.

The simple expressions in (105) are often modified by including a *statistical correction*, which is best explained by means of an example. Suppose that we have as catalyst a carboxylic acid $CH_3(CH_2)_nCO_2H$ whose catalytic effect is given correctly by Equation (105), and that we wish to compare it with a dicarboxylic acid $CO_2H(CH_2)_nCO_2H$, where $n$ is so great that the effect of the carboxyl groups upon one another is negligible. The tendency of the carboxyl groups to lose a proton will be essentially the same in the two acids, but the first dissociation constant of the dibasic acid $(K')$ will be twice as great as that of the monobasic acid $(K)$, since the ion $CO_2H(CH_2)_nCO_2^-$ can be formed by losing a proton from either end of the chain (see Chapter 6, p. 97). Similarly, the catalytic constant of the dibasic acid $(k'_A)$ will be twice that of the monobasic acid $(k_A)$, since in the former case the substrate can approach either end of the catalyst molecule. This is not, however, what is predicted by Equation (105) as it stands, which gives

$$k'_A/k_A = (K'/K)^\alpha = 2^\alpha.$$

The correct result is obtained if we reckon both the dissociation constant and the catalytic power *per carboxyl group*, giving

$$\tfrac{1}{2}k'_A = G_A(\tfrac{1}{2}K')^x = G_A K^x = k_A.$$

A similar problem arises if we compare the two acids $CO_2H(CH_2)_nCO_2^-$ (I) and $CO_2H(CH_2)_nCO_2CH_3$ (II), where $n$ is again large. In this case the tendency to lose a proton from the carboxyl group (and hence the catalytic power) will be the same for the two acids. On the other hand, the dissociation constant of (I) will be only half that of (II), since the conjugate base $CO_2^-(CH_2)_nCO_2^-$ has two equivalent groups to which a proton can be added, whereas $CO_2^-(CH_2)_nCO_2CH_3$ has only one such point. Here again the straightforward use of Equation (105) leads to an incorrect prediction, and in order to obtain the correct result it is necessary to multiply the observed dissociation constant of (I) by 2 before inserting it in the equation.

These arguments can easily be generalized. Thus if we have an acid-base pair A–B in which A has $p$ dissociable protons bound equally firmly and B has $q$ equivalent points at which a proton can be attached, then the catalytic power of A is related to its observed dissociation constant by the equation

$$k_A/p = G_A(qK/p)^x. \tag{110}$$

A similar treatment for base catalysis by B gives

$$k_B/q = G_B(p/qK)^\beta. \tag{111}$$

Analogous but more complicated equations can be developed[4] for the case in which the various protons or points of attachment are not all equivalent.

The statistical correction was first put forward by Brönsted and Pedersen,[3] though in an incomplete form, and later stated correctly by Brönsted.[5] Some authors have preferred to introduce the factor $p$ only when the protons are attached to $p$ different atoms (e.g., $p = 1$ for $NH_4^+$), but $p = 2$ for hydrogen peroxide HOOH or the hydrazinium ion $NH_3^+NH_3^+$), but there is no justification for this procedure. Recent derivations[6] use more sophisticated arguments involving the symmetry numbers

[4] F. H. Westheimer, *J. Org. Chem.*, **2**, 431 (1938).
[5] J. N. Brönsted, *Chem. Rev.*, **5**, 322 (1928).
[6] S. W. Benson, *J. Am. Chem. Soc.*, **80**, 5151 (1958); D. Rapp and R. E. Weston, *J. Chem. Phys.*, **36**, 2807 (1962); E. W. Schlag, *J. Chem. Phys.*, **38**, 2480 (1964); V. Gold, *Trans. Faraday Soc.*, **60**, 739 (1964); E. W. Schlag and G. L. Haller, *J. Chem. Phys.*, **42**, 584 (1965); D. M. Bishop and K. J. Laidler, *J. Chem. Phys.*, **42**, 1688 (1965); J. N. Murrell and K. J. Laidler, *Trans. Faraday Soc.*, **64**, 371 (1968).

of the reactants or the transition state. Care is often needed in assigning these numbers correctly, but the final conclusion is that the procedure proposed intuitively by Brönsted is the correct one. It is sometimes permissible to ignore statistical corrections altogether. This is obviously so when comparing a series of substances having the same values of $p$ and $q$ (e.g., a series of monocarboxylic acids). Moreover, since the effect on the predicted rate rarely exceeds a factor of two, it can be safely neglected when considering much larger effects, amounting perhaps to several powers of ten.*

The Brönsted relation is now in such general use that we shall not give any examples to demonstrate its validity. Provided that a closely similar series of acids or bases is studied, it has been found to hold with fair accuracy for all examples of general acid-base catalysis investigated, provided, of course, that no changes of mechanism or specific catalytic effects are involved. (We shall return later to the question of what constitutes 'closely similar' in this context.) When a very large range of catalytic power can be investigated, curvature is sometimes detectable in the plot of $\lg k$ against $pK$, always in the direction of a smaller $\alpha$ or $\beta$ at higher velocities. However, in order to obtain a sufficiently large range of velocities it is often necessary to use catalysts of different chemical or charge type, and there are few conventional studies of acid-base catalysis in which such curvature has been firmly established. One of the clearest cases is the study by Kreevoy and Konasewich[7] of the acid-catalysed decomposition of the diazoacetate ion. We have already seen (p. 170) that this reaction involves two consecutive steps of comparable rate, but the results can be analysed to yield velocity constants $k$ for the step $N_2CHCO_2^- + A \rightarrow N_2^+CH_2CO_2^- + B$ for a wide range of acids, and a plot of $\lg k$ against $\lg K_A$ gives a slope $\alpha = 0.74$ for four phenols, falling to $\alpha = 0.51$ for seven carboxylic acids.

In this last example an even greater curvature is apparent if we include catalysis by the species $H_2O$ and $H_3O^+$. The same is true for a number of other reactions, and also for base catalysis by $H_2O$ and $OH^-$, even when the rates of both the forward and reverse reactions are far below the diffusion-controlled limit. It is doubtful, however, how far general conclusions can be drawn from curvature based only on reaction with solvent

---

*An ambiguity arises for the pair $H_3O^+$—$H_2O$, for which it is usual to write $p = 3$, $q = 1$. As pointed out by Gold and Waterman [*J. Chem. Soc.*, B, 839 (1968)], $q = 2$ is more correct, since the pyramidal shape of $H_3O^+$ implies that $H_2O$ has two spatially different but equivalent basic sites.

[7] M. M. Kreevoy and D. E. Konasewich, *Adv. Chem. Phys.*, 21, 243 (1971).

species. In the first place the species $H_2O$, $H_3O^+$, and $OH^-$ are of different chemical type (and often also different charge type) from the other catalysts studied, and in catalysis by water it is necessary to divide the observed rate by $55.5 \text{ mol dm}^{-3}$ in order to obtain a second-order velocity constant comparable with catalytic constants for solute species. In the second place, the values $pK(H_2O) = 15.74$ and $pK(H_3O^+) = -1.74$, which are conventionally used, both involve putting $[H_2O] = 55.5$ and may not give a true representation of the acid-base strength of these species. However, it is of some practical importance that hydroxide ion (and to a lesser extent hydronium ion) frequently have a catalytic effect several powers of ten less than that predicted by the Brönsted relation for other catalysts, since it is easily shown that if they conformed to this relation it would be impossible to detect general acid-base catalysis in reactions for which $\alpha$ or $\beta$ is close to unity.

Since acid-base catalysis involves proton-transfer between the catalyst and the substrate, Brönsted-type relations should also apply to the variation of reaction velocity with the acid-base strength of the substrate, and the same statistical corrections should apply, as may be seen from the following general considerations.

Consider the general acid-base reaction

$$A_1 + B_2 \underset{\pi_{2,1}}{\overset{\pi_{1,2}}{\rightleftharpoons}} B_1 + A_2 \tag{112}$$

where $K_1$ and $K_2$ are the conventional strengths of $A_1$–$B_2$ and $A_2$–$B_2$, and the equilibrium constant of (112) is $K = K_1/K_2$. Suppose first that $A_2$–$B_2$ is kept constant and that we make a small variation in $A_1$–$B_1$, for example, by chemical substitution. If this variation increases the acid strength of $A_1$ (and hence decreases the basic strength of $B_1$), it is reasonable to suppose that $\pi_{1,2}$ will increase and $\pi_{2,1}$ will decrease. However, since $A_2$–$B_2$ remains the same, we must have

$$\{\partial \log(\pi_{1,2}/\pi_{2,1})\}_2 = (\partial \lg K)_2 = \alpha \, \mathrm{d} \lg K_1$$

which is satisfied by writing

$$(\partial \log \pi_{1,2})_2 = \alpha_1 \, \mathrm{d} \lg K_1,$$
$$(\partial \log \pi_{2,1})_2 = -(1-\alpha_1) \, \mathrm{d} \lg K_1, \tag{113}$$

where $\alpha_1$ is positive and less than unity. Exactly analogous arguments apply to the case in which $A_1$–$B_1$ is held constant and $A_2$–$B_2$ varies, and the general expressions covering variations of both pairs are

$$d \log \pi_{1,2} = \alpha_1 \, d \lg K_1 - (1 - \alpha_2) \, d \lg K_2,$$
$$d \log \pi_{2,1} = \alpha_2 \, d \lg K_2 - (1 - \alpha_1) \, d \lg K_1, \tag{114}$$

with both $\alpha_1$ and $\alpha_2$ positive and less than unity, but not necessarily equal. If $\alpha_1$ and $\alpha_2$ can be assumed constant over a certain range of velocities and equilibrium constants, then (114) can be integrated, giving

$$\pi_{1,2} = GK_1^{\alpha_1}(1/K_2)^{1-\alpha_2}, \qquad \pi_{2,1} = GK_2^{\alpha_2}(1/K_1)^{1-\alpha_1}, \tag{115}$$

equivalent to the usual form of relation (105). In this formulation the requirement that the variations should be among series of similar molecules is concealed in the assumption, implicit in (114), that the velocity is a unique function of the acid strength.

The dependence of the rate upon the acid-base strength of the substrate has proved less easy to establish, partly because substrates in catalytic reactions are frequently such weak acids or bases that their strength cannot be measured directly. Another factor is that the investigation of a series of substrates often involves substitution very close to the reaction site (as, for example, in compounds containing the group —CHXCO—), while series of catalysts usually involve a more remote substitution (as, for example, in a series of carboxylic acids or phenols); we shall see later that the latter type of substitution favours closer correlation between rates and equilibria. However, such correlations certainly exist for a series of similar substrates, as was first shown clearly by Pearson and Dillon[8] in a compilation of the rates of ionization of carbon acids in water. Provided that only chemically similar substances are considered (for example, ketones and keto-esters) there is close parallelism between p$K$ and rates of ionization, though the individual deviations are larger than those commonly found when the catalyst is varied. However, this parallelism breaks down badly if we consider carbon acids of different classes; for example, nitromethane dissociates in water about $10^5$ times more slowly than benzoylacetone, although their dissociation constants are very similar.

A striking example of a correlation between rate and the acid-base properties of the substrate is provided by recent work of Kresge[9] on tritium exchange between hydronium ion and seven hydroxy- and alkoxy-benzenes. The basic strengths of these aromatic species can be determined by spectroscopic measurements in aqueous perchloric acid, and by com-

[8] R. G. Pearson and R. L. Dillon, *J. Am. Chem. Soc.*, **75**, 2439 (1952).
[9] A. J. Kresge, H. J. Chen, L. E. Hakka, and J. E. Kouba, *J. Am. Chem. Soc.*, **93**, 6174 (1971); A. J. Kresge, S. G. Mylonakis, Y. Sato, and V. P. Vitullo, *J. Am. Chem. Soc.*, **93**, 6181 (1971).

bining the results with those of Long (Ref. 191 in Chapter 9) for four azulenes, a range of 25 p$K$ units can be covered. The plot of $K$ against p$K$ is a smooth curve with a slope which decreases from 0.9 to 0.5 as the reaction increases.

We have so far spoken separately of directly observed proton-transfer rates and of the effect of the acid-base strength of catalysts and substrates in catalysed reactions. The distinction between these topics is of course an artificial one, depending upon historical and experimental factors, and we shall now collect together all available data relating to aqueous solutions for reactions of the type $KH + B$, where $KH$ is a ketone, an ester, or a keto-ester, and B any base. These are taken from references 90–107 of Chapter 9, together with the results of Eigen and his collaborators for acetylacetone,[2] and a collection of some further data for aliphatic ketones.[10] Substances containing the group —$COCF_3$ have been omitted, since they exist in aqueous solution almost entirely in the form —$C(OH)_2CF_3$, and their observed acidities and rates of ionization cannot be simply interpreted.* Nitro-esters and nitro-ketones have also been omitted, since their behaviour as acids involves both nitro and keto groups, and we have also omitted reactions in which the base is a 2,4-disubstituted pyridine, since these are known to be subject to considerable steric hindrance. Finally, reactions with hydroxide ion have not been included since, as already noted, these often show anomalies.

The remaining data are shown in Figure 10, which gives a plot of $\lg(k/p_1 q_2)$ against $\lg(q_1 p_2 K/q_2 p_1)$, where $p_1$, $q_1$, and $p_2$, $q_2$ are statistical factors relating to the ketone and the base respectively, and $K$ is the equilibrium constant for the reaction. $K$ is equal to the ratio of the acidity constants of the ketone and the acid conjugate to B; the latter is always known directly, but the former must often be estimated either by analogy with other substances, or from the kinetic measurements themselves. Points corresponding to the latter procedure are designated separately in Figure 10; they confirm the internal consistency of the relation represented by the curve, but do not constitute independent evidence for it. Nevertheless, Figure 10 shows an excellent general correlation both for variation of ketone and for variation of base. The fact that both correlations can be

---

[10] R. P. Bell, G. R. Hillier, J. W. Mansfield, and D. G. Street, *J. Chem. Soc.*, B, 827 (1967).
* Strictly speaking, a correction for hydration ought to be applied to many of the other rates and equilibrium constants. However, for most of the compounds concerned the degree of hydration is unknown, and in those instances where it is known the correction is not large.

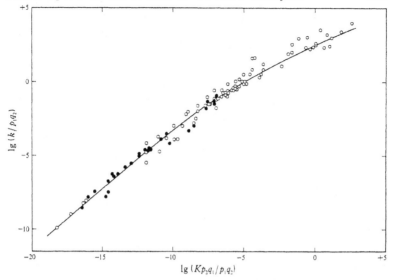

Fig. 10. Relation between rates and equilibrium constants for the reaction of ketones, esters, and keto-esters with bases. Open circles refer to systems in which $K$ is known by direct measurement, filled circles to estimated values of $K$. (Values from Refs. 90–107 of Chapter 9, and Ref. 10 of Chapter 10.) Reactions involving solvent species have been omitted, and also those in which steric effects or other specific interactions are suspected.

represented approximately by a single line shows that $\alpha_1$ and $\alpha_2$ in Equation (115) have similar magnitudes; this is not a thermodynamic necessity, though we shall see that it is understandable in molecular terms. The line in Figure 10 is noticeably curved, its slope changing by a factor of two over the whole range, but it is clear that departures from linearity would be barely detectable if only a limited range of rates (say five powers of ten) were studied. The curvature does however imply a fact first pointed out by Bell and Lidwell,[11] namely, that in base-catalysed reactions of ketonic substances the exponent of the Brönsted equation decreases markedly as the reactivity of the substrate increases.

The Brönsted relation was the first example of a *linear free energy relation* between rates and equilibrium constants, and would now be regarded as a particular example of this class. Such relations are now widely used for correlating the effects of substituents or change of solvent

[11] R. P. Bell and O. M. Lidwell, *Proc. Roy. Soc.*, A, **176**, 88 (1940).

upon the rates and equilibrium constants of organic reactions.[12] Nevertheless, the Brönsted relation has some particularly simple features, since it relates the rate and equilibrium constants *of the same reaction*.* Thus, although it applies only to the limited class of acid-base reactions, it usually covers a wider range of structural variation than most linear free energy relations; for example, both aromatic and aliphatic species are commonly covered by the same relation. It is therefore worthwhile to treat it as a special case and to seek a particular molecular interpretation applicable to proton-transfer reactions.

An interpretation in terms of *molecular potential—energy curves* was advanced almost simultaneously by Horiuti and Polanyi[13] and by Bell,[14] the two treatments corresponding respectively to a covalent and an ionic formulation. The former is probably closer to reality, and will be followed here.

A proton transfer is a special case of the general class of reaction $XZ + Y \rightarrow X + ZY$. The course of such a reaction is commonly depicted by an energy diagram such as the full line in Figure 11, in which the distance between the lowest and the highest points represents the activation energy of the reaction. However, this picture needs a number of qualifications and amplifications. Three co-ordinates are necessary to specify the distances between X, Y, and Z, so that the complete energy diagram is a surface in four dimensions. If we make the assumption (reasonable, but not necessarily correct) that we need consider only configurations in which X, Y, and Z are in a straight line, then the number of co-ordinates can be reduced by one and the energies now lie on a three-dimensional surface. The most probable path of the system during reaction can be represented by a line on the surface which follows the lowest possible energy contours between the initial and the final state, so that a two-dimensional diagram like Figure 11 represents a section (though not a plane section) of the energy surface.

[12] See particularly J. E. Leffler and E. Grunwald, *Rates and Equilibria of Organic Reactions*, Wiley, New York, 1963; S. Ehrenson, *Progr. Phys. Org. Chem.*, **2**, 195 (1964); C. D. Ritchie and W. F. Sager, *Progr. Phys. Org. Chem.*, **2**, 323 (1964); L. P. Hammett, *Physical Organic Chemistry*, 2nd edn., McGraw-Hill, New York, 1970, Ch. 11 and 12; P. R. Wells, *Linear Free Energy Relationships*, Academic Press, London, 1968.
*In many catalysed reactions the absolute value of the equilibrium constant is not accessible, but it is still true that changes in this equilibrium constant are directly proportional to changes in the acid-base strength of the catalyst.
[13] J. Horiuti and M. Polanyi, *Acta Physicochim. U.R.S.S.*, **2**, 505 (1935).
[14] R. P. Bell, *Proc. Roy. Soc.*, A, **154**, 414 (1936).

In general all three distances XZ, ZY, and XY will change during the important stages of the reaction, so that the abscissa in Figure 11 (the so-called 'reaction co-ordinate') is not related simply to the geometry of the system. However, in the particular case of proton transfers $(Z = H^+)$ the position is simpler. Since the proton has no attendant sheath of electrons, the repulsions between X and H and between Y and H can be neglected.

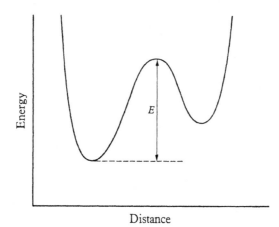

Fig. 11. Energy diagram for a proton-transfer reaction.

The only important repulsion term is that between X and Y, and since these are always considerably heavier than H it is a good approximation to picture the reaction process as the movement of the proton between two centres X and Y which remain stationary at a fixed distance apart. The original four-dimensional problem is thus reduced to a two-dimensional one, and the reaction co-ordinate has a simple significance, namely, the distance of the proton from either X or Y.

The energy diagram can now be amplified and made more specific, as shown in Figure 12.* The two heavy curves represent the energies of the two separate systems $XH + Y$ and $X + YH$ as a function of the distances X–H and Y–H. The relative positions of the two curves are chosen so that the vertical distance between the two minima is equal to the energy change in the reaction. The energy of the transition state is given approximately by the point of intersection of the two curves. Its true position is a little lower, as indicated by the broken line, because of resonance between the

* Zero-point energies have been omitted in this diagram and in Figures 12–14, as they do not affect the arguments of this chapter.

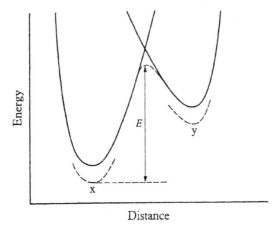

Fig. 12. More detailed energy diagram for a proton-transfer reaction.

two valency states, but there are theoretical reasons for believing that this resonance lowering is small in reactions involving a transfer of charge.[15] The point of intersection is lowered if the two curves are moved closer together horizontally, but this does not necessarily imply a reduction in the energy of activation, since energy must be expended against forces of repulsion in bringing X and Y together. The minimum activation energy will be achieved by some compromise between these two factors, and the minima of the full curves in Figure 12 represent energies somewhat higher than those corresponding to infinite separation in the systems $XH + Y$ and $X + HY$. The original energy levels are indicated in the figure by the broken curves x and y. If we take into account all these factors, the actual activation energy for the reaction from left to right is that given by the vertical line $E$ in the figure.

This kind of picture can be used to obtain a reasonable basis for the Brönsted relation, as shown in Figure 13, in which the resonance energy lowering has been omitted for the sake of clarity; the energy needed to overcome repulsion has also been neglected, since this will be the same for any series of similar reactions. As a concrete example, curve I may be taken to represent $SH + B$, where SH is a substrate and B a basic catalyst, while curve II represents the reaction products $S^- + BH^+$. The activation energy is then $E^0$, and the energy change in the reaction $\varepsilon^0$. Now suppose that the basic catalyst is modified slightly by the introduction of a sub-

[15] R. A. Ogg and M. Polanyi, *Trans. Faraday Soc.*, **31**, 604, 1375 (1935); M. G. Evans and M. Polanyi, *Trans. Faraday Soc.*, **34**, 11 (1938).

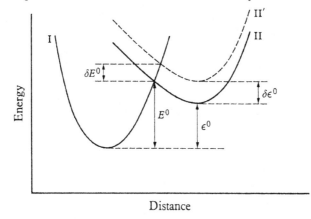

Distance

Fig. 13. Molecular basis for the Brönsted relation.

stituent, producing a slightly weaker base. The simplest way in which this can be represented in the diagram is to replace curve II by II′, which has the same shape and position along the distance axis but is displaced vertically in the direction of higher energy. The consequent changes in $E^0$ and $\varepsilon^0$ are shown in the diagram as $\delta E^0$ and $\delta\varepsilon^0$, and it is clear from the geometry of the figure that

$$\delta E^0 = \frac{s_1}{s_1 + s_2}\delta\varepsilon^0 = \beta\delta\varepsilon^0 \qquad (116)$$

where $s_1$ and $s_2$ are the slopes of the two curves at the point of intersection, both reckoned as positive.

Equation (116) is clearly closely related to (113) and (114). In fact, for the present example the latter could be written as

$$(\delta G^{\ddagger})_{SH} = \beta\delta G_{BH} \qquad (117)$$

where $G^{\ddagger}$ is the free energy of activation and $G_{BH}$ the standard free energy change for the acid-base pair $BH^+$–$B$ relative to any standard acid-base system. The problem of relating (116) and (117) is the same as was considered in Chapter 5 in connection with molecular models and thermodynamic quantities. Energy diagrams such as Figures 12 and 13 take no account of thermal motions and are strictly appropriate only at absolute zero. At a finite temperature, $\delta G$ and $\delta H$ will both differ from $\delta\varepsilon^0$. Similarly, $\delta G^{\ddagger}$ and $\delta H^{\ddagger}$ will both differ from $\delta E^0$. Thus although at first sight Equation (116) might seem to imply a relation between observed activation energies and enthalpy changes, it could equally well be regarded as relating observed velocity constants $(\exp - G^{\ddagger}/RT)$ to observed equilibrium constants $(\exp - G/RT)$. We saw in Chapter 5 that there was some justification for regarding free energies rather than enthalpies as being closely related to molecular models, and just the same arguments suggest that the relation between reaction velocities and equilibrium constants may be simpler than that between observed activation energies and enthalpy changes.[16] There is a certain amount of experimental support for this view,[17] but there are few reactions in which accurate activation energies are available for a range of catalysts, so that the comparison between rates and equilibrium constants at a given temperature is usually the only one possible in practice.

Figure 13 implies that a change in the strength of the catalyst or substrate can be represented by a vertical displacement of the energy curves, without any change of shape. It is likely that this assumption is too restrictive, even for a series of substances with very similar structures, since any change in the energy of a bond is usually accompanied by a change in the shape of the whole energy curve, including the curvature at the minimum (i.e., the fundamental frequency). However, as long as the shape is a unique and continuous function of the strength, it is easily shown that relations of the form of (114) will still hold, and that they can be integrated over a range of acid strengths to give (115). While (116) and (117) imply the same value for $\alpha_1$ and $\alpha_2$ in (114) and (115), both being related to the slopes of the energy curves at their point of intersection, this equality is no longer predicted if the curves undergo changes in shape as well as in position. In fact the curvature shown in Figures 9 and 10, and found in other cases,[2,7,9] is good evidence that changes in both shape and position must be assumed, since the slopes of appropriate energy curves do not vary sufficiently over the ranges of energy involved to account for the observed curvatures.

The shapes of the energy curves can also be made responsible for the large differences in rate which have already been noted for acid-base systems of widely differing structures, in particular the low rates observed for systems involving a large structural change on ionization (cf. Table 20 and Figure 10). For example, curves (a) and (b) in Figure 14 might represent respectively the loss of a proton by a nitroparaffin and a phenol of equal

---

[16] M. G. Evans and M. Polanyi, *Trans. Faraday Soc.*, **32**, 1333 (1936).
[17] K. J. Pedersen, *J. Phys. Chem.*, **38**, 501 (1934); G. F. Smith, *J. Chem. Soc.*, 1744 (1934); G. F. Smith and M. Smith, *J. Chem. Soc.*, 1413 (1937); E. C. Baughan and R. P. Bell, *Proc. Roy. Soc.*, A, **158**, 464 (1937).

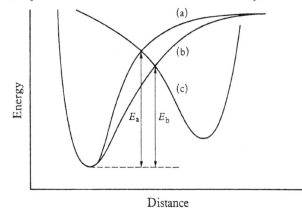

Fig. 14. Energy curves for proton transfer with and without structural change.

strengths. If curve (c) corresponds to the system BH, where B is the base to which the proton is being lost, then it is clear from the diagram that the activation energy for the nitroparaffin, $E_a$, will be greater than that for the phenol, $E_b$, corresponding to the observed difference in rates. The same is of course true for the reverse process, corresponding to the addition of a proton to the anions of the nitroparaffin and the phenol.

In this and similar cases it is easy to see why the energy curve for the pseudo-acid is initially the steeper of the two, although the two curves end up at the same level, corresponding to the equal strengths of the two acids. When a proton is lost by a nitroparaffin, the initial stage of the reaction is equivalent to the removal of a proton from the group

$$\overset{\diagdown}{\underset{\diagup}{-}}CH,$$

which is a very difficult process; hence the initial part of the energy curve is steep. As the distance between the proton and the anion increases, the negative charge shifts from the carbon to the oxygen of the nitro group with consequent stabilization of the anion, and the curve flattens off, as in Figure 14, curve (a).

The same kind of argument applies to any acid-base pair whose inter-conversion involves a considerable charge shift, and it is of interest to construct a simple electrostatic model of this situation. In this the anion consists of two conducting spheres of radii $a$ and $b$ connected by a conducting wire, as in Figure 15. It is convenient to assume that $a$ and $b$ are so far apart that the interaction between them can be neglected, though this assumption is not a necessary feature of the model. If the proton is close to sphere $a$, most of the negative charge will be resident on this sphere, but as

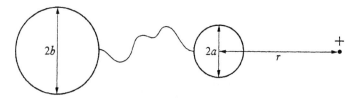

Fig. 15. Electrostatic model for proton transfer with charge shift.

it moves away the charge will pass gradually into sphere $b$, and when the proton is at an infinite distance the charge will be shared between the two in proportion to their capacities (i.e., their radii). If $r$ is the distance of the proton from the centre of sphere $a$ and $\phi$ the energy of the system corresponding to this distance, then simple electrostatics gives

$$\frac{\phi}{D} = \frac{2\rho + K}{\rho^2 (2 + K)} \tag{118}$$

where $\rho = r/a$, $K = b/a$, and $D$ is the energy needed to remove the proton from $r = a$ to $r = \infty$. Figure 16 shows how the energy varies with the

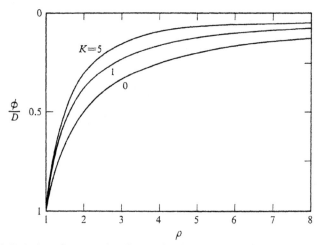

Fig. 16. Variation of energy with distance for the model in Figure 15.

distance for various values of $K$. $K = 0$ corresponds to a simple acid in which the charge remains on one atom, while high values of $K$ represent acids in which the negative charge moves to a more electronegative atom (e.g., from carbon to oxygen) as dissociation progresses. Figure 16 shows that high values of $K$ correspond to steep energy curves, and hence to slow proton-transfer reactions.

The behaviour just discussed applies to the reactions of several classes of carbon acids, notably the removal of a proton from nitro- and keto-compounds, and the addition of a proton to aromatic systems. In all of these systems there is good evidence that the reaction is accompanied by a drastic electronic rearrangement, and it is of interest to enquire why some other classes of carbon acids, notably the cyano-compounds and the disulphones, behave quite differently. Some of these compounds are strong enough acids for their dissociation constants to be measurable in aqueous

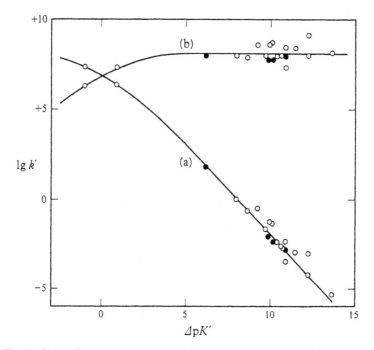

Fig. 17. Rates of proton transfers involving cyano-compounds, disulphones, and their anions. (Values from Refs. 18 and 19, statistically corrected.) Reactions involving solvent species or steric hindrance have been omitted. Curve (a), reaction of carbon acids with bases; curve (b), reaction of carbanions with acids; open circles, cyano-compounds; filled circles, $(EtSO_2)_2CHMe$.

solution, and since their rates of reaction with bases can be measured by halogenation, isotope exchange, or proton magnetic resonance line broadening, the velocity constants for both the forward and the reverse reactions can be determined. Figure 17 shows the results of Bell and Cox[18] for $(EtSO_2)_2CHMe$ and of Long and co-workers[19] for a number of cyano-compounds. The curves drawn correspond to Brönsted exponents of zero or unity over most of the range, and are closely similar to those shown in Figure 8 for the theoretical expression (40) of the theory for diffusion control, the only differences being that the limiting values of the velocity constants are $10^7-10^8$ dm$^3$ mol$^{-1}$ s$^{-1}$ rather than $10^9-10^{10}$ as predicted by diffusion theory; however, they show no trend with $\Delta pK$ over a considerable range.

The behaviour of the cyano-carbon acids and the disulphones suggests that (in contrast to other carbon acids) their ionization is not accompanied by any considerable shift of charge away from carbon, i.e., that their anions are true carbanions. This would make their kinetic behaviour resemble that of simple oxygen and nitrogen acids, corresponding to the lower curves in Figures 14 and 16. The acidifying power of the CN and SO$_2$ groups is frequently explained by a mesomeric effect, in which the charge of the anion is located largely on the nitrogen and oxygen atoms respectively, but it could alternatively be attributed to electrostatic stabilization of the anions in the structures

This kind of effect must be responsible for the high acidities of $CH_2CNCO_2H$ (pK 2.47) and $CH_3SO_2CH_2CO_2H$ (pK 2.36), in which the polar group is three atoms removed from the site of acidity, and there is no possibility of mesomerism. On the other hand, $CH_3COCH_2CO_2H$ (pK 3.58) is a relatively weak acid, and there is no doubt that the acidity of ketonic substances must be attributed to a shift of negative charge on to the oxygen atom; as we have already seen (p. 105), there is direct evidence for the charge distribution in the anions of these substances.

[18] R. P. Bell and B. G. Cox, *J. Chem. Soc.*, B, 654 (1971); B. G. Cox, F. G. Riddell, and D. A. R. Williams, *J. Chem. Soc.*, B, 859 (1970).
[19] E. A. Walters and F. A. Long, *J. Am. Chem. Soc.*, **91**, 3733 (1969); F. Hibbert, F. A. Long, and E. A. Walters, *J. Am. Chem. Soc.*, **93**, 2829 (1971); F. Hibbert and F. A. Long, *J. Am. Chem. Soc.*, **93**, 2836 (1971); **94**, 2647 (1972).

The nature of sulphur–oxygen links has been discussed in Chapter 6 in connection with the strengths of oxyacids and the effect of the group $—SO_3^-$ on the strengths of carboxylic acids and amines. The evidence was inconclusive, though it favoured a formulation in which little change of bonding takes place on the addition or removal of a proton. There seems to have been little speculation or evidence about the corresponding problem for the cyano group. The true state of affairs probably lies between the extreme electrostatic and mesomeric pictures, but there is no doubt that proton transfers for carbon acids containing these substituents differ drastically from those for most carbon acids.

If proton transfers involving nitrogen or oxygen acids are accompanied by considerable structural changes in other parts of the system we should again expect to observe low rates and fractional Brönsted exponents, since a considerable part of the activation energy will be associated with these changes. One such reaction is the decomposition of nitramide, discussed in Chapter 9. The rate-determining step is believed to be $HN{=}NO\cdot OH + B \rightarrow N{\equiv}N^+{—}O^- + OH^- + A$, and there will certainly be considerable changes in the transition state in the binding of the nitrogen and oxygen atoms. Correspondingly, general base catalysis of this reaction has a Brönsted $\beta$ in the range 0.6–0.8. The p$K$ of the NH group in $HN{=}NO\cdot OH$ is not experimentally accessible, but a plot of lg $k$ against the p$K$ of the catalyst suggests a limiting rate several powers of ten below the diffusion-controlled value.[1] Another nitrogen acid which reacts relatively slowly is the mono-anion of ethylenedinitramine, which reacts with ammonia according to the equation $^-O_2NHCH_2CH_2NHNO_2 + NH_3 \rightarrow {}^-O_2NNCH_2CH_2NNO_2 + NH_4^+$. Although this reaction is thermodynamically favourable, its rate constant is only about $10^5$ dm$^3$ mol$^{-1}$ s$^{-1}$,[20] which is considerably less than the diffusion-controlled value, but much greater than the value of $7 \times 10^{-3}$ dm$^3$ mol$^{-1}$ s$^{-1}$ found for the corresponding reaction of the analogous carbon acid nitroethane. This suggests that the anions of nitramines have structures intermediate between the two extremes,

Among oxygen acids, proton-transfers coupled with structural change elsewhere occur particularly in the addition reactions of the carbonyl

[20] R. P. Bell and R. G. Pearson, *J. Chem. Soc.*, 3443 (1953).

group, for example, the addition of water and alcohols and the mechanistically similar mutarotation of glucose, already discussed in the preceding chapter (pp. 183–190). These reactions involve the making and breaking of covalent links either with other molecules or (in the mutarotation reactions) with distant parts of the same molecule. The observed rates are low, and Brönsted reactions with exponents in the range 0.4–0.7 are valid over a wide range.

So far our discussion of slow proton-transfers and Brönsted exponents has been a qualitative one, apart from the crude electrostatic model represented by Figures 15 and 16. Recently considerable use has been made of a general equation relating the rate of a reaction to its standard free energy change, which was first derived by Marcus[21] for electron-transfer reactions, and later applied by him[22] and by others[7] to reactions involving the transfer of atoms or protons. For present purposes the Marcus equation can be written as*

$$\Delta G^{\ddagger} = W^{r} + \Lambda \left( 1 + \frac{\Delta G^{\circ}}{4\Lambda} \right)^{2} \tag{119}$$

where $\Delta G^{\circ}$ is the standard free energy change of the reaction and $\Delta G^{\ddagger}$ is the free energy of activation, defined in terms of the velocity constant $k$ by $\Delta G^{\ddagger} = -RT \ln(k/Z)$, where $Z$ is a collision number. For the sake of simplicity statistical factors have been omitted in both $\Delta G^{\ddagger}$ and $\Delta G^{\circ}$. $W^{r}$ is the work done in bringing the reactants together in a suitable position, orientation, and state of solvation for reaction to take place; it may include contributions from repulsion and from solvent reorganization, and corresponds to the vertical distance between the broken and solid curves in Figure 12. When $\Delta G^{\circ} = 0$, $\Delta G^{\ddagger} = W^{r} + \Lambda$, and $\Lambda$ is therefore termed the *intrinsic free energy barrier* for the series of reactions being considered. For proton-transfer reactions of the type $XH + Y^{-}$ it can be approximated by the mean of the corresponding quantities $\Lambda_{1}$ and $\Lambda_{2}$ for the symmetrical processes $XH + X^{-}$ and $YH + Y^{-}$.

Equation (119) follows directly from the energy diagram in Figure 13

---

[21] R. A. Marcus, *J. Chem. Phys.*, **24**, 966 (1956); *Disc. Faraday Soc.*, **29**, 21 (1960); *J. Phys. Chem.*, **67**, 853, 2889 (1963); *Ann. Rev. Phys. Chem.*, **15**, 155 (1964); *J. Chem. Phys.*, **43**, 679 (1965).
[22] R. A. Marcus, *J. Phys. Chem.*, **72**, 891 (1968); A. O. Cohen and R. A. Marcus, *J. Phys. Chem.*, **72**, 4249 (1968); R. A. Marcus, *J. Am. Chem. Soc.*, **91**, 7224 (1969).
*The nomenclature used by Marcus and others has varied in rather a confusing way. Equation (119) corresponds to the terminology of the last paper in Ref. 22, which, however, omits the term $W^{r}$. It should be noted that (119) applies only if $|\Delta G^{\circ}| < 4\Lambda$, since otherwise either the forward or the reverse reaction will become diffusion-controlled, and the picture of an activation barrier will not be valid.

if it is assumed that the relevant parts of the energy curves are parabolas of equal curvature. The picture of intersecting potential energy curves is particularly appropriate to electron-transfer reactions in which there is weak interaction between the initial and final states. On the other hand, it is certainly inappropriate to atom-transfer reactions, in which such interaction is strong. These last reactions have been treated, particularly by Johnston,[23] by what is termed the 'bond energy bond order' (BEBO) method, in which the energies of the two partial bonds during the atom-transfer process are related to their bond orders, these being in turn related to the extent to which the atom has been transferred between the two centres. By making certain simplifying assumptions Marcus has shown[22] that instead of (119) the appropriate expression is now

$$\Delta G^{\ddagger} = W^r + \Lambda(1 + \tfrac{1}{2}x) + \Lambda[\ln \cosh(\tfrac{1}{2}x \ln 2)]/\ln 2 \qquad (120)$$

where $x = \Delta G^{\circ}/4\Lambda$. However, in spite of the different appearances of (119) and (120), they lead to very similar predictions provided that $\Delta G^{\circ}/4\Lambda$ is not too great, as is likely to be the case in examples of practical interest. Since proton-transfers can be thought of as intermediate in character between electron-transfers and atom-transfers, it is reasonable to use the simpler Equation (119) in the present context.

It will be clear from the above that the Marcus Equation (119) does not rest on any clear-cut theoretical argument, and it is of interest that essentially the same equation can be derived from an apparently quite different set of assumptions.[24] Nevertheless, it constitutes a useful framework for rationalizing experimental results. If it can be assumed that $W^r$ and $\Lambda$ remain constant in a series of reactions, then the Brönsted exponent $\alpha$ (or $\beta$) is given by (119) as

$$\alpha = \frac{\partial \Delta G^{\ddagger}}{\partial \Delta G^{\circ}} = \frac{1}{2}\left(1 + \frac{\Delta G^{\circ}}{4\Lambda}\right). \qquad (121)$$

This predicts that $\alpha = \tfrac{1}{2}$ for reactions with zero overall free energy change, while $\alpha < \tfrac{1}{2}$ for 'downhill' reactions, and $\alpha > \tfrac{1}{2}$ for 'uphill' reactions. If $\Delta G^{\circ}$ can be measured or estimated, the observed value of $\alpha$ for a limited

[23] H. S. Johnston, *Gas Phase Reaction Rate Theory*, Ronald Press, New York, 1966, Ch. 11 and Appendix E.
[24] E. D. German, R. R. Dogonadze, A. M. Kuznetsov, V. G. Levich, and Yu. I. Kharkats, *J. Res. Inst. Catalysis Hokkaido Univ.*, **19**, 99, 115 (1971), and earlier papers quoted therein; also J. R. Murdoch, *J. Am. Chem. Soc.*, **94**, 4410 (1972). The last paper emphasizes that equations such as (119) apply only to the process (b) ⇌ (c) in a reaction scheme like (36), so that care is necessary in applying them to observed rates.

range of $\Delta G°$ (sufficiently removed from $\Delta G° = 0$) will yield a value for $\Lambda$, and any individual observed velocity constant than enables $W^r$ to be determined from (119). This kind of treatment does in fact fit the experimental data for a number of proton transfer reactions,[25] and the values obtained for $W^r$ can reasonably be attributed to the changes in solvation accompanying the coming together of the two reactants. However, there are certainly many cases in which it is not legitimate to assume that $W^r$ and $\Lambda$ remain constant in a series of reactions. This is particularly the case when compounds of different chemical type are being compared, or when considering substitution close to the seat of reaction. This has already been mentioned in connection with the variation of the substrate in catalysed reactions, and the same point arises when pseudo-acids (i.e., acids whose ionization involves a drastic structural change) are used as catalysts. Thus in the acid-catalysed hydration of acetaldehyde, nitroalkanes and di-ketones were found to be 10–100 times less effective as catalysts than would be predicted from the Brönsted relation which is valid for carboxylic acids and phenols.[26] A particularly striking anomaly has been studied by Bordwell and his collaborators,[27] who found that the rate of ionization of ring-substituted 1-aryl-nitroethanes and 1-aryl-2-nitropropanes varied in the opposite direction to the dissociation constants of these substances. This implies a negative value of $\alpha$ for the ionization process, and hence a value greater than unity for the reverse reaction. This behaviour is not consistent with (121) or with any simple interpretation of the Brönsted relation in terms of potential energy curves. It can formally be accounted for by supposing that substitution affects the quantities $W^r$ and $\Lambda$ in (119), but the molecular interpretation of such behaviour is still uncertain.

The quantity $\Delta G°$ in Equations (119)–(121) can be related qualitatively to the *asymmetry of the transition state*, and terms such as the *degree of proton transfer* have been widely used; highly asymmetric transition states are often described as being *reactant-like* or *product-like*. These concepts could also be expressed quantitatively in terms of the bond lengths, bond orders, or force constants of the two bonds in a transition state $X\cdots H\cdots Y$. However, the relation between these quantities is a matter for speculation, and $\Delta G°$ is the only one of them which can (in favourable cases) be

[25] A. O. Cohen and R. A. Marcus;[22] M. M. Kreevoy and D. E. Konasewich.[7]

[26] R. P. Bell and W. C. E. Higginson, *Proc. Roy. Soc.*, A, **197**, 141 (1949).

[27] F. G. Bordwell, W. J. Boyle, J. A. Hautala, and K. C. Yee, *J. Am. Chem. Soc.*, **91**, 4002 (1969); F. G. Bordwell, W. J. Boyle, and K. C. Yee, *J. Am. Chem. Soc.*, **92**, 5926 (1970); F. G. Bordwell and W. J. Boyle, *J. Am. Chem. Soc.*, **93**, 512 (1971); **94**, 3907 (1972). See also M. Fukuyama, P. W. K. Flanagan, F. T. Williams, L. Frainier, S. A. Miller, and H. Schechter, *J. Am. Chem. Soc.*, **92**, 4689 (1970).

measured directly. Much of the evidence about degrees of proton transfer derives from studies of kinetic isotope effects, and will be considered in this context in the last chapter of this book.

We shall now consider some examples of *specific deviations from the Brönsted equation*, especially those for which a reasonable molecular interpretation can be given. Much attention has been paid to the relative catalytic activities and basic strengths of *primary, secondary, and tertiary amines*. For the limited class of ring-substituted anilines there is an excellent relation between basic strength and catalytic power in the decomposition of nitramide in aqueous solution.[28] Similar relations hold for the same reaction in isopentanol[29] and in *m*-cresol,[30] still using the p$K$ values in water as a measure of basic strength. The catalytic constants of $N$-methylaniline and $N,N$-dimethylaniline in *m*-cresol deviate considerably from the relation holding for the primary anilines, though these deviations disappear if the basic strengths are measured in *m*-cresol instead of in water. Similarly, in the nitramide decomposition in water, catalysis by dimethylamine and trimethylamine does not conform to the equation for primary amines.[31] The same point is illustrated by the rates at which the series $NH_3$, $NH_2Me$, $NHMe_2$, and $NMe_3$ react with nitroethane to form the anion; these bear no relation to the basic strengths of the substances concerned.[32]

The above rather fragmentary observations suggest that substitution on the nitrogen atom is too drastic a structural variation for the validity of the Brönsted relation. The same point is brought out very clearly by systematic work on catalysis by amines in the decomposition of nitramide.[38] If lg $k_b$ is plotted against p$K$, the results in aqueous solution give two parallel straight lines, one for primary amines and the other for tertiary amines; i.e., a tertiary amine is about twice as effective a catalyst as a primary amine of the same p$K$. The same effect appears even more markedly when the kinetic data refer to anisole solution. In the plot lg $k_b$(anisole) against p$K$(water) there are now four parallel straight lines, represented by the equations

[28] J. N. Brönsted and H. C. Duus, *Z. Phys. Chem.*, **117**, 299 (1925).
[29] J. N. Brönsted and J. E. Vance, *Z. Phys. Chem.*, **163**, A, 240 (1933).
[30] J. N. Brönsted, A. L. Nicholson, and A. Delbanco, *Z. Phys. Chem.*, **169**, A, 379 (1934).
[31] H. L. Pfluger, *J. Am. Chem. Soc.*, **60**, 1513 (1938).
[32] R. G. Pearson, *J. Am. Chem. Soc.*, **70**, 204 (1948); R. G. Pearson and F. V. Williams, *J. Am. Chem. Soc.*, **76**, 1258 (1954); M. J. Gregory and T. C. Bruice, *J. Am. Chem. Soc.*, **89**, 2327 (1967).
[33] R. P. Bell and A. F. Trotman-Dickenson, *J. Chem. Soc.*, 1288 (1949); R. P. Bell and G. L. Wilson, *Trans. Faraday Soc.*, **46**, 407 (1950).

| Primary aromatic amines | $\lg k_b = -5.39 + 0.64 \, pK$ |
|---|---|
| Secondary aromatic amines | $\lg k_b = -5.04 + 0.64 \, pK$ |
| Tertiary aromatic amines | $\lg k_b = -4.42 + 0.64 \, pK$ |
| Tertiary heterocyclic amines | $\lg k_b = -3.49 + 0.64 \, pK,$ |

showing that there is a factor of nearly 100 between the catalytic power of a primary aromatic amine and a tertiary heterocyclic amine of the same $pK$.

Similar anomalies arise when we consider the $pK$ values of primary, secondary, and tertiary amines in different solvents. Thus on going from water to ethanol we find the following average changes of $pK$:[34]

|  | $pK(\text{EtOH}) - pK(\text{H}_2\text{O})$ |
|---|---|
| Primary amines | $1.0 \pm 0.2$ |
| Secondary amines | $0.6 \pm 0.2$ |
| Tertiary amines | $-0.3 \pm 0.3$ |

again showing a clear separation into three classes. Finally, there is the well-known anomaly in the effect of successive alkyl groups on the basic strengths of aliphatic amines in water; for example, we have $pK(\text{NH}_4^+) = 9.25$, $pK(\text{MeNH}_3^+) = 10.62$, $pK(\text{Me}_2\text{NH}_2^+) = 10.77$, and $pK(\text{Me}_3\text{NH}^+) = 9.80$, instead of the steady rise in basic strength which would be expected.

It seems certain that the main explanation of all these anomalies lies in the interaction of the cations with water, which will vary considerably in a series such as $\text{NH}_4^+$, $\text{MeNH}_3^+$, $\text{Me}_2\text{NH}_2^+$, and $\text{Me}_3\text{NH}^+$. This can be expressed either by saying that the degree of hydrogen bonding with the solvent depends upon the number of hydrogen atoms in the cation[35] or that the alkyl groups act by excluding water molecules from close inter-action with the positive charge.[36] In either case the effect of alkyl substitution is to decrease the stabilization of the cation by interaction with the solvent, thus acting in the opposite direction to the inductive effect. This could well produce the observed order of basic strengths. This view is supported by the observed entropy changes in the reaction

$$\underset{/}{\overset{\backslash}{-}}\text{NH}^+ + \text{H}_2\text{O} \rightarrow \underset{/}{\overset{\backslash}{-}}\text{N} + \text{H}_3\text{O}^+,$$

given in Table 5 (p. 75), which become more negative with increasing alkyl substitution, corresponding to decreasing orientation of water molecules round the cation. Even more convincing is the fact that when

[34] Data from H. Goldschmidt and E. Mathiesen, *Z. Phys. Chem.*, **119**, 439 (1926).
[35] A. F. Trotman-Dickenson, *J. Chem. Soc.*, 1293 (1949).
[36] A. G. Evans and S. D. Hamann, *Trans. Faraday Soc.*, **47**, 34 (1951).

the basic strengths of the alkylamines are investigated by indicator measurements in solvents with little or no power of hydrogen bonding we find the expected order

$$pK(RNH_3^+) > pK(R_2NH_2^+) > pK(R_3NH^+).^{37}$$

Further, the application of mass-spectrometric techniques gives just the same order for the proton affinities of amines in the gas phase,[38] and a considerable amount of quantitative information is now available.[39] It is of interest in this connection that just the same effects of alkyl substitution are observed on the tendency of amines to lose a proton,[40] i.e., in the absence of solvation effects alkyl groups have a similar stabilizing influence both on cations $\geqslant NH^+$ and on anions $> N^-$. This suggests that the effect of alkyl groups on the acid-base properties of amines, and also alcohols,[41] is best explained by the stabilization of a localized charge by a polarizable environment, and casts considerable doubt on the usual explanation in terms of the inductive effect.

It is now clear why the relative strengths in water are not adequate to account for the catalytic effects of primary, secondary, and tertiary amines in non-aqueous solvents, since the effect of hydration will not operate in the latter case. The position is similar even when both rates and equilibria are studied in aqueous solution, since the transfer of the proton is incomplete in the transition state, and the interaction with the solvent will be less than it is in the final state, in which the proton has been completely transferred. This picture agrees with the observation that in the plots of $\lg k_b$ against $pK(H_2O)$ in the decomposition of nitramide the separation between primary and tertiary amines is much greater for $k_b$(anisole) than for $k_b(H_2O)$.

Deviations from the Brönsted relation may appear when there are *specific interactions in the transition state*. The ordinary type of steric hindrance has little effect on acid-base equilibria on account of the small size of the proton, but it may of course affect the rate of an acid-base reaction, since the two species $A_1$ and $B_2$ must be close together in the transition state, and this may be hindered by the presence of bulky groups in one or both reactants. There are few clear examples of this effect in

[37] R. P. Bell and J. W. Bayles, *J. Chem. Soc.*, 1518 (1952); R. G. Pearson and D. C. Vogelsong, *J. Am. Chem. Soc.*, **80**, 1038 (1958); J. W. Bayles and A. Chetwyn, *J. Chem. Soc.*, 2328 (1958).
[38] M. S. B. Munson, *J. Am. Chem. Soc.*, **87**, 2332 (1965); J. I. Braumann, J. M. Riveros, and L. K. Blair, *J. Am. Chem. Soc.*, **93**, 3914 (1971).
[39] D. H. Aue, H. M. Webb, and M. T. Bowers, *J. Am. Chem. Soc.*, **94**, 4726 (1972).
[40] J. I. Braumann and L. K. Blair, *J. Am. Chem. Soc.*, **93**, 3911 (1971).
[41] J. I. Braumann and L. K. Blair, *J. Am. Chem. Soc.*, **92**, 5986 (1970).

reactions catalysed by acids or bases, though it appears in the catalytic effect of substituted pyridines and their cations in the hydration of acetaldehyde,[42] where the presence of alkyl substituents in the 2- and 6-positions causes a lowering of the catalytic power. Similar steric retardations are observed when alkyl-pyridines or their cations act as catalysts in the halogenation of ketones,[43] the mutarotation of glucose,[44] and the inversion of menthone.[44] An effect in the opposite direction has been observed in the anion-catalysed halogenation of various ketones and esters.[45] For most substrates and carboxylate anions the Brönsted relation is accurately obeyed, but if both the catalyst and the substrate contain a large group (alkyl, aryl, or bromine) near the seat of reaction, the observed velocity is greater than anticipated by up to 300%. This means that the proximity of the two large groups in the transition state must lower its energy. This probably depends not so much on any direct attraction between the groups as on the necessity of making a cavity in the solvent, thereby doing away with some of the attractions between the water molecules. When the two groups are close together, they will cause the separation of fewer water molecules than when they are apart, and this factor will tend to stabilize the transition state. The order of magnitude of this effect is illustrated by the work of Butler on the solubilities of homologous series of molecules in water. He found that the addition of each $CH_2$ group caused on the average an increase of 160 cal mol$^{-1}$ in the free energy of solution.[46] Since the effect may be increased by the direct attraction between the groups, it could easily account for the observed deviations from the Brönsted relation. This type of attraction, under the name of *hydrophobic bonding*, is commonly used to account for the configuration and association of macromolecules. It is also possible in principle for polar interactions between dipoles to stabilize the transition state, and there is some evidence for this,[47] though the resulting deviations from the Brönsted relation are not large.

Apparent deviations from the Brönsted relation can also occur if the equilibrium used to define the acid-base strength of the catalyst involves *subsidiary equilibria* in addition to the gain or loss of a proton, and such

[42] R. P. Bell, M. H. Rand, and K. M. A. Wynne-Jones, *Trans. Faraday Soc.*, **52**, 1093 (1956).

[43] J. A. Feather and V. Gold, *J. Chem. Soc.*, 1752 (1965).

[44] F. Covitz and F. H. Westheimer, *J. Am. Chem. Soc.*, **85**, 1773 (1963).

[45] R. P. Bell, E. Gelles, and E. Möller, *Proc. Roy. Soc.*, A, **198**, 308 (1949).

[46] J. A. V. Butler, *Trans. Faraday Soc.*, **33**, 229 (1937).

[47] A. J. Kresge, H. L. Chen, Y. Chiang, E. Murrill, M. A. Payne, and D. S. Sagatys, *J. Am. Chem. Soc.*, **93**, 413 (1971).

deviations can be used to obtain information about these subsidiary equilibria. This problem, like many related ones, was fully treated by Brönsted and Pedersen[3] in their first paper on the decomposition of nitramide, and their treatment is followed here. Consider the general equilibrium scheme

$$HX \rightleftharpoons X^- + H^+$$
$$\Updownarrow \qquad \Updownarrow \qquad \qquad (122)$$
$$HY \rightleftharpoons Y^- + H^+$$

and let $x_1$ and $x_2$ be the equilibrium fractions of the acid and its anion which are in the forms HY and $Y^-$ respectively. (The interconversion of HX and HY or of $X^-$ and $Y^-$ can involve the loss or gain of a solvent molecule without affecting the resulting equations, and in fact not all of these species need be acids or bases in terms of the usual definition of these terms, as will appear from the examples given later.) The 'true' dissociation constants of HX and HY are defined by $K_X = [H^+][X^-]/[HX]$ and $K_Y = [H^+][Y^-]/[HY]$, and are related to the apparent dissociation constant of the equilibrium mixture, $K_a$, by the equations

$$K_a = \frac{[H^+]([X^-]+[Y^-])}{[HX]+[HY]} = \frac{1-x_1}{1-x_2}K_X = \frac{x_1}{x_2}K_Y. \qquad (123)$$

If the anions $X^-$ and $Y^-$ can act separately as basic catalysts for a given reaction, with catalytic constants $k_X$ and $k_Y$ obeying the same Brönsted relation, then the observed catalytic constant for the equilibrium mixture is given by

$$k = (1-x_2)k_X + x_2 k_Y = GK_a^{-\beta}\{(1-x_1)^\beta(1-x_2)^{1-\beta} + x_1^\beta x_2^{1-\beta}\}. \quad (124)$$

A similar equation can be derived for acid catalysis by HX and HY. The expression in brackets in (124) represents the deviation from the simple Brönsted relation caused by the presence of the subsidiary equilibria in (122); if $x_1 = x_2$ no deviation will be observed, since we shall then have $K_X = K_Y = K_a$. If a deviation exists, it is clear that the separate values of $x_1$ and $x_2$ cannot be derived from observations on a single catalysed reaction, though in principle they can be obtained if the deviations are measured for two reactions with different values of $\beta$. However, a simpler result is obtained if, as is often the case, either $x_1$ or $x_2$ is close to zero. Thus if $x_2 = 0$, corresponding to an anion which exists in a single form, (124) becomes

$$k = G\{K_a/(1-x_1)\}^{-\beta} \qquad (125)$$

and $x_1$ can be determined by observations on a single reaction.

This aspect of the Brönsted relation has not been much exploited, and there are some ambiguities in its application. Brönsted and Pedersen[3] applied (125) to the unexpectedly low catalytic constants found for the nitramide ion $HN=NO_2^-$ and the nitrourethane ion $CO_2EtN=NO_2^-$ in the decomposition of nitramide, and arrived at values of 130 and 6 for the equilibrium ratios $[NH_2NO_2]/[NH=NO_2H]$ and $[CO_2EtNHNO_2]/[CO_2EtN=NO_2H]$ respectively.* It was also pointed out by Brönsted and Pedersen that basic catalysis by the bicarbonate ion is more properly related to the 'true' $pK$ of $H_2CO_3$ (3.89) rather than to the conventional value (6.35), which includes unhydrated carbon dioxide as well as $H_2CO_3$ (cf. p. 38). It was not practicable to investigate this point for the decomposition of nitramide, but it was shown later[48] that in the reaction $CO_2 + H_2O \rightarrow H_2CO_3$, which is catalysed by basic anions, the effect of bicarbonate ion is about 100 times smaller than would be expected on the basis of $pK = 6.35$, and is consistent with $pK = 3.89$.

A recent application of the same principle[49] deals with *o*-formylbenzoic and *o*-acetylbenzoic acids. These compounds exhibit ring–chain tautomerism, the possible form for *o*-formylbenzoic acid being as follows:

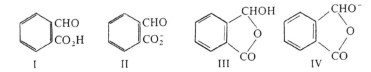

I                  II                 III                IV

There is good evidence from infrared and proton magnetic resonance spectra that in aqueous solution the acid exists predominantly in the cyclic form III, but that its anion is predominantly, if not exclusively, in the carboxylate form II.[50] When the anion is used as a catalyst for the decomposition of nitramide[49] application of (125) leads to the conclusion that 94% of the acid is in the cyclic form III. A similar result is obtained by

* Brönsted and Pedersen doubted this conclusion for catalysis by the nitramide anion, since they thought that the nitramide molecule might have the structure $HN=NO_2H$ or $HN\underset{O}{\overset{O}{\diagup}}NOH$. However, we have seen (p.160) that there is good evidence for the structure $NH_2NO_2$, and their conclusion can therefore stand.

[48] F. J. W. Roughton and V. H. Booth, *Biochem. J.*, **32**, 2049 (1938); cf. A. R. Olson and P. V. Youle, *J. Am. Chem. Soc.*, **62**, 1027 (1940).

[49] R. P. Bell, B. G. Cox, and B. A. Timimi, *J. Chem. Soc.*, B, 2247 (1971); R. P. Bell, B. G. Cox, and J. B. Henshall, *J. Chem. Soc., Perk. Trans.* II, 1232 (1972).

[50] D. D. Wheeler, D. C. Young, and D. S. Erley, *J. Org. Chem.*, **22**, 547 (1957); E. Bernatek, *Acta Chem. Scand.*, **14**, 785 (1960); J. Kagan, *J. Org. Chem.*, **32**, 4060 (1967).

studying the mutarotation of glucose, but is less accurate, since for this reaction the contribution of the water-catalysed reaction is considerable, the Brönsted exponent $\beta$ is smaller (0.40 as compared with 0.80 for the nitramide decomposition), and the points on the Brönsted plot are more scattered, perhaps because of steric effects. However, the agreement between the values obtained for the tautomeric equilibrium constant by studying two catalysed reactions with different values of $\beta$ is good evidence that there is no appreciable catalysis by the species IV.

The assumption that the catalytic powers of the isomeric species $X^-$, $Y^-$, HX, HY are related to their 'true' dissociation constants implies that proton transfers such as $X^- + SH \rightleftharpoons HX + S^-$ and $Y^- + SH \rightleftharpoons HY + S^-$ are much faster than reactions such as $X^- + SH \rightleftharpoons HY + S^-$ and $Y^- + SH \rightleftharpoons HX + S^-$, which also involve structural changes. This is likely to be generally true, and has been confirmed directly for the carbon dioxide system, and by relaxation studies on *o*-formylbenzoic acid[49] and on a number of keto–enol systems.[50] However, there may well be systems for which this assumption cannot be made. Thus there is good evidence (cf. p. 9) that the dissociation of boric acid in aqueous solution is represented by the equation $H_3BO_3 + 2H_2O \rightleftharpoons B(OH)_4^- + H_3O^+$, so that the conventional dissociation constant of boric acid would not be expected to give a good measure of the catalytic effect of boric acid or borate ion. In fact the rather scanty evidence[52] indicates that these species do not exhibit any marked deviations from the Brönsted relation, which might suggest that the loss or gain of the extra water molecule takes place synchronously with the proton transfer, without the intermediate formation of species such as $B(OH)_2O^-$ or $\overset{+}{B}(OH)_3OH_2$. On the other hand, relaxation measurements on borate solutions[53] reveal the presence of several relaxation processes, probably involving the above species. The whole problem illustrates the difficulty of distinguishing between successive and synchronous processes in solution, and further work is needed.

Finally, we shall consider some special problems connected with the catalytic effect of *hydrogen fluoride and related species*. The acid HF has a p$K$ of 3.23,[54] so that the species HF and $F^-$ should act as acid and basic catalysts respectively in suitable reactions. The fluoride ion does, however, differ from most basic anions in being monatomic, and the process of

[51] M. Eigen, G. Ilgenfritz, and W. Kruse, *Chem. Ber.*, **98**, 1623 (1965).
[52] For a summary see R. P. Bell, J. O. Edwards, and R. B. Jones, in *The Chemistry of Boron and its Compounds* (ed. E. L. Muetterties), Wiley, New York, 1966, pp. 209–221.
[53] Personal communication from Dr. J. G. Beetlestone, 1971.
[54] P. R. Patel, E. C. Moreno, and J. M. Patel, *J. Res. Nat. Bur. Stand.*, A, **75**, 205 (1971).

ionization involves no structural rearrangement or charge delocalization. The catalytic effect of HF has been measured for the iodination of two ketones[55] and the hydrolysis of four vinyl ethers,[56] and in each case has been found to exceed considerably (by a factor of 3–9) the catalytic effect of a carboxylic acid of the same p$K$. This is understandable in terms of what has been said earlier in this chapter (pp. 208–211) about the effect of charge delocalization upon rates of proton transfer, since the delocalization which occurs in the carboxylate ion is absent in the fluoride ion. The same explanation may account for the fact[26] that in the acid-catalysed hydration of acetaldehyde the oximes show considerable positive deviations from the Brönsted relation based upon the carboxylic acids and phenols, both of which owe their high acidity to delocalization of charge in the anion. From this point of view hydrogen fluoride and the oximes might be said to possess 'less pseudo character' than the carboxylic acids and the phenols.

On the other hand, when fluoride ion acts as a basic catalyst for the removal of a proton from an uncharged substrate, its catalytic activity is found to be very similar to that of a carboxylate anion of the same basic strength.[55] In this situation the effect of charge delocalization is much less marked, since the effect of the proton upon the charge distribution in the catalyst anion is largely counterbalanced by the effect of the negative charge borne by the deprotonated substrate.*

Aqueous fluoride solutions also contain appreciable concentrations of the ion $HF_2^-$. This species can act either as an acid or as a base by virtue of the hypothetical reactions $HF_2^- \rightarrow H^+ + 2F^-$ and $HF_2^- + H^+ \rightarrow 2HF$, although, as pointed out in Chapter 2 (p. 42), it is not possible to give a numerical measure of its acid-base strength on the usual scale. It was first reported[55] that in a number of reactions $HF_2^-$ and HF act as acid catalysts of comparable effectiveness, and in the first edition of this book several pages were devoted to explaining this observation. However, it was later shown that the apparent catalytic activity of $HF_2^-$ was due to an algebraic error, and that no such catalysis was detectable in any of the reactions studied.[56] It is in fact quite natural that $HF_2^-$ should be a poor acid catalyst, since it has a negative charge, and the proton is well buried between the two fluoride ions.

[55] R. P. Bell and J. C. McCoubrey, *Proc. Roy. Soc.*, A, **234**, 192 (1956). The interpretation of these results involved an algebraic error, which was corrected in Ref. 56.
[56] A. J. Kresge and Y. Chiang, *J. Am. Chem. Soc.*, **90**, 5309 (1968); **94**, 2814 (1972).
*I am indebted to Professor A. J. Kresge for pointing out the importance of charge type in this situation.

The latter half of this chapter has dealt with many speculative topics, and it is likely that some of the interpretations offered will eventually prove to be wrong. Nevertheless, it should be clear that a quantitative study of protolytic equilibria and kinetics still has much to offer as a tool for investigating molecular structures and the nature of kinetic processes in solution.

# 11 Isotope Effects in Proton-Transfer Equilibria

Since all the processes so far considered involve the transfer of a proton from one species to another, it is to be anticipated that the substitution of hydrogen by deuterium or tritium will affect both the rates and the equilibrium constants of these processes. There are in fact two reasons why isotope effects involving hydrogen will usually be much greater than those for any other elements. In the first place, the mass ratios $m_H : m_D : m_T = 1 : 2 : 3$ differ greatly from unity, while the corresponding ratios for other common elements are nearly always between unity and 1.1. In the second place, the low mass of these nuclides in itself favours large isotope effects, since these are essentially quantum effects, depending upon deviations from classical mechanics, and such deviations are greatest, other factors being equal, for particles of small mass. This last point will be justified in more detail in the subsequent discussion.

Large hydrogen isotope effects were found experimentally very soon after the discovery of deuterium in 1932, and there is now an extensive literature on the subject, recently supplemented by work with tritium. Much of this relates to the kinetics of reactions involving proton transfer, but there is also a large amount of information on acid-base equilibria. A recent review[1] estimated that about 300 papers on isotope effects are now published each year, and a considerable proportion of these relate to the isotopes of hydrogen. The present chapter will be devoted to the effect of isotope substitution on acid-base equilibrium constants, while Chapter 12 will deal with kinetic isotope effects in proton transfer reactions.

The magnitude of the effects to be expected can be estimated by considering the dissociation energies of the bonds X—H and X—D. To a very good approximation the electron distributions in these two bonds are the same for a given internuclear distance, and the curves relating energy to internuclear distance are therefore identical. The dissociation energies, however, are not the same, since according to the quantum theory the

---

[1] M. Wolfsberg, *Ann. Rev. Phys. Chem.*, **20**, 449 (1969).

lowest energy level is higher than the minimum point in the energy curve by the *vibrational zero-point energy*, which is different for the two isotopes. More specifically, the zero-point energy is given by $E_0 = \frac{1}{2}h\nu$, where $\nu$ is the vibration frequency, and the difference between the two dissociation energies is

$$\Delta E_0 = E_0 - E_0' = \frac{1}{2}h(\nu - \nu') \tag{126}$$

where $\nu$ and $\nu'$ refer respectively to the bonds X—H and X—D.* For a particular case the values of $\nu$ and $\nu'$ can be obtained from vibrational spectra; moreover, since the hydrogen or deuterium atom is usually attached to a much heavier atom, it is good approximation to write $\nu/\nu' = 2^{\frac{1}{2}}$, so that (126) becomes

$$\Delta E_0 = \frac{1}{2}h\nu(1 - 2^{-\frac{1}{2}}) = 0.146h\nu \tag{127}$$

where it is now only necessary to know the frequency in the hydrogen compound. For polyatomic molecules there will also be contributions to the zero-point energy from bending vibrations of the bond X—H, though these will be smaller because of the lower frequencies of bending vibrations.

Values of $\nu$ and $\Delta E_0$ for typical C—H, N—H, O—H, and S—H bonds are given in Table 23. Apart from any other effects, a difference $\Delta E_0$ in

Table 23. ZERO-POINT ENERGIES FOR STRETCHING VIBRATIONS IN HYDROGEN AND DEUTERIUM COMPOUNDS

| Bond | $\nu$ (cm$^{-1}$) | $\Delta E_0$ (cal mol$^{-1}$) | $\exp(\Delta E_0/RT)$ at 25°C |
|------|------|------|------|
| C—H | 2800 | 1150 | 6.9 |
| N—H | 3100 | 1270 | 8.5 |
| O—H | 3300 | 1400 | 10.6 |
| S—H | 2500 | 1060 | 6.0 |

bond energy will contribute a factor $\exp(\Delta E_0/RT)$ to the isotope effect in equilibria or velocities, and values of this factor at 25°C are also given in the table. They all differ very considerably from unity.

We shall now consider in detail the isotope effect in equilibria of the type

$$AH + B \rightleftharpoons A + HB \tag{128}$$

* Throughout this chapter unprimed and primed quantities refer respectively to the light and heavy isotopes, usually H and D.

where A and B are either atoms or groups. If the process concerned is the transfer of a hydrogen atom, all the species will bear the same charge, while if a proton is being transferred there will be a unit difference of charge between the pairs AH–A and B–HB. The treatment which follows applies to either case. If the equilibrium constant for reaction (128) is $K$, and $K'$ is the corresponding constant when H is replaced by one of the heavier isotopes D or T, then the ratio $K/K'$ is the equilibrium constant for the isotopic exchange reaction

$$AH + H'B \rightleftharpoons AH' + HB. \tag{129}$$

At absolute zero the decrease in internal energy $\Delta\varepsilon_0$ in reaction (129) is given by the difference of zero-point energies, i.e.,

$$\Delta\varepsilon_0 = E_0(AH) - E'_0(AH) - \{E_0(HB) - E'_0(HB)\} \tag{130}$$

and the ratio $K/K'$ is then given by standard statistical mechanics as

$$\frac{K}{K'} = \frac{Q_{HB} Q'_{AH}}{Q'_{HB} Q_{AH}} e^{\Delta\varepsilon_0/kT} \tag{131}$$

where the $Q$'s are internal partition functions for the species concerned.

Each $Q$ is a product of translational, rotational, and vibrational parts, $Q = Q_t Q_r Q_v$, and it is of interest to see how each of these parts depends on the masses of the atoms concerned. Omitting constants which are independent of the mass, the usual expressions give

$$Q_t \propto M^{\frac{3}{2}}, \qquad Q_r \propto (XYZ)^{\frac{1}{2}}$$

$$Q_v = \prod_{3n-6} (1 - e^{-hv_i/kT})^{-1} = \prod_{3n-6} (1 - e^{-u_i})^{-1} \tag{132}$$

where $M$ is the total mass of the species, $X$, $Y$ and $Z$ are its moments of inertia, $v_i$ is a vibrational frequency, and $n$ is the number of atoms which the species contains.* Since the quantities $M$, $Z$, $Y$, $X$, and $u_i$ are all affected in different ways by isotopic substitution, it might appear that no simple general expression could be obtained for $K/K'$. However, the position is greatly simplified by introducing the *product rules*, which have general validity provided that the different degrees of freedom are inde-

---

*Equation (132) must be modified for linear molecules, which have only two rotational degrees of freedom, and hence an extra vibration. Similar modifications occur in the product rule (133), and the final expression for $K/K'$ differs from (135) only in the number of vibrational terms. Analogous modifications for linear molecules (or transition states) must be made in the expressions given later for the kinetic isotope effect. These are not given separately, though the case in which A and B are atoms does of course involve linear molecules throughout.

pendent and the vibrations are harmonic.[2] The most convenient form of the rule for our present purpose states that if a single atom in any molecule is isotopically substituted, then the following relation holds between the properties of the substituted and the unsubstituted molecules,

$$\left(\frac{X'Y'Z'}{XYZ}\right)^{\frac{1}{2}}\left(\frac{M'}{M}\right)^{\frac{3}{2}}\left(\frac{m}{m'}\right)^{\frac{3}{2}}\prod_{3n-6}\frac{u_i}{u_i'} = 1, \tag{133}$$

where as before $u = hv/kT$. If this is combined with (132), we find for the ratio of the internal partition functions

$$\frac{Q}{Q'} = \left(\frac{m}{m'}\right)^{\frac{3}{2}}\prod_{3n-6}\frac{u_i(1-e^{-u_i'})}{u_i'(1-e^{-u_i})} \tag{134}$$

and hence for the ratio of equilibrium constants (131)

$$\frac{K}{K'} = \frac{f_{HB}}{f_{AH}}e^{\Delta\varepsilon_0/kT} \tag{135}$$

where the function $f$ is defined as

$$f = \prod_{3n-6}\frac{u_i(1-e^{-u_i'})}{u_i'(1-e^{-u_i})}. \tag{136}$$

Since

$$\Delta\varepsilon_0 = \tfrac{1}{2}\sum_{AH}(u_i - u_i') - \tfrac{1}{2}\sum_{HB}(u_i - u_i'),$$

this simple expression involves only the vibrational frequencies of the molecules, the masses and moments of inertia having disappeared.[3] It can also be written in the alternative form,

$$\frac{K}{K'} = \prod_{HB}\frac{u_i\sinh\tfrac{1}{2}u_i'}{u_i'\sinh\tfrac{1}{2}u_i}\bigg/\prod_{AH}\frac{u_i\sinh\tfrac{1}{2}u_i'}{u_i'\sinh\tfrac{1}{2}u_i}. \tag{137}$$

It may be noted that, if we let $h \to 0$ or $T \to \infty$, then $K/K' \to 1$; i.e., in the classical limit there is no discrimination between isotopes, and the discrimination which actually occurs is a purely quantal effect.*

[2] O. Redlich, *Z. Phys. Chem.*, **28**, B, 371 (1935).
[3] This simplification was first introduced by H. C. Urey, *J. Chem. Soc.*, 569 (1947), and by J. Bigeleisen and M. G. Mayer, *J. Chem. Phys.*, **15**, 261 (1947).
* Even in the classical case $K/K'$ is not necessarily unity if the molecules have certain symmetry properties; for example, the reaction $H_2 + D_2 \rightleftharpoons 2HD$ has a classical equilibrium constant of 4 because HD has a lower symmetry than either $H_2$ or $D_2$. For the sake of simplicity we have omitted these symmetry numbers, which are equivalent to the statistical corrections previously discussed in connection with dissociation constants (p. 97) and rates (p. 197).

Although the values of $\exp(\Delta E_0/RT)$ in Table 23 differ greatly from unity, Equation (130) shows that $\exp(\Delta\varepsilon_0/kT)$ involves a good deal of cancelling out and may be quite close to unity unless A and B differ greatly from one another. Similarly, inspection shows that the term $f_{HB}/f_{AH}$ will not be very different from unity. A single term in the product (136) tends to $u_i/u_i'$ for high frequencies or low temperatures (large $u_i$) and to unity for low frequencies or high temperatures (small $u_i$). In comparing hydrogen and deuterium we have approximately $u_i/u_i' = 2^{\frac{1}{2}}$ for any vibration directly involving the isotope in question. In practice the isotopic substitution of a single hydrogen will affect only a small number of frequencies, so that $f_{HB}/f_{AH}$ is probably closer to unity than is $\exp(\Delta\varepsilon_0/kT)$; i.e., the difference of zero-point energy will be the main factor causing isotopic discrimination.

Equations such as (135) contain only vibrational frequencies, which can be derived from spectroscopic observations for species of not too great complexity. It should thus be possible to predict the isotope effect on simple equilibria, and this has been done successfully for a number of reactions in the gas phase. The problem is more difficult for reactions in solution, and few such quantitative predictions can be made for acid-base equilibria, though existing experimental data can be rationalized to some extent.

Since most common solvents contain readily exchangeable hydrogen atoms, it is rarely possible to compare the acid strengths of two acids XH and XD (or XT) in solvents of the same isotopic composition. Bell and Crooks[4] investigated the effect of replacing hydrogen by deuterium on the reaction of 2,4-dinitrophenol with various amines in toluene or chlorobenzene solution, thus avoiding any isotope exchange with the solvent. There is no detectable isotope effect with triethlamine or piperidine, while for pyridine $K^H/K^D = 1.40 \pm 0.05$. As we have seen in Chapter 4, in solvents of such low dielectric constant (2.4 and 5.6 for toluene and chlorobenzene respectively) the main product of an acid-base reaction between uncharged reactants is an ion pair, and the equilibrium constants are thus of the form $K = [B_1^-, HB_2^+]/[HB_1][B_2]$. The main source of any isotope effect will be in a change of stretching frequency when O—H is converted into N—H; these frequencies are accessible from infrared spectra. Measurements on solutions of nitrophenols in benzene[5] give $v^H = 3240\ \text{cm}^{-1}$, $v^H/v^D = 1.33$ for these frequencies. The N—H frequencies in ion pairs should be close to those in solid trimethylammonium and

[4] R. P. Bell and J. E. Crooks, *J. Chem. Soc.*, 3513 (1962).
[5] R. Cardinaud, *Bull. Soc. Chim. France*, 34 (1960).

pyridinium salts, for which infrared spectra have been determined.[6] It was found that the N—H stretching frequencies fall into two classes, according to whether hydrogen bonding to the anion is present (as with halides) or absent (as with perchlorates and tetrafluoroborates); average values are given in Table 24. It will be seen that in the absence of hydrogen bonding the N—H frequency is very close to that found for the O—H frequency for nitrophenols. This suggests that there is no hydrogen bonding in the ion pairs formed by 2,4-dinitrophenol with the strong bases triethylamine and piperidine, for which no isotope effect was observed. On the other hand, for a hydrogen-bonded ion pair there is a considerable change of frequency, and Equation (135) or (137) gives $K^H/K^D = 1.37$ if the frequency for pyridinium salts is used, and it is assumed that no other frequencies are affected by isotopic substitution; this is close to the observed values of $1.40 \pm 0.05$. The experimental results can thus be accounted for if it is assumed that the ion pairs formed from dinitrophenol and pyridine are hydrogen bonded, while those from triethylamine and piperidine are not. This is consistent with the stronger binding of the proton to nitrogen by the two stronger bases.

Table 24. N—H STRETCHING FREQUENCIES $(cm^{-1})$ IN AMINE SALTS

|  |  | Trimethylammonium | Pyridinium |
|---|---|---|---|
| Hydrogen-bonded | $\nu^H$ | 2730 | 2700 |
| Not hydrogen-bonded | $\nu^H$ | 3180 | 3240 |
|  | $\nu^H/\nu^D$ | 1.33 | 1.32 |

It would be of interest to have more information about isotope effects for acid-base equilibria in non-exchanging solvents, especially those of higher dielectric constant such as dimethylformamide ($\varepsilon = 37$) and dimethyl sulphoxide ($\varepsilon = 49$), in which ion association could be neglected. There are, however, considerable difficulties in carrying out and interpreting such measurements.

Almost all the available information about hydrogen isotope effects in acid-base equilibria involves a comparison of the dissociation constants of a given acid-base pair in $H_2O$ and in $D_2O$. In the first edition of this book (1959) only sixteen examples of such a comparison could be found. A more recent compilation,[7] giving references up to 1967, quotes values

[6] R. H. Nuttall, D. W. A. Sharp, and T. C. Waddington, *J. Chem. Soc.*, 4965 (1960).
[7] P. M. Laughton and R. E. Robertson, in *Solute-Solvent Interactions* (ed. J. F. Coetzee and C. D. Ritchie), Dekker, New York, 1969.

for more than two hundred acids and bases, and many more values have been reported during the last few years, for example, for carbonic acid,[8,10] phosphoric acid,[9] sulphurous acid,[10] inorganic oxyacids,[11] cationic indicators,[12] thiol acids,[13] and various weak acids.[14] Some of the most accurate measurements have been carried out in connection with studies of $H_2O$–$D_2O$ mixtures, and will be mentioned later in this chapter. All the conventional techniques for measuring dissociation constants have been employed, though special care is necessary in interpreting the results of measurements with glass electrodes.[15]

It should be noted that there is sometimes an ambiguity in the concentration scales used; because of the difference in densities between $H_2O$ and $D_2O$ the ratio $K^H/K^D$ will have different values on the molarity and molality scales. It is probably most satisfactory to express all concentrations as moles per 55.51 moles of solvent; this has been termed the 'aquamolality' scale. Since $H_2O$ and $D_2O$ differ in molar volumes by only 0.36% the use of molarities will give virtually the same value for $K^H/K^D$.

There is thus abundant experimental material, but it cannot be said that we have any satisfactory quantitative understanding of the absolute values of $K^H/K^D$, or of the way in which they vary from one acid to another. This is essentially because the comparison involves a change of solvent from $H_2O$ to $D_2O$, both of which are highly structured solvents which interact strongly (and possibly differently) with solute species, and particularly with the ions $H_3O^+$, $D_3O^+$, $OH^-$, and $OD^-$. For an uncharged acid HX the two equilibrium constants which are measured correspond to the reactions

$$HX(H_2O) + H_2O(liq) \rightleftharpoons X^-(H_2O) + H_3O^+(H_2O)$$

$$DX(D_2O) + D_2O(liq) \rightleftharpoons X^-(D_2O) + D_3O^+(D_2O)$$
(138)

where $(H_2O)$ and $(D_2O)$ represent the solvents in which the species in question occur. It is obvious that the ratio $K^H/K^D$ cannot be strictly regarded as the equilibrium constant for a simple isotope exchange reaction such as (129). If we can neglect the different solvent effects of

[8] M. Paabo and R. G. Bates, *J. Phys. Chem.*, **73**, 3014 (1969).

[9] M. Paabo and R. G. Bates, *J. Phys. Chem.*, **74**, 706 (1970).

[10] P. Salomaa, A. Vesala, and S. Vesala, *Acta Chem. Scand.*, **23**, 2107 (1969).

[11] P. Salomaa, R. Hakala, S. Vesala, and T. Aalto, *Acta Chem. Scand.*, **23**, 2116 (1969).

[12] V. Gold and C. Tomlinson, *J. Chem. Soc.*, B, 1707 (1971).

[13] W. P. Jencks and K. Salvesen, *J. Am. Chem. Soc.*, **93**, 4433 (1971).

[14] R. A. Robinson, M. Paabo, and R. G. Bates, *J. Res. Nat. Bur. Stand.*, A, **73**, 299 (1969).

[15] See, e.g., A. K. Covington, M. Paabo, R. A. Robinson, and R. G. Bates, *Analyt. Chem.*, **40**, 700 (1968); H. Kakihana, *Bull. Chem. Soc. Japan*, **43**, 1377 (1970).

$H_2O$ and $D_2O$ (and in particular any differences in their interactions with $H_3O^+$ or $D_3O^+$), then $K^H/K^D$ may be regarded as the equilibrium constant for the process

$$HX + H_2O + D_3O^+ \rightleftharpoons DX + D_2O + H_3O^+. \qquad (139)$$

However, if we wish to use Equation (137) to calculate this equilibrium constant it is clear that the appropriate frequencies to use are not those of the isolated species, but rather those of the strongly hydrogen-bonded system which actually exists in aqueous solution. A quantitative treatment is thus very difficult, but some qualitative interpretations are possible.

We shall consider first the *ionic products* of $H_2O$ and $D_2O$. It was shown at an early stage[16] that $K_w^D$ is considerably smaller than $K_w^H$. Recent determinations[17] are not in complete accord, but a recent critical comparison[18] gives $K_w^H/K_w^D = 7.47 \pm 0.24$ (in molarity units) at 25°C. The ionic product of tritium oxide has also been determined recently,[19] giving $K_w^H/K_w^T = 16.4$ at 25°C. For $H_2O$ and $D_2O$ the relevant equilibria are

$$2H_2O \rightleftharpoons H_3O^+ + OH^-$$
$$2D_2O \rightleftharpoons D_3O^+ + OD^- \qquad (140)$$

again omitting any effects of further solvation, and any departure from unity in $K_w^H/K_w^D$ must be attributed to differences between the vibration frequencies of the species $H_3O^+$, $H_2O$, and $OH^-$ (or their isotopic analogues), with the stretching frequencies playing a major role. The vibrational spectrum of liquid water is not fully understood, but a recent analysis of Raman spectra[20] gives 3460 cm$^{-1}$ as the mean frequency of the stretching vibrations. For $OH^-$ the corresponding frequency is somewhat higher,[21] being about 3570 cm$^{-1}$. On the other hand, there is no doubt that $H_3O^+$ in solution has considerably lower vibrational frequencies. This statement was originally based upon the observation of a broad infrared absorption band centred at about 2900 cm$^{-1}$ in aqueous solutions of strong acids.[22] However, we have seen in Chapter 2 (p. 21) that it is

[16] E. Abel, E. Bratu, and O. Redlich, *Z. Phys. Chem.*, A, **173**, 353 (1935): W. F. K. Wynne-Jones, *Trans. Faraday Soc.*, **32**, 1397 (1936); G. Schwarzenbach, A. Epprecht, and H. Erlenmeyer, *Helv. Chim. Acta*, **19**, 1292 (1936).
[17] V. Gold and B. M. Lowe, *Proc. Chem. Soc.*, 140 (1963); *J. Chem. Soc.*, A, 936 (1967): A. K. Covington, R. A. Robinson, and R. G. Bates, *J. Phys. Chem.*, **70**, 3820 (1966); L. Pentz and E. R. Thornton, *J. Am. Chem. Soc.*, **89**, 6931 (1967).
[18] P. Salomaa, *Acta Chem. Scand.*, **25**, 367 (1971).
[19] M. Goldblatt and W. M. Jones, *J. Chem. Phys.*, **51**. 1881 (1969).
[20] G. E. Walrafen, *J. Chem. Phys.*, **36**, 1035 (1962); **40**, 3249 (1964); **44**, 1546 (1966); **47**, 114 (1967); **48**, 244 (1968).
[21] L. H. Jones, *J. Chem. Phys.*, **22**, 217 (1954).
[22] M. Falk and P. A. Giguère, *Canad. J. Chem.*, **35**, 1195 (1957); **36**, 1680 (1958).

doubtful whether this band can actually be attributed to the hydronium ion, and a more reasonable model for $H_3O^+$ in aqueous solution is provided by the $H_9O_4^+$ ion in solid $H_9O_4^+ \cdot Br^-$, for which the mean stretching frequency was found[23] to be 2350 cm$^{-1}$. The dissociation of water is thus accompanied by a decrease in the zero-point energy of the stretching vibrations, and since this decrease is greater for $H_2O$ than for $D_2O$, we should have $K_w^H > K_w^D$, as observed.

Similar qualitative considerations apply to oxyacids in $H_2O$ and $D_2O$, for which $K^H/K^D$ is usually in the range 2.5–4.5, since the O—H stretching frequencies are similar to those in $H_2O$. It is of interest that for thiol acids[24] the values of $K^H/K^D$ are lower, 2.0–2.5, corresponding to the fact that S—H stretching frequencies are considerably lower than those for O—H; for example MeSH has $v = 2550$ cm$^{-1}$, compared with 3640 for MeOH.

This picture in terms of stretching frequencies is of course a gross over-simplification, and the changes of zero-point energy involved are in general too small to account for the observed magnitude of the isotope effect. Various authors[25-27] have therefore attempted to include bending and librational modes, which are particularly important when hydrogen bonding to neighbouring water molecules is considered. It is possible to obtain reasonable agreement with experiment by this means, but only by making particular assumptions in the controversial field of water structure, and it is difficult to know how much weight to attach to this type of treatment.

Early results for a number of weak acids suggested[28] that the value of $K^H/K^D$ decreased regularly with increasing acid strength, but when more extensive experimental data[7] are considered there appears to be no real basis for this generalization, except perhaps as an ill-defined qualitative trend. There is some evidence that such a relation holds approximately for a closely related series such as the phenols and alcohols[29] or the thiols[24] but not for the carboxylic acids, where the available p$K$ range is rather small. From a theoretical point of view a relation between p$K^H$ – p$K^D$ and p$K^H$ was first attributed to a decrease in the H—X stretching frequency as the acid HX becomes stronger.[30] However, in the isolated acid

[23] J. Rudolph and H. Zimmermann, *Z. Phys. Chem.* (Frankfurt), **43**, 311 (1964).
[24] W. P. Jencks and K. Salvesen, *J. Am. Chem. Soc.*, **93**, 4433 (1971).
[25] C. A. Bunton and V. J. Shiner, *J. Am. Chem. Soc.*, **83**, 42, 3207, 3214 (1961).
[26] C. G. Swain and R. F. W. Bader, *Tetrahedron*, **10**, 182, 200 (1960).
[27] R. A. More O'Ferrall, G. W. Koeppl, and A. J. Kresge, *J. Am. Chem. Soc.*, **93**, 1 (1971).
[28] C. K. Rule and V. K. LaMer, *J. Am. Chem. Soc.*, **60**, 1974 (1938).
[29] R. P. Bell and A. T. Kuhn, *Trans. Faraday Soc.*, **59**, 1789 (1963).
[30] R. P. Bell, *The Proton in Chemistry*, Methuen, London, 1959.

molecules any such variations are much too small to lead to measurable variations in $K^H/K^D$, and a more plausible basis was suggested by Bunton and Shiner,[25] who pointed out that the strengths of hydrogen bonds formed between the solvent and acidic or basic species, as measured by the shift of infrared frequencies, is known to be a function of acid-base strength.[31] Other factors are undoubtedly involved, and it cannot be said that either the absolute magnitudes of $K^H/K^D$ or their dependence on structural factors is really understood.

The difference between the dissociation constants of HX and DX in the same solvent is termed a *primary isotope effect*, since isotopic substitution takes place at the bond which is being broken. The kind of comparison discussed in the last three paragraphs really involves both a primary effect and a *solvent isotope effect*, though the latter term is often used to describe the gross effect observed, since it cannot readily be separated into its components. In addition we may have *secondary isotope effects*, where isotopic substitution takes place in a part of the molecule not directly concerned in the reaction;* for example, the dissociation constants of $CD_3CO_2H$ and $CH_3CO_2H$ are not identical. These effects are much smaller than the primary ones, typical values being

$$K(HCO_2H)/K(DCO_2H) = 1.08,[32]$$

$$K(CH_3CO_2H)/K(CD_3CO_2H) = 1.03,[33]$$

$$K(CH_3NH_3^+)/K(CD_3NH_3^+) = 1.14,[34]$$

though when many atoms are substituted close to the reaction site quite large differences may be observed, for example,

$$K\{(CH_3)_3NH^+\}/K\{(CD_3)_3NH^+\} = 1.61.[35]$$

In examples like these there is very little tendency for the substituted groups to undergo isotopic exchange with the solvent, and the comparison can therefore be made *in the same solvent*, usually $H_2O$. Moreover, since C—H groups have very little tendency to hydrogen bonding, vibrational frequencies observed in the absence of solvent should be relevant to an

[31] W. Gordy and S. C. Stanford, *J. Chem. Phys.*, **9**, 204 (1941).
* Solvent isotope effects can also be formally considered as secondary effects, though this term is commonly reserved for isotopic substitution in one of the reacting solute molecules.
[32] R. P. Bell and W. B. T. Miller, *Trans. Faraday Soc.*, **59**, 1147 (1963).
[33] A. Streitwieser and H. S. Klein, *J. Am. Chem. Soc.*, **85**, 2759 (1963).
[34] W. Van der Linde and R. E. Robertson, *J. Am. Chem. Soc.*, **86**, 4504 (1964).
[35] D. Northcott and R. E. Robertson, *J. Phys. Chem.*, **73**, 1559 (1969).

interpretation of the equilibria in solution. Both of these circumstances suggest that secondary isotope effects for acid-base equilibria in aqueous solution might be simpler to interpret in principle than primary ones.

Since Equation (137) must apply to primary and secondary effects alike, the existence of a secondary effect implies that some of the isotopically sensitive frequencies must differ in the acidic and basic forms of the species concerned. Since many of the normal vibrations will involve coupling between different parts of the molecule, and the differences concerned are small, it may be necessary to include several frequencies in any quantitative treatment. The best documented case is a comparison of $HCO_2H$ and $DCO_2H$, for which $K^H/K^D$ is the equilibrium constant for the exchange reaction $HCO_2H + DCO_2^- \rightleftharpoons HCO_2^- + DCO_2H$ (all species in $H_2O$). The relevant frequencies for both isotopic species can be derived from the infrared and Raman spectra of the vapours of the acid and of solid formates, and it was found[36] that five frequencies undergo isotopic changes which are appreciably different in the acid and the anion, including some frequencies which involve primarily vibrations of the carboxyl group. When all these vibrations are taken into account excellent agreement is found with the observed value[32] of $K^H/K^D = 1.084 \pm 0.002$.

A detailed analysis of this kind is rarely possible, but it is found in general that replacement of hydrogen by deuterium in a secondary position has the effect of decreasing the acidity of the species, and that this effect increases with the number of hydrogens replaced and decreases with the distance between the site of replacement and the acidic proton. It is therefore sometimes stated[37] that the inductive effect of C—D is greater than that of C—H, or that the deuterium atom is more electronegative than the hydrogen atom. This description is misleading if it is taken to imply any appreciable difference between the electronic distributions in C—H and C—D bonds, since these must be very closely similar.[38] There will, however, be a small difference in the required sense between the dipole moments of C—H and C—D because of the different zero-point amplitudes of the two bonds,[39] amounting to $3 \times 10^{-13}$ to $4 \times 10^{-13}$ m. This difference in dipole moment can be measured quite accurately by spectroscopic techniques; for example, the moment of $CH_3D$ is given as $0.011274 \pm 0.000005$ D by electric resonance experiments on molecular beams,[40] while differences of similar magnitude can be deduced from the

[36] R. P. Bell and J. E. Crooks, *Trans. Faraday Soc.*, **58**, 1409 (1962).
[37] For example, E. A. Halevi, *Tetrahedron*, **1**, 74 (1957).
[38] R. E. Weston, *Tetrahedron*, **6**, 31 (1959).
[39] V. W. Laurie and D. R. Herschbach, *J. Chem. Phys.*, **37**, 1687 (1962).
[40] J. S. Muenter, M. Kaufman, and W. Klemperer, *J. Chem. Phys.*, **48**, 3338 (1968).

Stark effect in microwave spectra[41] for pairs of molecules such as $CH_3F$ and $CD_3F$, or $CHF_3$ and $CDF_3$. Although these differences are small, they are of right order of magnitude to account for the observed secondary isotopic effects in terms of charge–dipole interactions, for example in the species $CH_3CO_2^-$ and $CD_3CO_2^-$.

It is important to stress that this type of explanation is not alternative or additional to the treatment already given in terms of zero-point energies. The difference between the C—H and C—D dipole moments arises only because of the difference in zero-point amplitudes, which is a purely quantum effect; similarly, the observed difference in the isotopically sensitive frequencies in species such as $CH_3CO_2H$ and $CH_3CO_2^-$ (or $CD_3CO_2H$ and $CD_3CO_2^-$) implies some interaction between the C—H bonds and the groups —$CO_2^-$ or —$CO_2H$, leading to a difference in the force constants of the C—H bonds in the two species. To quote a statement by Thornton,[42] 'the qualitative argument that the smaller average dipole moment of C—D than of C—H gives rise to an isotopic energy difference is identical with the statement that a change in charge distribution in the molecule will give rise to a force constant change which is "felt" more by H than by D because H vibrates into regions of greater displacement, i.e., greater bond dipole moment'.*

We shall now consider *acid-base equilibria in $H_2O$–$D_2O$ mixtures*, in which isotopic exchange can occur between solute and solvent. Since it did not prove possible to give a quantitative interpretation of the changes in dissociation constants in going from pure $H_2O$ to pure $D_2O$, it might appear profitless to investigate mixtures. In fact, however, the way in which dissociation constants depend on isotopic composition has proved of considerable interest, and is a potential source of much information about interactions in solution, especially solvation. It was realized at an early stage that both equilibrium and rate constants frequently showed a non-linear dependence upon $x$, the atom fraction of deuterium in the solvent, and the same is true for the logarithms of these constants. The essential explanation of this was given almost immediately by Gross[43] and by

[41] J. S. Muenter and V. W. Laurie, *J. Chem. Phys.*, **45**, 855 (1966).
[42] E. R. Thornton, *Ann. Rev. Phys. Chem.*, **17**, 354 (1966).
* The discussion in the last two paragraphs is slightly over-simplified, since, if the vibrations are simple harmonic and the dipole moment a linear function of internuclear distance, the mean dipole moment is independent of the amplitude. The differences which actually occur are due partly to anharmonicity, which is particularly marked in bonds involving hydrogen, and partly to a non-linear variation of dipole moment with internuclear distance, sometimes called electrical anharmonicity.
[43] P. Gross and A. Wischin, *Trans. Faraday Soc.*, **32**, 879 (1936); P' Gross, H. Steiner, and H. Suess, *Trans Faraday Society.*, **32**, 883 (1936); P. Gross, *Z. Elektrochem.*, **44**, 299 (1938).

Butler.[44] Purlee[45] gave a critical review of the problem in 1959; this was followed by renewed activity in the field, and an authoritative article on the subject has recently been published by Gold.[46] The remainder of this chapter draws heavily on this last article.

It is convenient to have a symbol to denote any one of the three hydrogen isotopes, and L (cf. lyonium and lyate ions) is commonly used. In a mixture of $H_2O$ and $D_2O$ the solvent $L_2O$ thus consists of a mixture of the species $H_2O$, HDO, and $D_2O$, while the ion $L_3O^+$ may be either $H_3O^+$, $H_2DO^+$, $HD_2O^+$, or $D_3O^+$, and the ion $OL^-$ represents only the two species $OH^-$ and $OD^-$. The complications introduced by this multiplicity of species are greatly reduced by introducing the *rule of the geometric mean*, according to which isotopic disproportionation equilibria are determined purely by statistical factors, so that the corresponding equilibrium constants are given by the ratios of symmetry numbers. Thus for the equilibrium $H_2O + D_2O \rightleftharpoons 2HOD$ we have

$$K = \frac{x_{HOD}^2}{x_{H_2O} x_{D_2O}} = \frac{2.2}{1} = 4. \tag{141}$$

Similar equations apply to the $L_3O^+$ ions. It is usual to write these equilibria in a form involving only one molecule of the mixed isotopic species, i.e., as $\frac{2}{3}H_3O^+ + \frac{1}{3}D_3O^+ \rightleftharpoons H_2DO^+$ and $\frac{1}{3}H_3O^+ + \frac{2}{3}D_3O^+ \rightleftharpoons HD_2O^+$, and since the equilibrium constants, including (141), involve only dimensionless ratios of mole fractions, we can conveniently replace mole fractions by molarities, designated by square brackets. The resulting expressions are

$$[HOD] = 2[H_2O]^{\frac{1}{2}}[D_2O]^{\frac{1}{2}}$$

$$[H_2DO^+] = 3[H_3O^+]^{\frac{2}{3}}[D_3O^+]^{\frac{1}{3}} \tag{142}$$

$$[HD_2O^+] = 3[H_3O^+]^{\frac{1}{3}}[D_3O^+]^{\frac{2}{3}}$$

so that the concentrations of the mixed isotopic species can be expressed in terms of concentrations of species containing only hydrogen or deuterium. The rule of the geometric mean has been given a theoretical basis by Bigeleisen.[47] It is physically equivalent to the assumption that isotopic substitution at one position of a species $XL_m$ has no effect on the

[44] J. C. Horwel and J. A. V. Butler, *J. Chem. Soc.*, 1361 (1936); W. J. C. Orr and J. A. V. Butler, *J. Chem. Soc.*, 330 (1937); W. E. Nelson and J. A. V. Butler, *J. Chem. Soc.*, 958 (1938).
[45] E. L. Purlee, *J. Am. Chem. Soc.*, **81**, 263 (1959).
[46] V. Gold, *Adv. Phys. Org. Chem.*, **7**, 259 (1969).
[47] J. Bigeleisen, *J. Chem. Phys.*, **23**, 2264 (1955).

exchange equilibrium at any other position, i.e., that secondary isotope effects are negligible. Theory shows that the rule should hold with high accuracy for isotopes of heavier elements, though there may be appreciable deviations for hydrogen isotopes. We shall see later that such deviations certainly do exist, but that their effect is usually small; for the present the validity of the rule will be assumed.

No simple rule applies to isotopic exchange equilibria involving different chemical species, and in particular the isotopic abundance of deuterium in the $L_3O^+$ ion will differ from that in the water with which it is in equilibrium. This is expressed in terms of the *fractionation factor*, $l$, defined by

$$l = \frac{(D/H)_{L_3O}}{(D/H)_{L_2O}} = \frac{F_{L_3O}(1-x)}{(1-F_{L_3O})x} \tag{143}$$

where $F_{L_3O}$ represents the isotopic abundance of deuterium in $L_3O^+$, and $x$ that in the solvent. The value of $l$ can be determined experimentally by observations of proton magnetic resonance in $H_2O–D_2O$ mixtures,[48] and by measuring the isotopic composition of water vapour above solutions of perchloric acid;[49] both methods give $l = 0.69 \pm 0.02$ at 25°C. Somewhat different values had been reported earlier on the basis of e.m.f. measurements of cells containing $H_2O$ and $D_2O$, but the calculations involved some doubtful assumptions, and a reconsideration of the results[50] also gives $l = 0.69$. The isotopic abundances in (143) are by definition

$$x = \frac{[D_2O] + \frac{1}{2}[HDO]}{[D_2O] + [HDO] + [H_2O]}$$

$$F_{L_3O} = \frac{[D_3O^+] + \frac{2}{3}[HD_2O^+] + \frac{1}{3}[H_2DO^+]}{[D_3O^+] + [HD_2O^+] + [H_2DO^+] + [D_3O^+]}, \tag{144}$$

and by combining (142), (143), and (144) we obtain

$$l = \frac{[D_3O^+]^{\frac{1}{3}}[H_2O]^{\frac{1}{2}}}{[H_3O^+]^{\frac{1}{3}}[D_2O]^{\frac{1}{2}}}, \tag{145}$$

i.e., $l^6$ is the equilibrium constant for the reaction $3D_2O + 2H_3O^+ \rightleftharpoons 3H_2O + 2D_3O^+$, the mixed isotopic species having been again eliminated.

[48] V. Gold, *Proc. Chem. Soc.*, 141 (1963); A. J. Kresge and A. L. Allred, *J. Am. Chem. Soc.*, **85**, 1541 (1963); V. Gold and M. A. Kessick, *Disc. Faraday Soc.*, **39**, 84 (1965).
[49] K. Heinzinger and R. E. Weston, *J. Phys. Chem.*, **68**, 744, 2179 (1965); K. Heinzinger, *Z. Naturforsch.*, **20a**, 269 (1965).
[50] P. Salomaa and V. Aalto, *Acta Chem. Scand.*, **20**, 2035 (1966).

Similar fractionation factors, usually denoted by $\phi$, can be defined for other solute species containing exchangeable hydrogens. For a species $XL_m$, $\phi$ is defined by

$$\phi_{XL_m} = \frac{F_{XL_m}(1-x)}{(1-F_{XL_m})x} \qquad (146)$$

where $F$ again represents the fractional abundance of deuterium. For an acid LA with a single exchangeable hydrogen this becomes

$$\phi_{LA} = [DA](1-x)/[HA]x. \qquad (147)$$

When several exchangeable hydrogen atoms are present, equations analogous to (142) will relate the concentrations of the various isotopic species, and general expressions for the concentrations of species such as $XD_p H_{m-p}$ can be obtained.[51]

In regarding $l$ and $\phi$ as constants we are assuming that their values are independent both of the concentrations of the solutes, and of the isotopic composition of the solvent. The first assumption is a safe one for isotopic exchange reactions, since the ratios of activity coefficients for isotopic species are unlikely to be affected by concentration or by changing from $H_2O$ to $D_2O$. In developing a simple theory it is also convenient to omit activity coefficients in equilibrium constants such as $K_{HA} = [H_3O^+][A^-]/[HA][H_2O]$, involving protium species only. As long as we are dealing with ratios of such constants in solutions of low and equal ionic strength (but different isotopic compositions) this omission is justified for interionic activity coefficients, since the dielectric constants of $H_2O$ and $D_2O$ differ by only 0.5%. However, this is not necessarily valid for the so-called degenerate or medium activity coefficients, which reflect the free energy of transfer of a species from one solvent to another (cf. p. 68). For the present we shall assume that these activity coefficients also cancel out, returning later to the problem of transfer effects.

We can now derive an expression for the observed dissociation constant of a monobasic acid LA in a $D_2O$–$H_2O$ mixture in which the atomic fraction of deuterium is $x$; this will be denoted by $K^x$, and the corresponding value in pure $H_2O$ by $K^H$. Since we are assuming that $K^H = [H_3O^+][A^-]/[HA][H_2O]$ retains the same value in an isotopically mixed solvent, we can write

$$\frac{K^H}{K^x} = \frac{[H_3O^+][LA][L_2O]}{[L_3O^+][HA][H_2O]} \qquad (148)$$

[51] J. I. G. Cadogan, V. Gold, and D. P. N. Satchell, *J. Chem. Soc.*, 561 (1955); A. J. Kresge, *Pure Appl. Chem.*, **8**, 243 (1964).

where $L_3O^+$, etc., represent the sums of the concentrations of all isotopic species.*

By means of Equations (142), (145), and (147), the right-hand side of (148) can be expressed solely in terms of $x$ and the two fractionation factors $l$ and $\phi_{LA}$, and we find after some algebra

$$\frac{K^H}{K^x} = \frac{1 - x + x\phi_{LA}}{(1 - x + xl)^3} \tag{149}$$

Equation (149) as it stands is not very useful in practice, since there are few acids for which $\phi_{LA}$ can be measured independently with sufficient accuracy; for acetic acid the use of a measured $\phi_{LA}$ gives reasonable agreement with the observed values of $K^H/K^x$.[52] However, if $x = 1$, corresponding to pure $D_2O$, (149) becomes $K^H/K^D = \phi_{LA}/l^3$, and can therefore be written in the more useful form

$$\frac{K^H}{K^x} = \frac{1 - x + xl^3 K^H/K^D}{(1 - x + xl)^3} \tag{150}$$

which predicts the form of the relation between $K^H/K^x$ and $x$ provided that the value of $K^H/K^D$ is known.

Figure 18 shows the predictions of Equation (150) for $K^H/K^D = 0.5$, 1.0, 2.0, and 3.0 in the form of plots of $K^H/K^x$ against $x$, taking $l = 0.69$. It will be seen that the plots are far from linear, and that their curvature depends considerably upon the value of $K^H/K^D$. Since free energies rather than equilibrium constants would normally be related to solvent properties, it is more rational to plot $\lg(K^H/K^x)$ or $(pK^x - pK^H)$ against $x$, and these plots are shown in Figure 19 for the same values of $K^H/K^D$. This type of plot is almost linear when $K^H/K^D = 3$, and this is even truer for larger values of $K^H/K^D$.

There is not a great deal of experimental material for simple monobasic acids which can be used to test Equation (150). Apart from some early measurements on formic acid[53] and benzoic acid,[28] recent experimental data are available for acetic acid,[52,54] monochloroacetic acid,[55] hydrazoic

---

*Most workers on equilibria or rates in $H_2O$–$D_2O$ mixtures have expressed the results as $K^x/K^H$, which is logical since $K^H$ may be regarded as the standard value. We have preferred $K^H/K^x$ for the sake of uniformity with the expression $K^H/K^D$ (or $k^H/k^D$) commonly used to express primary isotope effects or solvent isotope effects in pure $H_2O$ or $D_2O$. The latter usage has the advantage that values greater than unity are most commonly encountered, so that a 'large isotope effect' is associated with a larger number.

[52] V. Gold and B. M. Lowe, *J. Chem. Soc.*, A, 1923 (1968).
[53] W. J. C. Orr and J. A. V. Butler, *J. Chem. Soc.*, 330 (1937).
[54] P. Salomaa, L. L. Schaleger, and F. A. Long, *J. Am. Chem. Soc.*, **86**, 1 (1964).
[55] P. Salomaa, L. L. Schaleger, and F. A. Long, *J. Phys. Chem.*, **68**, 410 (1964).

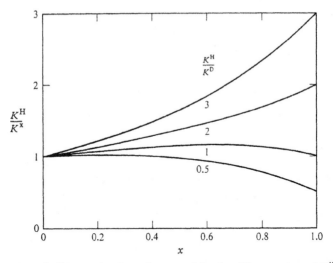

Fig. 18. Plots of $K^H/K^x$ against isotopic composition for different values of $K^H/K^D$.

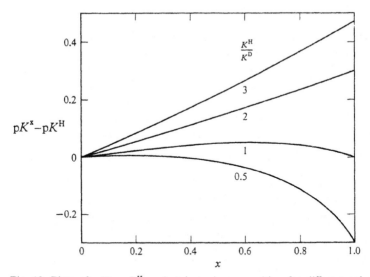

Fig. 19. Plots of $pK^x - pK^H$ against isotopic composition for different values of $K^H/K^D$.

acid,[55] 2-nitrophenol and 2,4-dinitrophenol.[56] Agreement with theory is satisfactory, though there are deviations which slightly exceed experimental error; moreover, the agreement obtained is not a very severe test

[56] L. Pentz and E. R. Thornton, *J. Am. Chem. Soc.*, **89**, 6931 (1967).

of the general theory, since the values of $K^H/K^D$ are all close to 3 for the acids cited. However, there seems little doubt that the treatment leading to (150) is essentially correct. In particular, the cubic form of the denominator arises from the assumption that the hydrogen ion contains three equivalent protons which differ appreciably in their isotopic fractionation factors from the average water protons. The agreement with experiment thus provides indirect evidence for the formulation of the hydrdonium ion as $H_3O^+$; at least it can be said that the assumption of any number of equivalent protons other than three often gives significantly poorer agreement with experiment. This statement does not deny the existence of any further hydration of $H_3O^+$, but merely implies that there is little isotopic fractionation between bulk water and any extra water molecules attached to $H_3O^+$. There is thus no contradiction with the evidence given in Chapter 2 for the existence in solution of a stable species $H_3O^+(H_2O)_3$, or $H_9O_4^+$.

Equations (149) and (150) apply to the dissociation of a monobasic acid containing only one exchangeable proton, giving a corresponding base with no exchangeable protons. For the general case of a dissociation

$$L_iA + L_2O \rightleftharpoons L_{i-1}A + L_3O^+$$

where the charges on $L_iA$ and $L_{i-1}A$ are not specified, but must of course differ by unity, an analogous treatment leads to the equation[54,57]

$$\frac{K^H}{K^x} = \frac{(1 - x + x\phi_{L_iA})^i}{(1 - x + xl)^3(1 - x + x\phi_{L_{i-1}A})^{i-1}} \tag{151}$$

When $x = 1$ this expression becomes

$$\frac{K^H}{K^D} = \frac{\phi_{L_iA}^i}{l^3\phi_{L_{i-1}A}^{i-1}} \tag{152}$$

so that the observed value of $K^H/K^D$ can be used to eliminate one (but not both) of the unknown fractionation factors. Equation (151) has been applied to a number of polybasic acids,* for example, phosphoric acid,[54] arsenic acid,[52] periodic acid,[58] and the ammonium ion.[55] Agreement with theory is satisfactory, but there is now an unknown fractionation factor, which can rarely be estimated independently and thus represents

[57] A. J. Kresge, *Pure Appl. Chem.*, **8**, 243 (1964).
*It is not necessary that more than one acidic proton should be dissociated appreciably under experimental conditions; for example, $NH_4^+$ corresponds to $i = 4$ in the above equations, since all four protons are exchangeable, although deprotonation proceeds only as far as $NH_3$.
[58] P. Salomaa and A. Vesala, *Acta Chem. Scand.*, **20**, 1414 (1966).

an adjustable parameter. Moreover, the calculated curves are rather insensitive to the form of the equation; for example, the results for the ammonium ion[55] agree about as well with the simple equation (150) as they do with (151) putting $i = 4$. Thus the experimental results are consistent with (151), but do not really demonstrate its correctness.

Special interest attaches to the ionic product of water in $H_2O-D_2O$ mixtures, for which there exist several reasonably concordant sets of data.[17] If the process is formulated as $2H_2O \rightleftharpoons H_3O^+ + OH^-$, we must put $i = 2$, $\phi_{LA} = 1$ in Equation (151), giving

$$\frac{K_w^H}{K_w^x} = \frac{(1-x+xl)^{-3}}{(1-x+x\phi_{OL})} \tag{153}$$

The fractionation factor for hydroxide ion has been measured[59] by isotopic analysis of water vapour above alkaline solutions, and if this value ($\phi_{OL} = 0.48$) is inserted in (153) fair agreement with experiment is obtained. The results can also be analysed without using this value, since for $x = 1$ (153) becomes $K_w^H/K_w^D = \phi_{OL}/l^3$, and can be rewritten as

$$\frac{K_w^H}{K_w^x} = \frac{(1-x+xl)^{-3}}{(1-x+xl^3 K_w^H/K_w^D)}. \tag{154}$$

This equation reproduces the experimental results to within 0.01 unit in $pK_w$, and it is noteworthy that for this system the representation of the hydronium ion as $H_3O^+$ is crucial, since any other formulation gives markedly poorer agreement with experiment. The use of (154) corresponds to $\phi_{OL} = 0.42$, which is fairly close to the directly determined value of 0.48.*

The application of the treatment outlined above presupposes a knowledge of the nature of the acidic and basic species in solution, in particular their degree of hydration. For example, if carboxylic acids existed in aqueous solution not as RCOOH but as $RC(OH)_3$, which is not inconceivable, especially for the stronger acids, then (150) would no longer apply. Measurements in $H_2O-D_2O$ mixtures could therefore be used in principle to obtain information about degrees of hydration, but in practice this procedure is limited both by experimental errors and by the assump-

[59] K. Heinzinger and R. E. Weston, *J. Phys. Chem.*, **68**, 2179 (1964).
*Attempts to determine $\phi_{OL}$ by measurements of proton magnetic resonance[60,61] at first appeared to give very different values, but the reasons for this apparent discrepancy have been recently cleared up.[62]
[60] A. K. Kresge and A. L. Allred, *J. Am. Chem. Soc.*, **85**, 1541 (1963).
[61] V. Gold, *Proc. Chem. Soc.*, 141 (1963).
[62] V. Gold and S. Grist, *J. Chem. Soc., Perk. Trans.* II, 89 (1972).

tions which underlie the simple theory. For example, the first dissociation of boric acid might be represented by one of the following schemes:

(a) $B(OH)_3 + H_2O \rightleftharpoons OB(OH)_2^- + H_3O^+$

(b) $B(OH)_3 + 2H_2O \rightleftharpoons B(OH)_4^- + H_3O^+$

(c) $H_2O^+ {}^- B(OH)_3 + H_2O \rightleftharpoons B(OH)_4^- + H_3O^+$

leading to different predictions for the variation of $K^x$ with $x$. Gold and Lowe[52] concluded that either (a) or (c) fitted the experimental results slightly better than (b), although we have seen (p. 9) that other evidence points clearly to (b). It must be concluded that it is dangerous to draw conclusions from small discrepancies with theory without examining in more detail the assumptions on which it is based.

In spite of the elegance and general success of the simple theory, it involves two assumptions which are only approximately valid. The first is the rule of the geometric mean, which is implicit in Equations (141) and (142), and has to be invoked again in deriving a more general expression such as (151). Theoretical considerations[47] suggest that this rule is only approximately correct for the isotopes of hydrogen. In fact, statistical thermodynamic calculations on the basis of observed spectroscopic frequencies[63] give the value 3.85 for the equilibrium constant $K = [HDO]^2/[H_2O][D_2O]$ rather than the value of 4 used hitherto; this is in reasonable agreement with two sets of direct mass-spectrometric measurements, which have given $K = 3.76 \pm 0.02$ and $K = 3.74 \pm 0.07$ respectively.[64,65] It is possible to carry through the calculation of $K^H/K^x$ using a value of $K$ less than 4, which has a small but detectable effect upon the shape of the curves relation $K^H/K^x$ to $x$.[56,66] However, the theory then loses its attractive simplicity, and it is doubtful whether this elaboration is justified, since the disproportionation equilibria involving the hydronium ion (cf. Equation 142) are also likely to show deviations in the same direction from simple statistical behaviour, though the magnitude of these deviations is unknown. One consequence of taking $K = 3.8$ rather than $K = 4.0$ is that the fractionation factor $\phi$ for a solute in equilibrium with $H_2O$–$D_2O$ mixtures should change by a few per cent in going from $x = 0$ to $x = 1$. There is some experimental evidence of such a trend,[66] but it has been shown[67] that very small errors are involved in using for the whole

[63] M. Wolfsberg, *J. Chem. Phys.*, **50**, 1484 (1969).
[64] L. Friedman and V. J. Shiner, *J. Chem. Phys.*, **44**, 4639 (1966).
[65] J. W. Pyper, R. S. Newbury, and G. W. Barton, *J. Chem. Phys.*, **46**, 2253 (1967).
[66] V. Gold, *Trans. Faraday Soc.*, **64**, 2770 (1968).
[67] W. J. Albery and M. H. Davies, *Trans. Faraday Soc.*, **65**, 1059 (1969).

range of isotopic compositions a constant value of $\phi$ based on measurements in the neighbourhood of $x = \frac{1}{2}$. The general conclusion is, therefore, that departures from the rule of the geometric mean are unlikely to introduce any serious errors into the predictions of the simple theory.

The second and much more serious assumption of the simple theory is that equilibrium constants such as $K^H$, $K^D$, $K_w^H$, and $K_w^D$ (involving only a single isotopic species) have values which are independent of the isotopic composition of the solvent. This is equivalent to assuming a cancellation of the degenerate activity coefficients (cf. p. 68), or of the free energies of transfer between $H_2O$ and $D_2O$ of the species involved in the equilibrium.* The possible importance of these transfer effects was realized at an early date by LaMer and his collaborators,[68] and they have been the subject of much discussion even since. For solutes containing exchangeable protons it is difficult or impossible to separate exchange and transfer effects, but the magnitude of such effects may be estimated by measurements on non-exchanging solutes, for which the standard free energy of transfer from $H_2O$ to $D_2O$ ($\Delta G_t^\circ$) can be determined from a variety of measurements, particularly of solubilities, partition coefficients, and electromotive force. A large amount of experimental material is now available, and a compilation by Arnett and McKelvey[69] covers work done up to 1967; more recent results have been given by various authors.[70-72] The results do not show any simple regularities, but if attention is confined to fairly small molecules and ions, it is true to say that $\Delta G_t^\circ$ is nearly always less than 100 cal mol$^{-1}$ for uncharged species, while it may reach 200–300 cal mol$^{-1}$ for uni-univalent salts. Since the free energy difference corresponding to $RT \ln(K^H/K^D)$ is commonly 400–600 cal mol$^{-1}$, the transfer terms are clearly likely to be important in many instances, though in considering an equilibrium there will frequently be some cancellation of terms corresponding to species on opposite sides of the equation.

The effect of including transfer terms in the equations of the simple theory has been carefully analysed by Gold,[46,52] especially for acetic

---

* It is important here to make a correct choice of standard states or concentration units, just as in making a comparison of $K^H$ and $K^D$. The most logical choice is mole fractions, and for dilute solutions almost equivalent results are obtained by using aquamolalities (moles solute per 55.51 moles of solvent).

[68] V. K. LaMer and E. Noonan, *J. Am. Chem. Soc.*, **61**, 1487 (1939); E. Noonan and V. K. LaMer, *J. Phys. Chem.*, **43**, 247 (1939).
[69] E. M. Arnett and D. R. McKelvey, in *Solute-Solvent Interactions* (ed. J. F. Coetzee and C. D. Ritchie), Dekker, New York and London, 1969.
[70] C. V. Krishnan, *J. Phys. Chem.*, **74**, 2356 (1970).
[71] P. Salomaa, *Acta Chem. Scand.*, **25**, 365 (1971).
[72] D. B. Dahlberg, *J. Phys. Chem.*, **76**, 2045 (1972).

acid. For this system Equation (149), with the directly measured value of $\phi_{LA}$, leads to predictions which differ from experiment by up to 0.05 p$K$ unit. Agreement with experiment to within 0.01 p$K$ unit can be secured if the values calculated from (149) are divided by a transfer activity factor $Y_{HA}^x$, it being assumed that $Y_{HA}^x = (Y_{HA}^D)^x$, where $Y_{HA}^D = 0.894$, the latter being the value for 100% $D_2O$ relative to $H_2O$.* The same degree of concordance with experiment can be obtained without considering any transfer effects by using Equation (150) with the experimental value of $K^H/K^D$; however, this corresponds to $\phi_{LA} = 1.05$, which is hardly compatible with the directly determined value of $0.96 \pm 0.02$. There is thus some evidence that the free energy of transfer is significant in this system, though it amounts to only about 10% of the total effect. Similarly, Salomaa[73] has recently concluded from distribution measurements that for the dissociation of picric acid the transfer effect contributes about 15% of the total free energy, though this involves the assumption that the free energy involved in transferring undissociated picric acid from $H_2O$ to $D_2O$ is the same as for 1,3,5-trinitrobenzene.

Salomaa[74] has used an ingenious procedure for separating the transfer effect in his studies of the first and second dissociation constants of carbonic and sulphurous acids in $H_2O–D_2O$ mixtures. We have already seen (p. 38) that only 0.3% of dissolved carbon dioxide is in the form of $H_2CO_3$, and there is some evidence[75] that the proportion of $H_2SO_3$ is also small. The product of the first and second dissociation constants thus represents the equilibrium constant for the process

$$XO_2 + 3L_2O \rightleftharpoons XO_3^{2-} + 2L_3O^+ \tag{155}$$

where X is either carbon or sulphur, and if transfer activity coefficients can be neglected the simple theory gives for this constant

$$K^H/K^D = l^{-6}, \qquad K^H/K^x = (1 - x + xl)^{-6} \tag{156}$$

since neither $XO_2$ or $XO_3^{2-}$ contains any exchangeable hydrogens.* In fact

---

*The assumption $Y_{HA}^x = (Y_{HA}^D)^x$ is equivalent to assuming that the free energy of transfer is a linear function of $x$.

[73] P. Salomaa, *Suomen Kem.*, B, **45**, 149 (1972).

[74] P. Salomaa, A. Vesala, and S. Vesala, *Acta Chem. Scand.*, **23**, 2107 (1969).

[75] M. Falk and P. A. Giguère, *Canad. J. Chem.*, **36**, 1121 (1958).

*The conclusions would not be affected if an appreciable fraction of $SO_2$ were present as $L_2SO_3$, since measurements were made at a constant pressure of $SO_2$ or $CO_2$ in the gas phase, and hence at constant activities of these species. Actually the high ratio of the solubilities of $SO_2$ in $H_2O$ and $D_2O$, 1.27 (compared with 1.006 for $CO_2$), suggests that it may be hydrated to a considerable extent. Direct evidence on this point is scanty,[75] but the high solubility of sulphur dioxide in water may be an indication in the same direction, and also the fact that the measured dissociation constant of sulphurous acid fits well into the classification of Table 11 (p. 92).

the use of Equation (156) with the well-established value $l = 0.69 \pm 0.01$ predicts equilibrium constants which differ from the experimental values by up to $0.16 \pm 0.04$ and $0.31 \pm 0.04$ pK unit for $CO_2$ and $SO_2$ respectively. There must therefore be considerable contributions from transfer effects in the equilibrium (155), amounting to 15–25% of the total free energy.

We have so far treated fractionation (or exchange) effects and transfer effects as though they were distinct phenomena, but in fact this distinction is an artificial one. According to the general theory of isotope effects, all such effects arise because of changes which the reaction produces in the environment of the isotopic atoms. In dealing with fractionation factors we normally include only those exhangeable atoms which appear in the chemical equation as usually written, so that solvent molecules not directly involved in the reaction are only included when there is good evidence for a very stable hydrate. Thus the hydrogen ion is written as $H_3O^+$ rather than $H^+$, and periodic acid as $H_5IO_6$ rather than $HIO_4$, but in general the effect of any species on more loosely held water molecules is classified as a transfer effect. Clearly, if such water molecules were included in writing the chemical equation, their part in determining isotope effects would logically appear in the guise of more fractionation factors. For example, the autoprotolysis of water could reasonably be written as

$$8H_2O \rightleftharpoons [OH(H_2O)_3]^- + [H_3O(H_2O)_3]^+$$

since there is good evidence for the further hydration of the hydronium and hydroxide ions. The disadvantage of this kind of approach is that it introduces a multiplicity of new fractionation factors, whose values can only be guessed at, so that they come close to being extra adjustable parameters. Thus Gold and Grist[62] write the hydroxide ion as

where the protons marked a, b, and c will have different fractionation factors, estimated by them as $\phi_a = 1.2\text{–}1.5$, $\phi_b = 0.65\text{–}0.70$, $\phi_c = 1$. On the other hand, Walters and Long[76] prefer to treat the same problem in terms of a single fractionation factor plus a transfer effect. These differences are to some extent matters of taste, but on the whole it seems likely that trans-

[76] E. A. Walters and F. A. Long, *J. Phys. Chem.*, **76**, 362 (1972).

fer effects will be gradually replaced by fractionation factors as our quantitative understanding of solvation increases. It is interesting to note that when there are a considerable number of protons with fractionation factors close to unity the resulting dependence of $K^H/K^x$ upon $x$ becomes very close to the logarithmic relation $K^H/K^x = (K^H/K^D)^x$. In fact if there are $n$ such protons each with a fractionation factor $\phi$, then if $\phi^n = F$, and is kept constant, it is easy to show that

$$\lim_{n \to \infty} (1 - x + x\phi)^n = \lim_{n \to \infty} (1 - x + xF^{1/n})^n$$

$$= F^x = \phi^{nx}. \tag{157}$$

Actually the limit is reached rapidly for moderate values of $n$; thus if $F = 1.5$ and $x = \frac{1}{2}$, the expression $(1 - x + xF^{1/n})^n$ has the values 1.233, 1.229, and 1.225 for $n = 3$, 6, and $\infty$. The shape of the plot of $K^x$ against $x$ is thus of little value in discriminating between the two interpretations.

The subject of isotope effects in $H_2O$–$D_2O$ mixtures has been treated in some detail because kinetic studies in these solvents have been recently used extensively to obtain detailed information about the nature of transition states in proton-transfer reactions. The problems involved are essentially the same as arise for equilibria, in particular with respect to transfer effects, and there is the added difficulty that fractionation factors and transfer activity coefficients for the transition state must either be guessed at by analogy with stable species, or derived from the kinetic measurements themselves. On the other hand, the numerical value of $k^H/k^D$ is often more favourable for distinguishing between different possibilities than are the values of $K^H/K^D$ commonly met with in equilibrium systems, and there is also another piece of experimental information, the so-called product isotope effect, which is sometimes helpful. These kinetic problems will be discussed briefly in the next chapter.

# 12 Kinetic Isotope Effects in Proton-Transfer Reactions

Kinetic isotope effects in general have now become an everyday tool of the mechanistic organic chemist, and this is particularly true of hydrogen isotope effects, partly because hydrogen is involved in so many reactions, and partly because such effects are much larger for hydrogen than for the isotopes of heavier atoms. It is interesting to note that the rate differences between hydrogen and deuterium compounds are sometimes so large that the use of deuterium compounds has been proposed as a practical expedient for slowing down harmful reactions, e.g., the deterioration of lubricants by oxidation.[1] Since the publication of the first edition of this book a number of books and review articles have appeared on the general subject of kinetic isotope effects.[2-5] The present chapter will therefore be confined almost entirely to isotope effects in proton-transfer reactions, though some reference will be made to the closely allied problem of reactions involving the transfer of hydrogen atoms, especially in connection with the tunnel effect. On the other hand, no reference will be made to the increasing use of secondary hydrogen isotope effects for obtaining information about neighbouring group participation, especially in solvolytic reactions,[6] since these do not normally involve proton transfers. Even in the field of proton-transfer reactions only a selection of the available material has been covered.

It is convenient to begin by considering what predictions can be made theoretically, first in the simplest possible terms and then using a more sophisticated model. The energy curve for a proton-transfer reaction has already been discussed, and the full curve in Figure 20 has the same

[1] P. Krumbiegel, Z. Chem., **8**, 328 (1968).
[2] L. Melander, Isotope Effects on Reaction Rates, Ronald Press, New York, 1960.
[3] Isotope Mass Effects in Chemistry and Biology, Butterworths, London, 1964; also in Pure Appl. Chem., **8**, Nos. 3 and 4 (1964).
[4] W. H. Sanders, Survey Progr. Chem., **3**, 109 (1966).
[5] Isotope Effects in Chemical Reactions (ed. C. J. Collins and N. S. Bowman), Van Nostrand Reinhold, New York, 1970.
[6] For recent reviews of this subject, see the articles by V. J. Shiner and by D. E. Sunko and S. Borcic in Ref. 5.

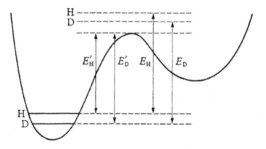

Fig. 20. Zero-point energies and hydrogen isotope effect for a proton-transfer reaction.

significance as those in Figures 11–14 of Chapter 10. Since intermolecular and interatomic forces are almost unaffected by a change in nuclear mass, it is a very good approximation to use the same curve for any of the hydrogen isotopes. On the other hand, the lowest energy levels of the initial and final states will depend on the isotopic mass, as illustrated by the zero-point energies for H and D indicated in the figure. The natural assumption is that the corresponding activation energies will differ by the same amount; i.e., for the reaction from left to right they will be $E'_H$ and $E'_D$ in the figure. The corresponding ratio of velocity constants would then be $k^H/k^D = \exp(\Delta E_0/RT)$, typical values of which have been given in Table 23, and for any particular reaction $k^H/k^D$ would be predictable in principle from spectroscopic data for the reacting species. Since the reaction velocity in one direction is unaffected by the properties of the products, there is no partial cancellation of zero-point energy differences of the kind met with in equilibria (cf. Equation 130), and the isotope effect on reaction velocities should be much greater than on equilibrium constants, as is indeed found to be the case.

There are, however, two reasons why the simple view of the situation is inadequate. The first of these is the neglect of *zero-point energy in the transition state*. There are of course many vibrations both in the initial and in the transition state which are not shown in Figure 20, but many of these are not affected by a change in the mass of the atom being transferred. However, in a reaction of the type $AH + B \rightarrow A + HB$ the transition state contains a bending vibration whose frequency is sensitive to the mass of the hydrogen atom. This can be represented as

$$A \cdots \overset{\uparrow}{\underset{\downarrow}{H}} \cdots \underset{\downarrow}{B}.$$

and is doubly degenerate, since a vibration of identical frequency could take place in a plane at right angles to the paper. If A and B are atoms, it is easy to count the degrees of freedom involved. The molecule AH has 3 translational, 2 rotational, and 1 vibrational degree of freedom $(3T + 2R + V)$, so that the system $AH + B$ can be represented by $6T + 2R + V$. The transition state (assumed linear) has $3T$ and $2R$, leaving four degrees of freedom to be accounted for. One of these is the 'internal translation' of zero (or imaginary) frequency corresponding to motion along the reaction co-ordinate and represented by

$$\overset{\leftarrow}{A} \cdots \overset{\rightarrow}{H} \cdots \overset{\leftarrow}{B}.$$

Since A and B are much heavier than H, they will be almost stationary in this vibration. A second vibration has the form

$$\overset{\leftarrow}{A} \cdots \overset{(?)}{H} \cdots \overset{\rightarrow}{B},$$

where the hydrogen will not be stationary except in the symmetrical case. This vibration will be somewhat affected by the isotopic mass of the hydrogen, but since its frequency will be low the contribution to the observed isotope effect may be small. The remaining two degrees of freedom are represented by the degenerate bending vibration already mentioned, so that the total degrees of freedom of the transition state are $3T + T^* + 2R + 3V$, where $T^*$ represents the internal translation. The formation of the transition state from $AH + B$ thus involves an increase of two in the vibrational degrees of freedom, at least one of which is sensitive to isotopic replacement. The position is somewhat different when, as is usually the case, A and B are not atoms. If they contain respectively $m$ and $n$ atoms and are not linear, then for the change $AH + B \rightarrow A\;H\;B$ the change in degrees of freedom is $6T + 6R + (3m + 3n - 9)V \rightarrow 3T + T^* + 3R + (3m + 3n - 4)V$. The formation of the transition state is thus attended by an increase of five in the number of vibrations (or four if the transition state contains an internal rotation not present initially), but not all of these will be sensitive to isotopic changes in the hydrogen atom. We may regard the bending vibrations in the transition state as being derived from two bending modes in the initial species AH, which of course do not exist if A is an atom.

The idea of bending vibrations in the transition state may seem an unfamiliar one, but it has a close parallel in the description of the reaction process given by the collision theory. This supposes that in the reaction between AH and B a particular direction of approach (usually collinear) gives a lower activation energy than any other direction. This means that the energy of a bent configuration

$$
\begin{array}{ccc}
 & \mathrm{H} & \\
\text{\tiny .·´} & & \text{\tiny `·.} \\
\mathrm{A} & & \mathrm{B}
\end{array}
$$

is higher than that of the straight configuration, which is equivalent to saying that the vibration

$$
\mathrm{A}\cdots\overset{\uparrow}{\underset{\downarrow}{\mathrm{H}}}\cdots\underset{\downarrow}{\mathrm{B}}
$$

has a finite frequency, as assumed above.

Returning to Figure 20, we see that the energy of the transition state is not given correctly by the maximum of the full curve but will be higher by an amount corresponding to the zero-point energy, and hence different for the two isotopes. The zero-point energy will of course change continuously as AH approaches B. The full line in the figure represents a section along the path of the particle but gives no information about the energy contours in other directions. The true energy paths therefore lie somewhat above the full line and will differ slightly for the two isotopes, the activation energies being now $E^{\mathrm{H}}$ and $E^{\mathrm{D}}$. The difference in activation energies is now $E^{\mathrm{D}} - E^{\mathrm{H}} = \Delta E_0 - \Delta E_0^{\ddagger}$, where $\ddagger$ refers as usual to the transition state. Unlike $\Delta E_0$, $\Delta E_0^{\ddagger}$ is not calculable from spectroscopic data, and its evaluation depends on a detailed knowledge of the frequencies of the transition state, which are not accessible experimentally and which cannot usually be calculated theoretically by present methods.

A possible model for the bending moles of the transition state is the bending vibration of the symmetrical ion $\mathrm{HF}_2^-$, which has $\nu = 1225\,\mathrm{cm}^{-1}$.[7] Since, in the transition state, the proton is still close to two centres, it seems likely that in general the bending frequency will be at least as great as that in a normal molecule, and we shall see later in this chapter that this expectation is borne out by some model calculations. A further contribution to $E_0$ comes from the 'symmetrical' stretching mode

$$
\overset{(?)}{\overset{\leftarrow}{\mathrm{A}}\cdots\mathrm{H}\cdots\overset{\rightarrow}{\mathrm{B}},}
$$

so that the figures in the last column of Table 23 should represent the *maximum* H/D isotope effects for reactions involving these bonds.

---

[7] G. L. Coté and H. W. Thompson, *Proc. Roy. Soc.*, A, **210**, 206 (1951).

The second modification to the simple expression

$$k^H/k^D = \exp(\Delta E_0/RT)$$

arises when we take into account the translational, rotational, and higher vibrational states of the reactants and transition state. This is best seen by applying the transition-state method to the isotope effect on the velocity of the reaction $AH + B \rightarrow A + HB$, just as in Equations (131)–(137) we have considered the equilibrium isotope effect. In this case the reaction co-ordinate $x$ can be closely represented by the position of the proton between the two centres A and B, and if we define the transition state as including all systems in which $x$ lies in a small arbitrary range $\delta$ in the neighbourhood of the energy maximum, then the velocity constant is given by

$$k = \kappa(K/\delta)(\mathbf{k}T/2\pi m_{\ddagger})^{\frac{1}{2}}. \tag{158}$$

In this equation $m_{\ddagger}$ is the reduced mass for motion along the reaction co-ordinate (in this case close to the mass of the hydrogen isotope in question), $\kappa$ is a transmission coefficient which will be referred to later, and $K$ is the complete equilibrium constant for the formation of the transition state from $AH + B$.* As before, the equilibrium constant is expressed in terms of partition functions and the expression for the effect of isotopic substitution simplified by using a product rule, but the special nature of the transition state introduces some modifications. In place of

$$Q_v^{\ddagger} = \prod_{3n-6} (1 - e^{-u_i})^{-1}$$

---

*An alternative form of (158) uses an incomplete expression for the equilibrium constant, omitting the degree of freedom corresponding to the reaction co-ordinate; the velocity constant is then obtained by multiplying this equilibrium constant by $\mathbf{k}T/\mathbf{h}$. The same result is obtained finally, but the form given in (158) shows a closer analogy to the statistical treatment of equilibria.

Although the quantity $\delta$ cancels out in the final result, the treatment outlined above suffers from a logical inconsistency. On the one hand, $\delta$ must be kept very small, so that the potential energy will vary inappreciably within the transition state, and hence motion through it can be treated as a translation. On the other hand, a very small value of $\delta$ raises difficulties in defining the transition state because of the uncertainty principle, or (which amounts to the same thing) because the quasi-classical expression for the translational partition function is no longer applicable. The second difficulty is particularly serious for the motion of light particles such as protons, for which it is impossible to choose a value of $\delta$ which satisfies both conditions even approximately. It has been shown[8] that these difficulties can be partly overcome if the transition state is defined in terms of a finite region within which the potential energy curve can be approximated by a parabola; the same final result is obtained.

[8] R. P. Bell, *Trans. Faraday Soc.*, **66**, 2770 (1970).

(Equation 132) we must write

$$Q_v^{\ddagger} = (2\pi m_{\ddagger} kT)^{\frac{1}{2}} (\delta/\mathbf{h}) \prod_{3n-7} (1 - e^{-u_i})^{-1} \tag{159}$$

since one vibration is replaced by an internal translation along the reaction co-ordinate. Similarly, in the product rule (133)

$$\prod_{3n-6} (u_i/u_i')$$

is replaced by

$$(m_{\ddagger}/m_{\ddagger}')^{\frac{1}{2}} \prod_{3n-7} (u_i/u_i').$$

This gives finally for the ratio of the velocity constants,

$$\frac{k}{k'} = \left(\frac{m_{\ddagger}'}{m_{\ddagger}}\right)^{\frac{1}{2}} \frac{f_{\ddagger}}{f_{AH}} e^{\Delta\varepsilon_0/kT} \tag{160}$$

in which

$$f_{\ddagger} = \prod_{3n-7} \frac{u_i(1 - e^{-u_i'})}{u_i'(1 - e^{-u_i})}, \qquad f_{AH} = \prod_{3n-6} \frac{u_i(1 - e^{-u_i'})}{u_i'(1 - e^{-u_i})} \tag{161}$$

$$\Delta\varepsilon_0 = \Delta E_0 - \Delta E_0^{\ddagger} \tag{162}$$

and the transmission coefficient $\kappa$ is assumed to be the same for the two isotopes. In Equation (161) the values of $n$ for $f_{\ddagger}$ and $f_{AH}$ are of course different, being equal to the numbers of atoms in the transition state and AH respectively. Similarly, the values of $u_i$ in each expression refer to the appropriate species. Since

$$\Delta\varepsilon_0/kT = \tfrac{1}{2}\sum_{AH} (u_i - u_i') - \tfrac{1}{2}\sum_{\ddagger} (u_i - u_i'),$$

(160) can also be written as

$$\frac{k}{k'} = \left(\frac{m_{\ddagger}'}{m_{\ddagger}}\right)^{\frac{1}{2}} \prod_{\ddagger} \frac{u_i \sinh \tfrac{1}{2} u_i'}{u_i' \sinh \tfrac{1}{2} u_i} \bigg/ \prod_{AH} \frac{u_i \sinh \tfrac{1}{2} u_i'}{u_i' \sinh \tfrac{1}{2} u_i}. \tag{163}$$

It is of interest to see what happens to (160) or (163) in the classical limit, i.e., when $\mathbf{h} \to 0$ or $T \to \infty$. $u$ then tends to zero and each factor $u/(1 - e^{-u})$ to unity, so that $k/k' \to (m_{\ddagger}'/m_{\ddagger})^{\frac{1}{2}}$. In contrast to the equilibrium expression (137) there is still an isotope effect under these conditions, and the factor $(m_{\ddagger}'/m_{\ddagger})^{\frac{1}{2}}$ represents the ratio of the vibration frequencies. For hydrogen and deuterium this factor has a maximum value of $2^{\frac{1}{2}}$, and since $k^H/k^D$ often exceeds this value in practice it is clear that quantal effects are also operating.

The factor $(m_{\ddagger}/m'_{\ddagger})^{\frac{1}{2}}(f_{\ddagger}/f_{AH})$ in equation (160) is not directly identifiable with the ratio of the experimental Arrhenius pre-exponential factors, $A/A'$. This is because, as shown by (161), $f_{\ddagger}$ and $f_{AH}$ are temperature-dependent. For the same reason, the observed difference of activation energies, $E' - E$, is not exactly equal to the difference of zero-point energies $\Delta\varepsilon_0$, though in practice these two quantities may differ only slightly.

Equation (163) is of very general validity, being dependent only on the following assumptions, of which the first two are also implicit in equation (137) for equilibria:

(a) The vibrations have been assumed simple harmonic. This will not cause any serious error unless the temperature is high, and such errors will partially cancel when comparing the two isotopes.

(b) The limiting value has been taken for the rotational partition functions (cf. Equation 132). This assumption will fail only at temperatures far below room temperature, and then only for molecules with low moments of inertia, i.e., simple hydrides.

(c) The transmission coefficient $\kappa$ has been assumed to be the same for the two isotopes. This is a classical effect depending on the shape of the energy surface and is not to be confused with the 'tunnel effect' treated later in this chapter. It cannot be calculated without a knowledge of the energy surface but will not be much less than unity for simple reactions. Although few calculations have been made, it seems intuitively probable that it will have closely the same value for the hydrogen isotopes, since these are considerably lighter than the atoms to which they are attached.

(d) The passage of the proton through the transition state has been treated by classical mechanics. This is much more doubtful than the three preceding assumptions, and we shall return to it in connection with the tunnel effect. However, we shall follow common practice in retaining this assumption for the present.

The above treatment was first given by Bigeleisen.[9] For heavier isotopes, such as those of carbon, it is convenient to write $m' = m + \Delta m$ and to expand the expressions in powers of $\Delta m/m$, but this is rarely advantageous for the isotopes of hydrogen.

In the problem considered here many of the vibrations of AH or the transition state will be little affected by the mass of the atom transferred, and the corresponding terms in (161) or (163) will be equal to unity. A realistic assumption is that the only changes involved in the isotope effect

[9] J. Bigeleisen, *J. Chem. Phys.*, **17**, 675 (1949).

are the disappearance of the A—H stretching frequency and a change in the frequencies of two bending vibrations. Equation (160) then becomes

$$-\frac{k}{k'} = \left(\frac{m'_{\ddagger}}{m_{\ddagger}}\right)^{\frac{1}{2}} \prod_3 \frac{u'(1-e^{-u})}{u(1-e^{-u'})} \prod_2 \frac{u_{\ddagger}(1-e^{-u_{\ddagger}^{\ddagger}})}{u'_{\ddagger}(1-e^{-u_{\ddagger}^{\ddagger}})} e^{\Delta\varepsilon_0/kT} \tag{164}$$

where $u$ and $u_{\ddagger}$ refer to the initial state and the transition state respectively. In comparing hydrogen and deuterium we have very nearly $(m'_{\ddagger}/m_{\ddagger})^{\frac{1}{2}} = u/u' = u_{\ddagger}/u'_{\ddagger}$. If it can be assumed that $e^{-u} \ll 1$ throughout, then (164) reduces to the simple form

$$k/k' = e^{\Delta\varepsilon_0/kT}. \tag{165}$$

This assumption is certainly justified for the initial state near room temperature and probably also applies to the transition state. However, since we have no real knowledge of the latter, it should be noted that the extreme assumption $u_{\ddagger} \ll 1$ lead to $k^{H}/k^{D} = \frac{1}{2}e^{\Delta\varepsilon_0/kT}$. Intermediate values are obviously also possible for the ratio of pre-exponential factors. At sufficiently high temperatures we have $u \ll 1$, $u_{\ddagger} \ll 1$, giving a limiting value of $k^{H}/k^{D} = (m'_{\ddagger}/m_{\ddagger})^{\frac{1}{2}} = 2^{\frac{1}{2}}$, though this is not really relevant for proton-transfer reactions in solution.

Essentially the same simplified model has been used by Swain and his collaborators to relate the primary isotope effects caused by deuterium and tritium substitution.[10] They derive the expression

$$k^{H}/k^{T} = (k^{H}/k^{D})^{1.422} \tag{166}$$

which is often referred to as the Swain or Swain–Schaad relation. Consideration of a more general model[11] gives a very similar result; the exponent in (166) was found to have the extreme values 1.33 and 1.58, but to be close to 1.44 for large primary effects at ordinary temperatures.

It is rarely possible to apply the full theoretical expressions such as (160) and (163) to actual reactions, since the required transition-state frequencies are not accessible experimentally, and can be calculated theoretically only for the very simplest systems, such as those involving three hydrogen atoms. However, much valuable information has been obtained by applying these equations to plausible model systems in which force constants and configurations can be varied systematically. Since computer programmes are available[12] for calculating vibration frequencies even in

[10] C. G. Swain, E. C. Stivers, J. F. Reuwer, and L. J. Schaad, *J. Am. Chem. Soc.*, **80**, 5885 (1958).
[11] J. Bigeleisen, *Tritium in the Physical and Biological Sciences*, I.A.E.H. Vienna, **1**, 161 (1962).
[12] J. H. Schachtschneider and R. G. Snyder, *Spectrochim. Acta*, **19**, 117 (1963).

relatively complicated systems, a wide variety of 'computer experiments' can be carried out, in particular to test the validity of various approximations, and this has been done especially by Wolfsberg and M. J. Stern and their collaborators.[13-18] Only a few of their conclusions will be summarized here.

A conclusion of general importance is embodied in the so-called *cut-off procedure*,[13,14] in which it was shown that no appreciable errors are involved in omitting from the calculation parts of molecules separated by more than two bonds from positions of isotopic substitution at which force-constant changes are occurring; this justifies the application of simple models even to reactions involving complicated molecules. Moreover, it turns out that several of the simplifications introduced above are essentially correct for reactions involving transfer of protons or hydrogen atoms at ordinary temperatures. Thus although the temperature variation of $k/k'$ may show anomalies (inflections, maxima and minima, and changes from values greater than unity to values less than unity) for small isotope effects involving pairs like $^{13}C-^{14}C$ or $^{16}O-^{18}O$,[19-21] these anomalies are absent at all temperatures for primary hydrogen isotope effects for which $k^H/k^D$ is greater than about 2.7 at 300 K, the observed activation energy being primarily determined by the change in zero-point energy. Similarly, consideration of a variety of model reactions[18] showed that the ratio of observed pre-exponential factors $A^H/A^D$ was always between 0.7 and 1.2 for a temperature range 20–2000 K, and had an absolute minimum value of 0.5; this is consistent with the limits $0.5 < A^H/A^D < 2^{\frac{1}{2}}$ derived from very simple considerations.

It is convenient to summarize here the rather meagre general results of a theoretical consideration of the reaction $AH + B \rightleftharpoons A + HB$, as follows:

(1) The major contribution to the kinetic isotope effect lies in the loss of the zero-point energy associated with the original A—H stretching frequency, since the contributions of bending frequencies tend to cancel

[13] M. Wolfsberg and M. J. Stern, *Pure Appl. Chem.*, **8**, 225, 325 (1964).
[14] M. J. Stern and M. Wolfsberg, *J. Chem. Phys.*, **39**, 2776 (1963); **45**, 2618, 4105 (1966); *J. Pharm. Sci.*, **54**, 849 (1965).
[15] M. J. Stern, M. E. Schneider, and P. C. Vogel, *J. Chem. Phys.*, **55**, 4286 (1971).
[16] P. C. Vogel and M. J. Stern, *J. Chem. Phys.*, **54**, 779 (1971).
[17] M. J. Stern and P. C. Vogel, *J. Am. Chem. Soc.*, **93**, 4664 (1971).
[18] M. E. Schneider and M. J. Stern, *J. Am. Chem. Soc.*, **94**, 1517 (1972).
[19] M. J. Stern, W. Spindel, and E. V. Monse, *J. Chem. Phys.*, **48**, 2908 (1968).
[20] E. V. Monse, W. Spindel, and M. J. Stern, *Adv. Chem. Ser.*, **89**, 148 (1969).
[21] T. T. S. Huang, W. J. Kass, W. E. Buddenbaum, and P. E. Yankwich, *J. Phys. Chem.*, **72**, 4431 (1968).

out as between the initial and transition states, while stretching frequencies in the transition state are likely to have lower and less isotopically sensitive frequencies.

(2) Most of the isotope effect should reside in a difference in activation energies rather than in the pre-exponential factor $A$. Limits for $A^H/A^D$ are 0.5 and $2^{\ddagger}$, and values much closer to unity are to be expected.

(3) For many reactions $k^H/k^D$ will be approximately equal to the values of $\exp(\Delta E_0/RT)$ in Table 23; however, these should be close to the maximum values for a given type of bond, and considerably lower values may be found for some reactions.

(4) When all three isotopes of hydrogen can be compared, the relation $k^H/k^T = (k^H/k^D)^{1.442}$ should be approximately valid.

We shall now consider some of the experimental material available for comparison with theory. Just as for equilibria, we can distinguish between *primary*, *secondary*, and *solvent isotope effects*, of which only the first is directly relevant to the theoretical treatment given above. In contrast to proton-transfer equilibria, it is often possible to study the pure primary kinetic effect for a proton transfer even in hydroxylic solvents such as water. This is because we are often concerned with proton transfer from a group such as C—H, which does not exchange rapidly with the solvent; thus it is quite simple to compare the rates of the reactions $CHMe_2NO_2 + OH^- \rightleftharpoons [CMe_2NO_2]^- + H_2O$ and $CDMe_2NO_2 + OH^- \rightleftharpoons [CMe_2NO_2]^- + HDO$, both in $H_2O$ as a solvent.* However, this experimental separation of primary and solvent effects is not always feasible. Thus rapid exchange will obviously make it impossible to compare $CH_3CO_2H$ and $CH_3CO_2D$ as acid catalysts in the same aqueous medium; similarly, the decomposition of nitramide-$d_2$, $ND_2NO_2$, cannot be studied in $H_2O$, since both hydrogen atoms exchange with the solvent by a reaction which is much faster than that leading to decomposition

---

* The species HDO will of course exchange rapidly with the solvent or its ions, so that when the second reaction has proceeded to an appreciable extent in $H_2O$ we shall also have the possibility of the reaction $CDMe_2NO_2 + OD^- \rightarrow [CMe_2NO_2]^- + D_2O$. However, provided that the solutions are reasonably dilute, this last reaction can be neglected because the species $H_2O$ is present in very large excess. Even in dilute solutions the situation is simple only if the reaction being studied is effectively irreversible. Thus if we are attempting to study the reaction $CDMe_2NO_2 + H_2O \rightarrow [CMe_2NO_2]^- + H_2DO^+$ in $H_2O$, there will be a progressive build-up of the species $CHMe_2NO_2$ because of the reverse reaction $[CMe_2NO_2]^- + H_3O^+ \rightarrow CHMe_2NO_2 + H_2O$, so that if the reaction is being followed by monitoring the concentration of the anion, the observed rate will include an increasing contribution from the reaction of $CHMe_2NO_2$ with $H_2O$. It is therefore often necessary to render the reaction irreversible by adding a reagent (commonly called a 'scavenger') which will remove the primary product as soon as it is formed.

(cf. p. 161 ). The same limitation applies to catalysed reactions of the type $>C{=}O + ROH \rightleftharpoons >C(OH)OR$, including the mutarotation of glucose and similar substances (cf. Chapter 9, pp. 183–190), since all the hydrogens attached to oxygen exchange rapidly.

There is little direct information about the magnitude of secondary kinetic isotope effects in proton-transfer reactions. Model calculations[13,14] predict that in the absence of changes in force constants the effect of deuterium substitution on atoms adjacent to the reaction site should not exceed 1–2%. Moderately large effects have been observed when several atoms are substituted. Thus the transfer of a proton from the ketones

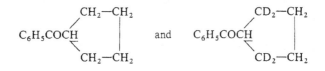

to acetate ions in 90% acetic acid has $k^H/k^D = 1.24$, although the same proton (underlined above) is being removed in each case.[22] Similarly, a comparison of the rates of the reactions $(CH_3)_2CHNO_2 + OH^-$ and $(CD_3)_2CHNO_2 + OH^-$ in $H_2O$ at 25°C gives $k^H/k^D = 1.14$.[23] A particular kind of secondary isotope effect arises when several hydrogens are attached to the same atom. For example, the removal of $D^+$ from $>CD_2$ is not strictly equivalent to its removal from $>CHD$, so that a comparison of rates of ionization of $>CH_2$ and $>CD_2$ involves both primary and secondary effects. A secondary effect of this kind is equivalent to a departure from the rule of the geometric mean (p. 238 ), and should therefore be small. Experimental values of 1.15 and 1.18 have been given for the ionization of toluene[24] and of 2-nitropropane,[25] though these values are subject to considerable uncertainty. Since these secondary effects are considerably smaller than primary ones, it is customary to neglect them when considering the latter.

The simplest reactions for comparing the above conclusions with experiment are naturally those involving a single rate-determining proton transfer, but for reactions involving two consecutive steps interesting information can sometimes be obtained by studying the effect of reaction conditions upon the isotope effect, quite apart from the absolute magnitude

[22] W. D. Emmons and M. F. Hawthorne, *J. Am. Chem. Soc.*, **78**, 5593 (1956).
[23] M. H. Davies, to be published (1972).
[24] A. Streitwieser and D. E. van Sickle, *J. Am. Chem. Soc.*, **84**, 254 (1962).
[25] R. P. Bell and D. M. Goodall, *Proc. Roy. Soc.*, A, **294**, 273 (1966).

of the latter. One example has already been quoted (p.192 ); the diazo-coupling reaction between 1-naphthol-3-sulphonate and *p*-chlorodiazo-benzene.[26] This reaction exhibits general catalysis by bases (e.g., pyridine), but there is a non-linear dependence of the rate upon base concentration, and the isotope effect $k^H/k^D$ decreases markedly with increasing concentration of base. If the coupling reaction is written schematically as

$$ArL + X^+ \underset{k_1}{\overset{k_1}{\rightleftharpoons}} XArL^+$$

$$XArL^+ + B \underset{k_2}{\rightarrow} ArX + BL^+$$

where L may be either hydrogen or deuterium, reference to Equation (101) shows that a decrease of $k^H/k^D$ with increasing [B] implies that $k_{-1}$ and $k_2$[B] must be of similar magnitudes, since of the three velocity constants only $k_2$ will show any appreciable isotope effect. A similar situation arises in the iodination of *m*-nitrophenol[27] and the bromination of the *p*-methoxybenzenesulphonate ion.[28] In these reactions it is not practicable to vary the nature or concentration of the base, but if we write the reactions as

$$ArL + X_2 \rightleftharpoons XArL^+ + X^-$$

$$XArL^+ + H_2O \rightarrow ArX + H_2LO^+$$

where $X_2$ is a halogen, and any charge on ArL has been omitted, kinetic analysis shows that if the intermediate $XArL^+$ reverts to ArL and reacts to form ArX at comparable rates, then the isotope effect $k^H/k^D$ should increase with increasing concentration of the halide ion $X^-$. This is in fact observed for both reactions.* Other examples could be given, and a review by Zollinger[29] gives a general account of hydrogen isotope effects in aromatic substitution reactions.

Returning to single-stage proton-transfer reactions, we have seen in Chapter 9 that *proton transfer from carbon acids to bases* is the rate-determining step in many reactions of these compounds, and during the last fifteen years or so many studies of isotope effects have been carried

[26] M. Christen and H. Zollinger, *Helv. Chim Acta*, **45**, 2057 (1962).
[27] E. Grovenstein and N. S. Aprahamian, *J. Am. Chem. Soc.*, **84**, 212 (1962).
[28] B. T. Baliga and A. N. Bourns, *Canad. J. Chem.*, **44**, 379 (1966).
*The above description omits two further complications: (a) the dissociation of *m*-nitrophenol, which varies with pH, and (b) the formation of trihalide ions in the equilibria $X_2 + X^- \rightleftharpoons X_3^-$. Both of these are of course taken into account in the original treatments.
[29] H. Zollinger, *Adv. Phys. Org. Chem.*, **2**, 163 (1964).

out, especially by F. A. Long, J. R. Jones, R. P. Bell, and their collaborators. The experimental material is very extensive, and only a few references will be given here. In some instances it is possible to study the ionization process directly, for example, in the reaction of $CH_3NO_2$ and $CD_3NO_2$ with hydroxide ions, but more frequently the rate of formation of the anion is determined indirectly, for example, by using a reactive species such as a halogen to remove it as soon as it is formed, or by studying the racemization or mutarotation of optically active compounds. Rates of isotope exchange may also be used, most conveniently by using proton magnetic resonance measurements for monitoring protium species, or radioactive counting for those containing tritium.

Most of the results are consistent with theoretical expectations, in that $k^H/k^D$ at ordinary temperatures usually lies between 3 and 7 (or the corresponding limits of about 7 and 12 for $k^H/k^T$). However, on closer examination two rather surprising features emerge. In the first place, it is fairly common to find values of $k^H/k^D$ of around 10 at 298 K, i.e., considerably greater than the supposedly maximum value of 7 given in Table 23. In a few instances even larger effects are encountered, the highest being $k^H/k^D = 23$ for proton transfer from 2-nitropropane to 2,4,6-trimethylpyridine,[30] together with the corresponding value $k^H/k^T = 79$. In the second place, there are unexpectedly large variations in $k^H/k^D$ for rather small structural changes in either the base or the acid. These two 'anomalies' may have a common cause, and we shall begin by considering the second one.

Table 25 illustrates the variation in isotope effect produced by variations in the nature of the base in proton abstraction from three carbon acids.[31,32] Most of the rates were measured by using bromine or iodine to scavenge the anion as soon as it is formed, but in the reactions of ethyl nitroacetate with the four strongest bases the appearance of anion could be followed directly by its ultraviolet absorption, since these reactions are thermodynamically favourable. For each carbon acid there is a large variation in $k^H/k^D$ as the base is varied, and for the first two compounds this is in the sense of an increasing isotope effect as the base becomes stronger. For these last reactions the anion of the carbon acid is nearly always stronger than the base with which it reacts, so that the reactions studied are all 'uphill'. This is conveniently expressed in terms of the quantity

[30] E. S. Lewis and L. H. Funderburk, *J. Am. Chem. Soc.*, **89**, 2322 (1967); E. S. Lewis and J. K. Robinson, *J. Am. Chem. Soc.*, **90**, 4337 (1968).
[31] R. P. Bell and J. E. Crooks, *Proc. Roy. Soc.*, A, **286**, 285 (1965).
[32] D. J. Barnes and R. P. Bell, *Proc. Roy. Soc.*, A, **318**, 421 (1970).

Table 25. ISOTOPE EFFECTS IN THE IONIZATION OF CARBON ACIDS IN WATER AT 25°C

$k_B^H$ and $k_B^D$ = velocity constants for transfer of protons and deuterons respectively, $dm^3 \, mol^{-1} \, s^{-1}$

$\Delta pK = pK_{SH} - pK_{BH}$
$\Delta pK'$ = statistically corrected value of $\Delta pK$

| Base | $k_B^H$ | $k_B^H/k_B^D$ | $\Delta pK'$ |
|------|---------|---------------|--------------|
| *Ethyl α-methylacetoacetate*, $pK_{SH} = 12.7$ | | | |
| Water | $1.14 \times 10^{-5}/55.5$ | 3.8 | 14.3 |
| Dichloroacetate | $3.57 \times 10^{-5}$ | 3.9 | 11.7 |
| Monochloroacetate | $3.00 \times 10^{-4}$ | 5.2 | 10.1 |
| β-Chloropropionate | $1.76 \times 10^{-3}$ | 5.7 | 8.9 |
| Acetate | $3.23 \times 10^{-3}$ | 5.9 | 8.3 |
| Trimethylacetate | $5.75 \times 10^{-3}$ | 6.4 | 8.0 |
| Hydrogen phosphate | $3.95 \times 10^{-2}$ | 6.3 | 5.6 |
| | | | |
| *Sodium propan-2-one-1-sulphonate*, $pK_{SH} = 13.6$ | | | |
| Water | $2.01 \times 10^{-5}/55.5$ | 2.5 | 16.0 |
| Monochloroacetate | $1.39 \times 10^{-5}$ | 2.6 | 11.5 |
| Acetate | $4.48 \times 10^{-4}$ | 3.8 | 9.6 |
| Trimethylacetate | $4.78 \times 10^{-4}$ | 4.4 | 9.4 |
| 2,6-Lutidine | $3.99 \times 10^{-3}$ | 7.3 | 7.4 |
| Hydroxide | $2.60 \times 10^{+2}$ | 7.4 | $-2.0$ |
| | | | |
| *Ethyl nitroacetate*, $pK_{SH} = 5.8$ | | | |
| Water | $1.57 \times 10^{-2}/5.55$ | 3.6 | 8.0 |
| Monochloroacetate | 0.92 | 6.6 | 3.5 |
| Acetate | 13.5 | 7.7 | 1.6 |
| 2-Picoline | 78 | 9.6 | 0.1 |
| 4-Picoline | 129 | 9.1 | 0.1 |
| 2.6-Lutidine | 207 | 9.9 | $-0.6$ |
| 2-Chlorophenoxide | $8.7 \times 10^3$ | 8.1 | $-2.3$ |
| Phenoxide | $2.9 \times 10^4$ | 6.7 | $-3.9$ |
| Hydroxide | $1.5 \times 10^5$ | 4.6 | $-10.0$ |

$\Delta pK = pK_{SH} - pK_{BH}$, where SH is the carbon acid and B the base with which it reacts; $\Delta pK$ is positive for all the reactions listed of ethyl α-methylacetoacetate and sodium propan-2-one-1-sulphonate (except for the reaction of the latter with hydroxide ion), and $k^H/k^D$ increases with decreasing $\Delta pK$. However, a new factor emerges if we consider the much stronger carbon acid ethyl nitroacetate, for which both positive and negative values of $\Delta pK$ (i.e., both uphill and downhill reactions) are accessible. Here there is clear evidence of a maximum isotope effect in the neighbourhood of $\Delta pK = 0$, with a decrease for reactions which are either markedly uphill or markedly downhill.

All the reactions in Table 25 refer to the solvent $H_2O$, so that no allowance need be made for solvent isotope effects. No correction has been applied for secondary isotope effects, since we have seen that these are small; on the other hand, a statistical correction has been applied to the value of $\Delta pK$. It is unlikely that $k^H/k^D$ will be affected by changes in free energy which depend only on statistical differences, and it is more logical to relate it to a statistically corrected $\Delta pK$ defined by

$$\Delta pK' = \Delta pK + \lg(p_{SH}q_B/p_{BH}q_S), \tag{167}$$

where $p$ and $q$ are the statistical factors already considered in connection with relations between rates and equilibrium constants (p.198 ).

Many other examples could be cited which show a smooth variation of $k^H/k^D$ with the strength of the base, and in a few cases some indication of a maximum isotope effect when $\Delta pK$ (and hence the standard free energy change of the reaction) is close to zero. Such a maximum may well be concealed in the results for sodium propan-2-one-1-sulphonate in Table 25, since $k^H/k^D$ has almost the same value for reaction with 2,6-lutidine and hydroxide, for which $\Delta pK$ is $+7.4$ and $-2.0$ respectively. One way of changing the value of $\Delta pK$ is to modify the nature of the solvent, and in particular the addition of dimethyl sulphoxide to aqueous solutions containing hydroxide ions will displace the equilibrium $SH + OH^- \rightleftharpoons S^- + H_2O$ to the right because the solvation of $OH^-$ is reduced more than that of $S^-$. It is thus of interest that when the hydroxide ion-catalysed inversion of $(-)$-menthone is studied in mixtures of water and dimethyl sulphoxide, the isotope effect has a maximum value in a solvent containing 30–40 mol% of dimethyl sulphoxide;[33] indicator measurements suggest that the hydroxide ion and the anion of menthone have similar basic strengths in solvents of this composition. Similarly, for the reaction of nitroethane with hydroxide ions the value of $k^H/k^D$ falls from 9.3 to 5.9 as the mole fraction of dimethyl sulphoxide increases from zero to 0.58, though in this case the reaction is thermodynamically favourable in all the solvents studied, and no maximum is observed.[34]

The position is less clear-cut as regards the effect on $k^H/k^D$ of varying the nature of the carbon acid. The same situation was encountered in Chapter 10 in connection with relations between rate and equilibrium constants, and it was pointed out there that different carbon acids often contain different mesomeric systems, and in any case involve substitution

[33] R. P. Bell and B. G. Cox, *J. Chem. Soc.*, B, 194 (1970).
[34] R. P. Bell and B. G. Cox, *J. Chem. Soc.*, B, 783 (1971).

close to the site of reaction. Nevertheless, an overall pattern appears when all the results are considered together. Figure 21 shows all the available results at 25° for carbonyl and nitro-compounds whose p$K$ values are known with reasonable certainty reacting with a variety of bases in aqueous solution.* In spite of the large scatter there is clear evidence of a maximum in $k^H/k^D$ in the neighbourhood of $\Delta pK' = 0$.

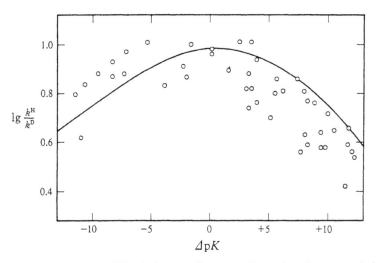

Fig. 21. Dependence of kinetic isotope effect on $\Delta pK$ (statistically corrected) for reactions of carbon acids with bases. (Values from Refs. 31–36.)

We have seen in Chapter 10 that for a series of similar bases (and to a lesser extent for a series of similar carbon acids) there is a relation between the reaction rate and $\Delta pK'$, so that $k^H/k^D$ should also vary smoothly with the observed velocity in a series of similar reactions. The velocities are always accessible, even when the values of $\Delta pK'$ are not, and this type of correlation has often been attempted. It does, however, break down badly when carbon acids of very different structures are compared; for example, when acetate ions remove protons from tricarbomethoxymethane and 2-nitropropane both $\Delta pK'$ and $k^H/k^D$ have similar values, but the actual velocities differ by five powers of ten. It was also shown in Chapter 10 that $\beta$, the exponent of the Brönsted relation, often correlates with the

*For sources, see Refs. 31–36. The line drawn in Figure 21 represents theoretical predictions which will be discussed later.
[35] R. P. Bell and D. M. Goodall, *Proc. Roy. Soc.*, A, **294**, 273 (1966).
[36] R. P. Bell, W. H. Sachs, and R. L. Tranter, *Trans. Faraday Soc.*, **67**, 1995 (1971).

free energy change in the reaction and might therefore be related to the magnitude of the isotope effect; in particular, the Marcus equation (121) predicts that $\beta = \frac{1}{2}$ when $\Delta G° = 0$. There is some evidence[37] of a correlation between $k^H/k^D$ and $\beta$, but reliable experimental information is sparse, and it appears that $\Delta pK'$ (or $\Delta G°$, statistically corrected) is the most useful experimental reaction parameter for comparing with isotope effects.

We now turn to the theoretical interpretation of these variable isotope effects, and in particular the occurrence of a maximum effect when $\Delta G°$ is close to zero. Since this last statement implies some kind of symmetry in the system, all theoretical treatments of the problem have involved the *degree of symmetry* or the *extent of proton transfer* in the transition state, concepts which have already been mentioned (p. 216). The most commonly accepted explanation was first put forward by Westheimer[38] and has been elaborated by a number of authors. Most of these[39-41] adopt a linear three-centre model $A\cdots H\cdots B$ for the transition state and consider only motion along the line of centres. For small displacements the potential energy of this system can be written

$$2\Delta V = k_1(\Delta r_{AH})^2 + k_2(\Delta r_{BH})^2 + 2k_{12}\,\Delta r_{AH}\Delta r_{BH} \tag{168}$$

where $\Delta$ signifies the departure of each quantity from its equilibrium value. The usual treatment for normal vibrations yields a quadratic equation for the characteristic vibration frequencies $v_1$ and $v_3$* which are given by

$$16\pi^4 v_1^2 v_3^2 = \frac{m_A + m_H + m_B}{m_A m_H m_B}(k_1 k_2 - k_{12}^2)$$

$$4\pi^2(v_1^2 + v_3^2) = \frac{k_1}{m_B} + \frac{k_2}{m_A} + \frac{k_1 + k_2 - 2k_{12}}{m_H}. \tag{169}$$

If $A\cdots H\cdots B$ were a stable triatomic molecule, both $v_1$ and $v_3$ would have real values, corresponding respectively to the 'symmetrical' and 'unsymmetrical' stretching vibrations. However, for a transition state the

[37] R. P. Bell, *Disc. Faraday Soc.*, **39**, 16 (1966).
[38] F. H. Westheimer, *Chem. Rev.*, **61**, 265 (1961).
[39] J. Bigeleisen, *Pure Appl. Chem.*, **8**, 217 (1964).
[40] A. V. Willi and M. Wolfsberg, *Chem. and Ind.*, 2097 (1964).
[41] W. J. Albery, *Trans. Faraday Soc.*, **63**, 200 (1967).
*These are termed $v_1$ and $v_3$ so as to reserve $v_2$ for the bending frequency in conformity with the usual spectroscopic nomenclature.

unsymmetrical motion $\overleftarrow{A}\cdots\overrightarrow{H}\cdots\overleftarrow{B}$ represents the reaction co-ordinate, corresponding to a maximum rather than a minimum in potential energy, and the corresponding root of (169) would represent an imaginary frequency, usually written as $iv_3$. It is clear from (169) that the condition for a transition state rather than a stable (hydrogen-bonded) system is $k_{12} > (k_1 k_2)^{\frac{1}{2}}$.

The relevance of this treatment to the problem of isotope effects lies in the sensitivity of the 'symmetrical' frequency to the mass of the central atom. It follows from the equations (and is intuitively obvious) that in a completely symmetrical situation with $k_1 = k_2$ and $m_A = m_B$ the value of $v_1$ will be independent of $m_H$, since the central atom will remain stationary. In a less symmetrical situation the central atom will move, $v_1$ will be dependent upon $m_H$, and the corresponding zero-point energy will be less for deuterium than for hydrogen (cf. Figure 20, p. 251), thus reducing the observed isotope effect below the value found in the symmetrical case.

This analysis is undoubtedly qualitatively correct, but it remains open to question whether the contributions of $v_1$ to the isotope effect are large enough to explain the considerable variations in $k^H/k^D$ which are observed for moderate changes in the free energy of reaction (cf. Table 25). Westheimer[38] calculated that a system in which $k_1 = 10k_2$ (or $k_2 = 10k_1$) would show an isotope effect much lower than the maximum value, and similar conclusions have been reached in more sophisticated calculations.[39,40] Apart from the fact that a ratio of 10 is rather extreme, all these calculations made the simplifying assumption $k_{12} = (k_1 k_2)^{\frac{1}{2}}$; this gives $v_3 = 0$, corresponding to zero curvature of the energy surface along the reaction path as it passes through the transition state. This is not physically plausible, and if we make the more realistic assumption that the energy surface has similar curvatures (positive or negative) in different directions, a different picture emerges:[37] even when $k_1 = 10k_2$ the value of $v_1$ is low and only slightly affected by isotopic substitution, so that the observed isotope effect is only slightly lower than for the symmetrical case. The interplay of the factors $k_1$, $k_2/k_1$, and $k_{12}/(k_1 k_2)^{\frac{1}{2}}$ is well shown in an analysis by Albery,[41] in which the isotope effect is calculated for independent variations of these quantities.

In all the papers quoted so far, the force constants $k_1$, $k_2$, and $k_{12}$ (or their ratios) are regarded as disposable parameters, and no attempt is made to relate them to observable or calculable properties of the initial or final states, such as the free energy change in the reaction. The analysis has been carried further in several respects by More O'Ferrall and his collaborators.[42,43] In the first place both linear and non-linear transition states were considered. Secondly, the three-centre model was extended to one of four or five centres, thus making it possible to include on a rational basis the bending vibrations of both the initial and transition states. Thirdly, the force constants in the transition state were correlated with those of the initial state through bond lengths and bond orders with the help of arbitrary but reasonable assumptions about the relations between these three quantities and the assumption that the total order of the bonds to hydrogen is unity; similar correlations of the imaginary frequency $iv_3$ with bond order led to values for $k_{12}$. This treatment again predicts a maximum isotope effect for an approximately symmetrical transition state, but the disposable parameters are now the bond orders, which were not related to any observable characteristics of the reactions. In particular, it might be doubted whether the range of velocities accessible in practice really corresponds to the rather large variations in bond order required to account for the observed variations in $k^H/k^D$.* In a somewhat similar treatment Willi[45] has attempted to relate the bond orders to estimates of the energy required to remove a proton completely from SH or BH. This involves a number of questionable assumptions, but it is interesting to note that his conclusions support the last statement. For 14 reactions of carbon acids with bases, including several of those listed in Table 25, he deduces bond orders varying between 0.48 and 0.88, but the predicted isotope effects vary only between $k^H/k^D = 7.0$ and $k^H/k^D = 7.3$, compared with the observed variations between 3.5 and 10.3.

It is therefore of interest to calculate isotope effects for proton-transfer reactions in terms of a model which will also yield other observable characteristics of the system. In principle such a model should yield the following properties of the initial, final and transition states: internuclear distances, energies, and frequencies of both stretching and bending variations. (For the transition state it will also give the imaginary frequency $iv_3$, or the curvature of the energy barrier, which will be shown later to be important in connection with the tunnel effect.) The frequencies

[42] R. A. More O'Ferrall and J. Kouba, *J. Chem. Soc.*, B, 985 (1967).
[43] R. A. More O'Ferrall, *J. Chem. Soc.*, B, 785 (1970).
* It might be argued that the concept of bond orders is more appropriate to reactions involving transfer of hydrogen atoms rather than of protons. It is therefore of interest that a study of 17 reactions between free radicals and thiols[44] does suggest a maximum isotope effect for thermoneutral reactions, though this is less conclusively shown than for proton transfers.
[44] W. A. Pryor and K. G. Kneipp, *J. Am. Chem. Soc.*, **93**, 5584 (1971).
[45] A. V. Willi, *Helv. Chim. Acta*, **54**, 1220 (1971).

can then be used to calculate the isotope effect, while the energies give the energy change in the reaction and the activation energy, or at least the variation of these quantities with some parameter of the model.

A fundamental quantum-theoretical treatment is intractable, and has been attempted only for very simple systems.[46] Several electrostatic models have been proposed, and these may be particularly suitable for treatment the transfer of a proton. The simplest model[47] treated the motion of a proton between two negative point charges of varying magnitude. In order to represent the stretching vibrations of the transition state it is necessary to add a repulsive potential between the two centres, which was taken as $V = Ar_{AB}^{-m}$, with $m$ between 8 and 12. This model predicts[37] that even for extreme asymmetry such as $k_1/k_2 = 10$ the 'symmetrical' stretching vibration of the transition state makes a negligible contribution to the isotope effect; on the other hand, the contribution of the bending vibration is considerable, but varies very little with the symmetry of the transition state. Thus according to this model neither the stretching nor the bending vibrations of the transition state can account for the observed variations in $k^H/k^D$; however, it is not possible to relate calculated isotope effects to any properties of the reactants or products, since the point-charge model is not consistent with any equilibrium positions for the proton.

This last defect could be removed, somewhat artificially, by introducing a distance of closest approach of the proton to the negative charges, but it seems better to use a slightly more elaborate model which does not imply that the electron density at the proton is always zero. It has been shown by various authors[48] that the internuclear distances and vibration frequencies of diatomic hydrides are predicted surprisingly well by a model in which the proton moves in a rigid spherical electron distribution surrounding the nucleus. This model has been extended by Bader,[49] who treated hydrogen bonds and the transition states of proton-transfer reactions in terms of the motion of a proton in the superimposed electron distributions of two negative ions. In treating transition states Bader concentrates attention upon the bending vibrations and considers the negative charge clouds to be fixed relative to one another, so that no information is obtained about the 'symmetrical' stretching frequency. He also makes the un-

[46] C. D. Ritchie and H. F. King, *J. Am. Chem. Soc.*, **90**, 825, 833, 838 (1968).

[47] R. P. Bell, *Trans. Faraday Soc.*, **57**, 961 (1961).

[48] J. R. Platt, *J. Chem. Phys.*, **18**, 932 (1950); H. C. Longuet-Higgins and D. A. Brown, *J. Inorg. Nucl. Chem.*, **1**, 60 (1955); L. Salem, *J. Chem. Phys.*, **38**, 1227 (1963).

[49] R. F. W. Bader, *Canad. J. Chem.*, **42**, 1822 (1964).

realistic simplifying assumption $k_{12} = (k_1 k_2)^{\frac{1}{2}}$, and arrives at the conclusion that the isotope effect should have a *minimum* value for a symmetrical transition state, in disagreement with experiment. However, a more general treatment of the charge-cloud model[50] leads to different results. In this treatment a repulsive potential $V = Ar_{AB}^{-12}$ is added, which makes it possible to calculate both bending and stretching force constants, while the basic strength of one of the charge clouds is varied by varying its effective number of electrons, corresponding to calculable changes in $\Delta pK$. The stretching frequencies $v_1$ are in the range 500–700 cm$^{-1}$ (according to the details of the model), but they vary only slightly for a change of some 40 units in $\Delta pK$, and, most significantly, are unaffected (to within 1 cm$^{-1}$) by isotopic substitution; they can therefore make no detectable contribution to the observed isotope effect. The bending frequencies, $v_2$, are in the range 800–1100 cm$^{-1}$ for the hydrogen compounds, i.e., not far from 1225 cm$^{-1}$ observed[7] for the bending vibration of the ion $HF_2^-$. This frequency is considerably reduced by deuterium substitution, and therefore contributes to reducing the observed isotope effect. However, it varies only by a few per cent for a change of 40 units in $\Delta pK$, and the resulting variation in $k^H/k^D$ also amounts to only a few per cent, with no indication of a maximum near $\Delta pK = 0$. The smallness of these variations does of course reflect very small changes in the symmetry of the transition state, and in fact the change of 40 units in $\Delta pK$ corresponds to variations of only 15 pm in the position of the proton, and of only a few per cent in $k_1/k_2$, in contrast to the ten-fold variation of the latter quantity which is often postulated. It is thus clear that the explanation of variable isotope effects in terms of the zero-point energy of the transition state, although qualitatively plausible, is quite inadequate to account quantitatively for the experimental facts; this conclusion is strictly valid only for the model just described, but seems likely to stand for any reasonable description of the system.

It is natural, therefore, to look for some aspect of the problem which has been entirely omitted so far. This is almost certainly to be found in the so-called *tunnel effect*, which is in fact a logical consequence of the quantum theory, and has a bearing on a number of phenomena other than isotope effects.* The essence of the tunnel effect can be illustrated by

[50] R. P. Bell, W. H. Sachs, and R. L. Tranter, *Trans. Faraday Soc.*, **67**, 1995 (1971).
* The term 'tunnel effect' is somewhat misleading, since it seems to suggest a separate and special effect outside the framework of standard quantum theory, and from this point of view 'tunnel correction' would be a happier description. However, 'tunnel effect' has become a commonly used term, and we shall retain it in this chapter.

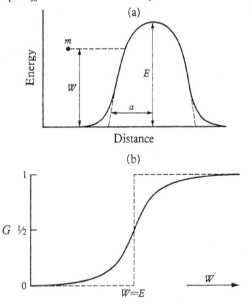

Fig. 22. Barrier permeabilities in classical and in quantum theory.

considering the behaviour of a particle of mass $m$ and energy $W$ moving towards an energy barrier of height $E$ as in Figure 22(a). If $G$ is the probability that the particle shall cross the barrier (often described as the *permeability* of the barrier), then classical mechanics predicts $G = 0$ for $W < E$ and $G = 1$ for $W > E$, corresponding to the broken line in Figure 22(b). According to quantum theory, on the other hand, the permeability is a continuous function of $W$, $G(W)$, as shown by the full line in Figure 22(b). The most striking contrast with classical behaviour lies in the finite probability predicted for $W < E$, which in classical mechanics would correspond to negative kinetic energies (or imaginary velocities) near the centre of the barrier, and it is this feature which has led to the name 'tunnel effect'. It is, however, equally inconsistent with the classical picture that $G < 1$ for values of $W$ a little greater than $E$, i.e., that some of the particles in this energy range should be reflected back.

The quantal result depends upon the wave-particle duality of matter, and there is a close optical analogue. If a ray of light inside a piece of glass strikes the surface at an angle greater than the critical angle, there will be total internal reflection. However, the position is altered if a second piece of glass is brought close to the first so that the width of the air gap between them is not very great compared with the wavelength of the light being used. As illustrated in Figure 23, there will then be a transmitted ray as well as a reflected one, and the intensity of the transmitted ray increases exponentially as the width of the gap decreases. This phenomenon is incomprehensible in terms of ray optics, or a corpuscular theory of light, but is predicted quantitatively by the wave theory.

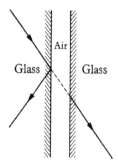

Fig. 23. Optical analogue of the tunnel effect.

The de Broglie relation assigns a wavelength $\lambda = h/mv$ to a particle of mass $m$ and velocity $v$, and we shall therefore expect the largest deviations from classical behaviour for particles of low mass. It is generally accepted that the motion of electrons on a molecular scale cannot be treated even approximately by classical mechanics, and one of the earliest applications of the tunnel effect was to the emission of electrons from metals in strong electric fields. It is found that the energies of the electrons emitted are much lower than the maximum potential energy in the region through which they have passed, and a quantitative theoretical treatment[51] predicts correctly the dependence of emission upon field strength. A similar treatment can be applied to the emission of $\alpha$-particles from radioactive nuclei. Although the mass of the $\alpha$-particle is much greater than that of the electron, corresponding to a shorter wavelength, the energy barrier surrounding the nucleus is a very narrow one (about $10^{-12}$ cm) so that the permeabilities are appreciable. A quantitative treatment leads to a relation between the half-life of the nucleus and the energy of the $\alpha$-particle emitted, which was originally established empirically under the name of the Geiger–Nuttall relation.[52]

In chemical reactions the passage of an electron through an energy barrier may sometimes be important in oxidation-reduction processes

[51] R. H. Fowler and L. Nordheim, *Proc. Roy. Soc.*, A, **119**, 173 (1928); L. Nordheim, *Proc. Roy. Soc.*, A, **121**, 626 (1928).
[52] See, e.g., G. Gamow, *Structure of Atomic Nuclei and Nuclear Transformations*, Oxford, 1937, Ch. 5.

which can be formulated as electron transfers,[53] but in most types of reaction the motion of nuclei is also involved. It is clear that appreciable deviations from classical behaviour can only be expected for light nuclei, notably protons, and calculation shows that for protons moving with thermal velocities at ordinary temperatures the wavelength $\lambda = \mathbf{h}/mv$ has values in the range $10^{-8}$–$10^{-9}$ cm. Since the barriers appropriate to chemical reactions have a total width of a few Ångström units, we may expect the tunnel effect to be of some importance in at least some proton-transfer reactions, especially at low temperatures. This was suggested at an early date by a number of authors,[54] though without any experimental evidence. More recently experimental evidence has been forthcoming, and there has been renewed interest in the part played by the tunnel effect in the transfer of protons or hydrogen atoms and in other phenomena. Reviews have been published by several authors,[55-57] and a series of papers by Christov[58] deal with various aspects of the subject.

It is, however, misleading to regard the tunnel effect as an optional or additional effect outside the framework of the usual treatment of reaction kinetics, since it has in fact just the same logical status as zero-point energy. Both of these phenomena depend on the quantum theory and can be related to the operation of the uncertainty principle for motion along one co-ordinate, the difference being that for the tunnel effect the co-ordinate is one in which the potential energy passes through a maximum, whereas for zero-point energy it passes through a minimum. It might even be anticipated that the two effects would be of the same order of magnitude in the transition state, and we shall see later that this is in fact the case. These considerations do not of course apply to the initial or final states of the system, in which the energy is at a minimum for any

[53] J. Weiss, *Proc. Roy. Soc.*, A, **222**, 128 (1954); R. J. Marcus, B. J. Zwolinski, and H. Eyring, *J. Phys. Chem.*, **58**, 432 (1954).
[54] F. Hund, *Z. Physik.*, **43**, 805 (1927); D. G. Bourgin, *Proc. Nat. Acad. Sci.*, **15**, 357 (1929); R. M. Langer, *Phys. Rev.*, **34**, 92 (1929); S. Roginsky and L. Rosenkewitsch, *Z. Phys. Chem.*, B, **10**, 47 (1930); E. Wigner, *Z. Phys. Chem.*, B, **19**, 203 (1932); R. P. Bell, *Proc. Roy. Soc.*, A, **139**, 466 (1933); C. E. H. Bawn and G. Ogden, *Trans. Faraday Soc.*, **30**, 434 (1934).
[55] H. S. Johnston, *Adv. Chem. Phys.*, **3**, 131 (1961).
[56] E. F. Caldin, *Chem. Rev.*, **69**, 135 (1969).
[57] M. D. Harmony, *Chem. Soc. Rev.*, **1**, 211 (1972).
[58] S. G. Christov, *Ann. Univ. Sofia Fac. Phys. Math.*, **42**, 69 (1945–1946); *C. R. Acad. Bulg. Sci.*, **1**, 43 (1948); *Z. Elektrochem.*, **62**, 567 (1958); **64**, 840 (1960); *Dokl. Akad. Nauk SSSR*, **125**, 141 (1959); **136**, 663 (1960); *Z. Physik. Chem.* (Leipzig), **212**, 40 (1959); **214**, 40 (1960); *Ber. Bunsengesell. Phys. Chem.*, **67**, 117 (1963); **76**, 507 (1972); *Electrochim. Acta*, **4**, 194, 306 (1961); **9**, 575 (1963); *Ann. Phys.*, **12**, 20 (1963); **15**, 87 (1965); *Disc. Faraday Soc.*, **39**, 60, 254, 263 (1965); *J. Res. Inst. Catalysis, Hokkaido Univ.*, **16**, 169 (1968); *Croat. Chem. Acta*, **44**, 67 (1972).

type of displacement, and in general the tunnel effect has no relevance for any equilibrium problem. However, in any kinetic treatment which is sufficiently refined to take into account the zero-point energy of the transition state it is not justifiable to neglect the tunnel effect.

In carrying out a quantitative treatment of the tunnel effect it is usual to regard it as a one-dimensional problem, as illustrated by Figure 22(a). There is no strict justification for thus separating out the reaction co-ordinate from other types of motion, but the error is probably similar to that involved in ignoring the interaction between different vibrational modes. For a particle of given mass and energy the value of the permeability depends not only on the height of the barrier, but also on its shape, and in particular on its curvature at the top. Since we have no detailed knowledge of the shape of the energy surface in the neighbourhood of the transition state, it is natural to approximate the true barrier by means of a parabola, as shown by the broken curve in Figure 22(a). The discrepancy at the base of the curve will not be important unless the main contribution to the reaction velocity is made by particles of low energy, and we shall see that this is not likely to be the case in chemical problems, except perhaps at very low temperatures. The assumption of a parabolic barrier is analogous to the harmonic oscillator approximation for vibrations and gives correspondingly simple results.

The evaluation of the permeability is simple in principle, depending upon the conditions of continuity for the wave function in passing from one part of the energy diagram to another, and it is frequently unnecessary to obtain an explicit expression for the wave function. It is convenient to define the curvature of the barrier at the top by a frequency $v_t$ given by

$$v_t = E^{\frac{1}{2}}/\pi a(2m)^{\frac{1}{2}} \tag{170}$$

where $E$ and $a$ are the height and half-width of the parabola as shown in Figure 22(a). ($v_t$ is actually the frequency with which a particle of mass $m$ would vibrate in a parabolic potential well having the same curvature as the barrier, and this is often expressed symbolically by saying that motion in the reaction co-ordinate corresponds to an imaginary frequency $iv_t$.) The permeability of the barrier for a particle of energy $W$ is then given by the expression[59]

[59] R. P. Bell, *Trans. Faraday Soc.*, **55**, 1 (1959). In this paper expression (171) was regarded as an approximate one, since it was based on the Brillouin–Wentzel–Kramers (B.W.K.) approximate solution of the wave equation. However, in this particular case the result is exact [E. C. Kemble, *Fundamental Principles of Quantum Mechanics*, McGraw-Hill, New York, 1937, Ch. 3; D. L. Hill and J. A. Wheeler, *Phys. Rev.*, **89**, 1140 (1953)].

$$G(W) = \{1 + \exp[2\pi(E - W)/h\nu_t]\}^{-1}. \qquad (171)$$

Equation (171) for $G(W)$ represents a curve like that in Figure 22(b), which is symmetrical about the point $W = E$, $G(W) = \frac{1}{2}$.

Equation (171) expresses the reaction probability for particles of a given energy. In a chemical reaction we are dealing with systems having a thermal distribution of energies, and in order to obtain an expression for the reaction velocity it is necessary to average $G(W)$ appropriately over all possible energies. The most convenient assumption (strictly true when the energy can be expressed as two classical square terms) gives a simple Boltzmann distribution for the energy. If the motion of the proton is treated classically, this leads to the expression $\exp(-E/kT)$ for the integrated reaction probability, and we can thus formulate $Q_t$, the tunnel-effect correction to the reaction velocity, as

$$Q_t = \exp(E/kT) \int_0^\infty \frac{1}{kT} \exp(-W/kT)G(W)dW \qquad (172)$$

with $G(W)$ given by (171). Equation (172) can be evaluated exactly, giving

$$Q_t = \frac{\frac{1}{2}u_t}{\sin \frac{1}{2}u_t} - u_t \exp\left(\frac{E}{kT}\right)\left(\frac{y}{2\pi - u_t} - \frac{y^2}{4\pi - u_t} + \frac{y^3}{6\pi - u_t} - \cdots\right) \qquad (173)$$

where $u_t = h\nu_t/kT$ and $y = \exp(-2\pi E/u_t kT)$
$$= \exp(-2\pi E/h\nu_t)*$$

Provided that $y \exp(E/kT) \ll 1$, which is often the case for chemical reactions at ordinary temperatures, only the first term of (173) need be retained, giving the simple result*

$$Q_t = \frac{1}{2}u_t/\sin \frac{1}{2}u_t \qquad (u_t = h\nu_t/kT). \qquad (174)$$

Equation (174) has several points of interest. It bears a remarkable formal resemblance to the quantum correction to the partition function for a real harmonic frequency in the transition state, which is

$$Q = \frac{1}{2}u/\sinh \frac{1}{2}u$$

---

*It was originally stated[59] that (173) is valid only if $u_t < 2\pi$, but it can be shown that it is in fact valid for all values of $u_t$. If $u_t$ is an integral multiple of $2\pi$ each of the terms of (173) becomes infinite, but their sum remains finite, so that $Q_t$ is always a continuous function of $u_t$. (Personal communication from Dr. I. Shavitt.)

*It should be noted that the mass $m$ which occurs in (170) and subsequent equations represents the reduced mass for motion along the reaction co-ordinate. It will, therefore, not always be equal to the mass of one of the hydrogen isotopes, but may also depend upon the masses of the other atoms and on how the reaction co-ordinate is formulated.

(cf. Equation 163), and can in fact be derived from it by replacing the real frequency $v$ by the imaginary one $iv_t$. If (174) is expanded in powers of $u_t$ we obtain

$$Q_t = 1 + \frac{u_t^2}{24} + \frac{7u_t^4}{5760} + \cdots \qquad (u_t < 2\pi) \qquad (175)$$

$$\ln Q_t = \frac{u_t^2}{24} + \frac{u_t^4}{2880} + \cdots \qquad (u_t < 2\pi) \qquad (176)$$

which are identical with the corresponding expansions for a harmonic oscillator[60] except that all the terms are positive instead of being alternately positive and negative. The term $u_t^2/24$ in (175) was derived by Wigner[54] on very general grounds as the first correction, assumed small, for the tunnel effect. It differs only in sign from the corresponding correction for zero-point energy in the transition state, and to this approximation the two corrections are of the same order of magnitude provided that the curvature of the energy surface in the direction of the reaction co-ordinate does not differ greatly from the curvatures in other directions, though this is no longer true when the corrections are not small.

There are some reservations about the strict applicability of the treatment outlined above to actual reactions. In the first place a parabolic energy barrier becomes unrealistic for configurations far removed from the transition state, and a more appropriate type of barrier is shown by the full curve in Figure 22(a), in which the broken curve represents a parabola. There is one potential energy function of this kind for which an explicit expression for the permeability can be obtained, commonly known as the Eckart barrier.[61] For a symmetrical barrier ($\Delta H = 0$) the equations are

$$V(x) = E/\cosh^2(\pi x/l) \qquad (177)$$

$$G(W) = \frac{\sinh^2(2\pi l \sqrt{2m\,W}/h)}{\sinh^2(2\pi l \sqrt{2m\,W}/h) + \cosh^2\left[\frac{1}{2}\pi\{(32ml^2 E/h^2) - 1\}^{\frac{1}{2}}\right]} \qquad (178)$$

where the barrier extends effectively between $x = -1$ and $x = +1$, with somewhat more complicated expressions for unsymmetrical barriers. It is not possible to integrate Equation (172) in terms of known functions, but numerical integrations have been carried out in several applications,[55,62,63] useful tables have been published,[64] and an approximate

[60] J. Bigeleisen, *Proceedings of International Symposium on Isotope Separation*, Amsterdam, 1958, p. 148.
[61] C. Eckart, *Phys. Rev.*, **35**, 1303 (1930).
[62] H. S. Johnston and D. Rapp, *J. Am. Chem. Soc.*, **83**, 1 (1961).
[63] T. E. Sharp and H. S. Johnston, *J. Chem. Phys.*, **37**, 1541 (1962).
[64] H. S. Johnston and J. Heicklen, *J. Phys. Chem.*, **66**, 532 (1962).

expression which is adequate for many chemical applications has been derived.[65] There is no reason to suppose that the Eckart equation will accurately express real energy barrier, and it is therefore of interest that numerical methods are available[66] for computing the permeability of a one-dimensional barrier of arbitrary form; these methods are valuable in the rare instances when the energy surface can be predicted theoretically, or for testing approximate solutions for model barriers. In the usual situation in which the true energy profile is unknown (which certainly includes all proton-transfer reactions) it is probably adequate for moderate tunnel corrections to use Equation (173) or (174) for a parabolic barrier, and to treat the value of $v_t$ as a parameter to be derived from experiment. This procedure is supported by the fact that motion along the reaction co-ordinate is only strictly separable from other motions of the system over the range in which the energy varies parabolically; this is analogous to the restriction that real vibrations can be strictly resolved into separable normal vibrations only to the harmonic approximation.

A more fundamental criticism of the simple treatment arises whenever the tunnel correction becomes large, i.e., whenever it is necessary to consider tunnelling through parts of the barrier which are appreciably below the energy maximum. Since the configuration of the reacting system needs at least two co-ordinates to describe it (for example, the distances $A\cdots H$ and $H\cdots B$ in a linear system $A\cdots H\cdots B$), the energy diagram must be a surface in at least three dimensions, and a curve such as that in Figure 20 or 22(a) represents only one possible route from reactants to products. When tunnelling is neglected these other possible routes are taken care of by the real stretching and bending vibrations of the transition state, but the problem of calculating tunnel corrections for more than one geometrical co-ordinate is a difficult one, even if we assume that the complete shape of the energy surface is known. There have been a number of attempts to estimate the error involved in using the one-dimensional model[62,67-69] but these do not agree even as to the sign of the error. In view of our lack of detailed knowledge of energy surfaces, even for the simplest reactions, it seems justifiable at present to continue to use the equations for tunnel corrections in one dimension.

[65] H. Shin, *J. Chem. Phys.*, **39**, 2934 (1963).
[66] R. J. Le Roy, K. A. Quickert, and D. J. Le Roy, *Trans. Faraday Soc.*, **66**, 2997 (1970).
[67] E. M. Mortensen and K. S. Pitzer, *Chem. Soc. Special Publ. No. 16*, 57 (1962).
[68] E. M. Mortensen, *J. Chem. Phys.*, **48**, 4029 (1968); **49**, 3526 (1968).
[69] D. G. Truhlar and A. Kuppermann, *J. Chem. Phys.*, **52**, 3841 (1970); *Chem. Phys. Letters*, **9**, 269 (1971).

We return now to the bearing of tunnel corrections on the problem of variable hydrogen isotope effects in proton-transfer reactions. The point-charge electrostatic model[47] predicts that the imaginary frequency $iv_3$ is related to the bending frequency $v_2$ by $v_3/v_2 = 2^{\frac{1}{2}}$. The more realistic charge-cloud model[50] predicts somewhat smaller values of $v_3$, but it remains larger than $v_2$ and varies only slightly with $\Delta G$ for the reaction; it is of course sensitive to isotopic substitution because of the occurrence of $m$ in (170). When tunnel corrections computed from (173) are included in the calculation of $k^H/k^D$ a very interesting result emerges. The isotope effect now varies considerably with $\Delta G$ (i.e., with $\Delta pK$) and passes through a clear maximum close to $\Delta G = 0$; the curve drawn in Figure 21 corresponds to that calculated from the model, and it is clear that the theory is well able to reproduce the kind of variation of $k^H/k^D$ observed in practice. This behaviour is not due to the small variations in $v_3$, but to the second term of (173), in which the barrier height $E$ refers only to that part of the barrier which lies above both the initial and final states. This is illustrated in Figure 24, which shows that the 'area' available for tunnelling is greater for a thermoneutral reaction than for either exothermic or endothermic ones. It is physically obvious that this kind of dependence

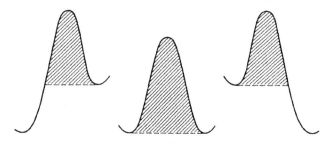

Fig. 24. Tunnelling regions as a function of the free energy change in the reaction. (The curves have been drawn so as to satisfy the Marcus equation (119), and the shaded areas show the region available for tunnelling.)

upon the symmetry of the reaction is bound to arise whenever the tunnel correction becomes considerable, independent of the exact model or the procedure used to calculate this correction. The present interpretation of variable isotope effects among chemically similar systems therefore seems preferable to any explanation in terms of real stretching frequencies of the transition state, which we have seen to be inadequate in its quantitative aspects.

Apart from the above problem, the introduction of tunnel corrections of appreciable magnitude has a number of other consequences for the theory of isotope effects and for the general interpretation of reactions involving the motion of light nuclei. These will now be described briefly, and it is useful to introduce the nomenclature $(X)_s$ to denote the semi-classical value of a quantity $X$, i.e., the value which would be predicted if the tunnel correction is neglected; in particular, the quantity $Q_t$ already introduced in equations (172)–(176) is equal to $k/(k)_s$. Much useful information comes from considering the temperature coefficients of reaction velocity, and this is conveniently expressed in terms of the Arrhenius activation energy and pre-exponential factor, $E_A$ and $A_A$ defined operationally by the equations

$$E_A = \mathbf{k}T^2 \, d \ln k/dT = -\mathbf{k} \, d \ln k/d(1/T) \tag{179}$$

$$\ln A_A = \ln k + E_A/\mathbf{k}T \tag{180}$$

from which it follows that

$$(E_A)_s - E_A = \mathbf{k}T^2 \, d \ln\{(k)_s/k\}/dT \tag{181}$$

$$\ln A - \ln(A)_s = \ln\{k/(k)_s\} + \{E_A - (E_a)_s\}/\mathbf{k}T. \tag{182}$$

Any theoretical expression or calculated numerical values for $Q_t = k/(k)_s$ can be inserted in the last two equations; thus the simple expression (174) yields

$$(E_A)_s - E_A = \mathbf{k}T(\tfrac{1}{2}u_t \cot \tfrac{1}{2}u_t - 1) \tag{183}$$

$$\ln A - \ln(A)_s = \ln(\tfrac{1}{2}u_t/\sin \tfrac{1}{2}u_t) + \tfrac{1}{2}u_t \cot \tfrac{1}{2}u_t - 1. \tag{184}$$

Equations (174), (183), and (184) lead to the following qualitative predictions, which are also valid for any theoretical treatment of moderate tunnel corrections:

(1) $E_A < (E_A)_s$, with the difference increasing with decreasing temperature. $(E_A)_s$ cannot be estimated independently, but since it should be almost independent of temperature (and close to the height of the energy barrier), the tunnel correction will appear as positive deviations at low temperatures from the Arrhenius equation applicable at higher temperatures. Although there are other causes which may contribute to such deviations,[70] in particular changes in mechanism, there are now several examples of low-temperature deviations for proton-transfer reactions which can almost certainly be attributed to the tunnel correction. Thus

[70] J. R. Hulett, *Chem. Soc. Quart. Rev.*, **18**, 227 (1964).

Caldin and his collaborators[71,72] studied the reaction of the trinitrobenzyl anion $[C_6H_2(NO_2)_3CH_2]^-$ with various weak acids in ethanol at temperatures between $-114°C$ and $+20°C$. The reaction with acetic acid showed appreciable positive deviations below $-90°$, reaching 45% at $-114°$, while hydrofluoric acid deviates appreciably at all temperatures below $-20°$, the reaction rate at $-90°$ being more than twice that extrapolated from measurements above $-20°$. The reaction of ethoxide ion with 4-nitrobenzyl cyanide in a mixture of ethanol and ether[73] shows deviations at $-124°C$ of similar magnitude, and we shall see later that an interpretation in terms of a tunnel correction is supported by the deuterium isotope effect. This method of detecting the tunnel effect is not often applicable to proton-transfer reactions, since it demands accurate measurements over a large temperature range extending to low temperatures.

(2) $A_A < (A_A)_s$, the difference again increasing with decreasing temperature. It seems at first sight unreasonable that the tunnel correction should lead to low rather than high values of $A$. The reason for this is illustrated in Figure 25, which shows the usual plot of $\lg k$ against $1/T$ with and without the tunnel correction. The value of $\lg A$ is equal to the extrapolated intercept at $1/T = 0$, and it will be seen that when this extrapolation is

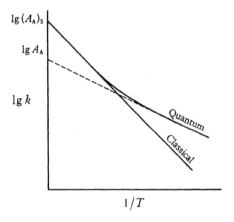

Fig. 25. Temperature variation of reaction velocity with and without the tunnel correction.

[71] E. F. Caldin and E. Harbron, *J. Chem. Soc.*, 3454 (1962).
[72] E. F. Caldin and M. Kasparian, *Disc. Faraday Soc.*, **39**, 25 (1965).
[73] E. F. Caldin, M. Kasparian, and G. Tomalin, *Trans. Faraday Soc.*, **64**, 2823 (1968).

made from measurements over a finite temperature range it will lead to $A_A > (A_A)_s$ because of the curvature of the plot. Although it is rarely feasible to make reliable theoretical calculations of $(A_A)$, it should be possible in principle to set approximate limits to its value for a given type of reaction, and the observation at low temperatures of values considerably below these limits would constitute good evidence for a considerable tunnel correction. So far, however, there is no example of a proton-transfer reaction in solution for which this has been demonstrated.

Corresponding predictions can be made of the hydrogen isotope effect on these quantities, and these provide more useful criteria for assessing the importance of tunnelling corrections, since many of the uncertainties cancel out in comparing behaviour of two isotopes. The following predictions (and corresponding ones for tritium) follow from the Equations (174), (183), and (184), or from any other expressions for moderate tunnelling:

(1) $k^H/k^D > (k^H/k^D)_s$. The semi-classical value is given by equations such as (160), (163), and (164). Although these cannot be evaluated exactly without some assumptions about the isotopically sensitive frequencies in the transition state, we have seen that the major contribution comes from the difference in zero-point energies between the initial and transition states, as expressed in (165). Moreover, since there will be a considerable cancellation between the bending frequencies in these two states, the maximum value of $(k^H/k^D)_s$ is unlikely to be much greater than $[\{E_0(AH) - E_0(AD)\}/kT]$ where $E_0(AH)$ and $E_0(AD)$ are the zero-point energies of the stretching vibration in the initial state. Values of this expression for several bonds have been given in Table 23 (p. 227). For the common example of C—H it has the value 6.9 at 25°C, and any values of $k^H/k^D$ greater than about 10 at this temperature may be reasonably attributed to an appreciable tunnelling correction; the corresponding value for $k^H/k^T$ is $10^{1.442} = 28$.

There is only one class of acid-base reaction for which conspicuously high values of $k^H/k^D$ have been reported, namely the reaction of 2-nitropropane with pyridine bases.[30,35] For pyridine itself and for methylpyridines with no substituent adjacent to the nitrogen atom, 'normal' (though rather high) isotope effects of about 10 were observed. However, for the hindered bases 2,6-dimethylpyridine and 2,4,6-trimethylpyridine, $k^H/k^D$ in alcohol–water mixtures at 25° was 24, while 2,6-dimethylpyridine gave the slightly lower value $k^H/k^D = 19$ in aqueous solution.[35] Similarly, the reaction between 2-nitropropane and 2,4,6-trimethylpyridine gave the abnormally high tritium isotope effect of $k^H/k^T = 79$.[30] All of these values are impossible to account for without invoking tunnelling; they suggest a connection between steric hindrance and tunnelling corrections, to which we shall return.

(2) $E_A^D - E_A^H > (E_A^D)_s - (E_A^H)_s$. Although the second of these differences is not accessible experimentally, we have seen that it will be close to the difference of zero-point energies, limits for which can be estimated from spectroscopic frequencies. Thus for proton transfer from C—H acids, if we again assume an approximate cancellation of bending frequencies between initial and transition states, Table 23 shows that any value of $E_A^D - E_A^H$ considerably above 1.2 kcal mol$^{-1}$ is suggestive of tunnelling.

(3) $A_A^D/A_A^H > (A_A^D)_s/(A_A^H)_s$. As already mentioned (p. 258), model calculations predict that the semi-classical value of this ratio will normally be close to unity, with absolute limits of $2^{-\frac{1}{2}}$ and 2. Any observed value of $A_A^D/A_A^H$ appreciably greater than unity is thus suggestive of tunnelling, and values greater than 2 provide strong evidence that the tunnelling correction is considerable. As shown by Equations (179)–(182), $E_A$ and $A_A$ both derive from measurements of the variation of reaction velocity with temperature, and the experimental tests of (2) and (3) are therefore best considered together.

The earliest investigation of this kind was by Bell, Fendley, and Hulett,[74] who studied the rate of abstraction of protons or deuterons from 2-ethoxycarbonylcyclopentanone by the bases $D_2O$, $CH_2ClCO_2^-$, and $F^-$. The values found for $E_A^D - E_A^H$ are $1.21 \pm 0.08$, $1.45 \pm 0.08$, and $2.44 \pm 0.10$ kcal mol$^{-1}$ respectively, while the corresponding values of $A_A^D/A_A^H$ are $2.3 \pm 0.3$, $2.9 \pm 0.4$, and $24 \pm 4$. For $D_2O$ the differences observed might just be accommodated by a semi-classical model, but for the other two bases, and especially for fluoride ion, there is strong evidence of tunnelling. A recent investigation of the tritium isotope effect[75] for the same reaction strengthens this interpretation, though there are some unexplained features of the results.

Interesting results were also obtained for the elimination reaction $PhMeCHCH_2Br + EtO^- \rightarrow PhMeC{=}CH_2 + Br^- + EtOH$, in which the rate-determining step involves the loss of a proton from carbon; both protium–deuterium[76] and deuterium–tritium[77] comparisons were made. The former gave $E_A^D - E_A^H = 1.77 \pm 0.11$ kcal mol$^{-1}$ and $A_A^D/A_A^H = 2.5 \pm 0.3$, both of which are considerably greater than the expected semi-classical

[74] R. P. Bell, J. A. Fendley, and J. R. Hulett, *Proc. Roy. Soc.*, A, **235**, 453 (1956).
[75] J. R. Jones, *Trans. Faraday Soc.*, **65**, 2430 (1969).
[76] V. J. Shiner and M. L. Smith, *J. Am. Chem. Soc.*, **83**, 593 (1961).
[77] V. J. Shiner and B. Martin, *Pure Appl. Chem.*, **8**, 371 (1964).

values. The latter gave $E_A^T - E_A^D = 0.68 \pm 0.03$ kcal mol$^{-1}$, $A_A^T/A_A^D = 1.19 \pm 0.04$, which may be compared with the difference of the zero-point energies of the initial stretching vibrations, estimated at 0.52 kcal mol$^{-1}$, and the semi-classical value of unity for the ratio of the $A$-values. The results thus agree with the expectation that the tunnel correction will be considerable for protium, small for deuterium, and even smaller for tritium.

Larger deviations from semi-classical behaviour have been observed in the reaction of $o$-methylacetophenone and $p$-methoxyacetophenone with hydroxide ions, in which the protium compound was studied by bromination and the tritium compound by detritiation. For these two reactions $E_A^T - E_A^H$ had values of $3.6 \pm 0.4$ and $3.3 \pm 0.9$ kcal mol$^{-1}$ respectively, compared with the expected value of 1.7 kcal mol$^{-1}$, while the values of $\lg(A_A^T/A_A^H)$ were $1.5 \pm 0.3$ and $1.1 \pm 0.6$. Although the precision of these last values is not high, they certainly lie outside the semi-classical limits.[78]

It is satisfactory to find that the large value of $k^H/k^D$ found for the reaction of 2-nitropropane with 2,4,6-trimethylpyridine is accompanied by large values of $E_A^D - E_A^H$ (3.0 kcal mol$^{-1}$) and $A_A^D/A_A^H$ (about 7), though these values are subject to some uncertainty.[30] Similarly, the reaction of 4-nitrobenzyl cyanide, which exhibits deviations from the Arrhenius equation at low temperatures,[73] also gives the high values $E_A^D - E_A^H = 1.85 \pm 0.2$ kcal mol$^{-1}$, $A_A^D/A_A^H \sim 5$.[79] The correspondence between two criteria for the same reaction increases our confidence that the explanation in terms of the tunnel correction is the right one.

If it is assumed that deviations from a simple Arrhenius law or values of $A_A^D/A_A^H$ differing from unity can be attributed solely to a tunnel correction, then the experimental results can be used to calculate the parameters characterizing the barrier from the theoretical equations. This has been done particularly by Caldin and his collaborators,[56,79] and some of his results for proton-transfer reactions are given in Table 26. Since $\Delta G^\circ$ is unknown for most of the reactions, a symmetrical parabolic barrier was assumed throughout, and the unknowns are then the true barrier heights $E^H$ and $E^D$ or $E^T$ (which differ because of the zero-point energies of the initial and transition states) and a parameter characterizing the curvature of the barrier, which may be either $2a$, its width at the base, or $v_3^H$, the numerical value of the imaginary frequency for the light isotope. The last two parameters are related by Equation (170).

Several points of interest emerge from Table 26. All the values of $2a$ are of the right order of magnitude, since Figure 22(a) shows that they should

[78] J. R. Jones, R. E. Marks, and S. C. Subba Rao, *Trans. Faraday Soc.*, **63**, 993 (1967).
[79] E. F. Caldin and G. Tomalin, *Trans. Faraday Soc.*, **64**, 2814, 2823 (1968).

Table 26. BARRIER DIMENSIONS FOR PROTON-TRANSFER REACTIONS
CALCULATED FROM ISOTOPE EFFECTS

Values at 25°C unless otherwise stated

| Reaction | $k^H/k^D$ | $2a$, pm | $v_3^H$, cm$^{-1}$ | $E^D - E^H$, kcal mol$^{-1}$ | $Q_t^H/Q_t^D$ | $E_A^H/E^H$ |
|---|---|---|---|---|---|---|
| I | 3.4 | 1.26 | 878 | 0.39 | 1.71 | 0.90 |
| II | 3.9 | 1.17 | 924 | 0.42 | 1.79 | 0.88 |
| III | 2.6 | 1.17 | 1106 | 0.06 | 2.45 | 0.81 |
| IV | 18.1[a] | 1.22 | 1001 | 1.2[a] | 2.46[a] | 0.87 |
| V | 19.8[a] | 1.27 | 908 | 1.4[a] | 1.92[a] | 0.84 |
| VI | 24 | 1.14 | 965 | 1.3 | 2.76 | 0.88 |
| VII[b] | 7.8 | 1.59 | 908 | 0.9 | 1.74 | 0.95 |
| VIII[c] | 29 | 1.63 | 600 | 1.0 | 1.89 | 0.89 |

(a) Values for tritium rather than deuterium; (b) Values for tritium yield identical barrier dimensions for this reaction; (c) Values at $-90°C$; deviations from the Arrhenius equation yield barrier dimensions agreeing well with those in the table.

*Key to reactions*

| | |
|---|---|
| I | 2-Ethoxycarbonylcyclopentanone + $D_2O$ |
| II | 2-Ethoxycarbonylcyclopentanone + $CH_2ClCO_2^-$ |
| III | 2-Ethoxycarbonylcyclopentanone + $F^-$ |
| IV | $p$-Methoxyacetophenone + $OH^-$ |
| V | $o$-Methylacetophenone + $OH^-$ |
| VI | $Me_2CHNO_2$ + 2,4,6-trimethylpyridine |
| VII | $PhMeCHCH_2Br$ + $OEt^-$ |
| VIII | 4-Nitrobenzyl cyanide + $OEt^-$ |

be somewhat smaller than the distance through which the proton moves. Similarly, the values of $v_3$ are close to those derived from models for proton-transfer reactions.[42,43,47,50] The difference between the true barrier heights, $E^D - E^H$, comes out to be less than the difference of the initial zero-point energies, as required by the theory. The whole picture is thus physically self-consistent, but too much weight should not be attached to any quantitative interpretation of the barrier parameters, since they are derived from the simplified assumption of a symmetrical parabolic barrier. In particular, the small range of values found for $2a$, $v_3$, and $Q_t^H/Q_t^D$ may be illusory, since when the tunnel correction is considerable it does in fact depend appreciably on the symmetry of the barrier, i.e., on $\Delta G$.[50]

The quantity $E_A^H$ is a measure of the average excess energy of the systems which react, and Table 26 shows that for the systems considered it amounts to 81–95% of $E^H$, the true barrier height. This implies that our molecular picture of proton-transfer reactions at ordinary temperatures is not usually modified in any major way by the inclusion of tunnel corrections. For example, the semi-quantitative use of hydrogen isotope effects for drawing conclusions about reaction mechanisms remains valid, since it is still true to say that a large isotope effect implies a considerable loosening of hydrogen in the transition state. It can also be shown theoretically[80,81] that the usual type of Brönsted relation between rates and equilibrium constants will still hold over a considerable range even when there are large tunnel corrections. However, it is certainly necessary to include tunnel corrections in any quantitative treatment of hydrogen isotope effects, and this is particularly true if activation energies and pre-exponential factors are being considered. Since the tunnel correction is sensitive to the exact dimensions of the barrier it may be responsible for individual deviations shown by particular systems. For example, there is some evidence that reactions involving steric hindrance may exhibit abnormally large isotope effects, which can be understood in terms of a large tunnel correction arising from a higher and steeper energy barrier.

It might be thought that the presence of considerable tunnelling would cause deviations from the Swain relation (Equation 166), which was derived on the basis of a semi-classical model. However, quantitative examination[17,30] shows that this is an insensitive criterion for detecting tunnelling, though a large tunnel correction should give an exponent $r$ in the relation $k^H/k^T = (k^H/k^D)^r$ which is smaller than the value 1.442 in Equation (166). There is only one proton-transfer for which a markedly low value has been reported, namely, the reaction of acetone with hydroxide ions,[82] for which $r = 1.12$.

Most of our general considerations about hydrogen isotope effects should be equally valid for the transfer of hydride ions or hydrogen atoms, and in fact these classes of reaction have provided much of the evidence for tunnel corrections. Some of it will be summarized briefly here, beginning with solution reactions. Many oxidation reactions are believed to involve hydride ion transfer, and large isotope effects have been reported in the oxidation of 1-phenyl-2,2,2-trifluoroethanol by alkaline permanganate[30,83] ($k^H/k^D = 16$, $k^H/k^T = 57$, both at 25°C). The activation energies for this reaction are not known accurately, but the reported values lead to $E_A^D - E_A^H = 2.3$ kcal mol$^{-1}$, $A_A^D/A_A^H = 3.0$. The chromic acid oxidation of this and related substances in aqueous acetic acid also exhibits rather

[80] R. P. Bell, *Proc. Roy. Soc.*, A, **148**, 241 (1935).
[81] J. J. Weiss, *J. Chem. Phys.*, **41**, 1120 (1964).
[82] J. R. Jones, *Trans. Faraday Soc.*, **65**, 2138 (1969).
[83] R. Stewart and R. van der Linden, *Disc. Faraday Soc.*, **29**, 211 (1960).

large isotope effects, the largest being $k^H/k^D = 12.9$ for the 3,5-dinitro-phenyl compound.[84] The oxidation by permanganate of $CF_3CH(OH)O^-$ also gave results[85] which suggest tunnelling, namely $k^H/k^D = 14$ at 25°C, $E_A^D - E_A^H = 1.9$ kcal mol$^{-1}$ and $A_A^D/A_A^H = 2.3$. The most striking results for this class of reaction were obtained for the oxidation by 2,3,5,6-tetra-chloro-*p*-benzoquinone (choranil) of $(p\text{-}Me_2NC_6H_4)_3CH$ (leuco Crystal Violet) in acetonitrile solution.[86] The value of $k^H/k^D$ is only moderately high (11.7 at 25°C), but $E_A^D - E_A^H$ (3.36 kcal mol$^{-1}$) and $A_A^D/A_A^H$ (24) are strikingly so; this appears to be another instance in which a sterically hindered system leads to a large tunnel correction.

Reactions involving transfer of hydrogen atoms necessarily also involve free radicals, which may lead to chain reactions. When long chains are present it is possible for a relatively small isotope effect on the component reactions to influence the chain length and thus produce a large overall isotope effect, as has been observed in various autoxidation processes.[87,88] In order to obtain isotope effects for simple H-transfers it may therefore be necessary to isolate the individual processes, either by adding chain inhibitors or by using competitive methods. When this is done there are several solution reactions which exhibit abnormally large isotope effects. The abstraction of hydrogen from $CH_3CH_2OH$ and $CH_3CD_2OH$ by hydrogen atoms (produced by the action of X-rays on water)[89] has $k^H/k^D = 17 \pm 1$ at 25°C. Abstraction by various radicals from the —OH or —OD groups of phenols is associated with several high values,[90] for example, $k^H/k^D = 17$ and 30 at 30°C, and 19 at 50°C for different systems. In the reaction of methyl radicals with 2,4,6-tri-t-butylphenol in heptane solution at 60–90°C, $k^H/k^T$ is about 50, i.e., about three times the value which might be expected on the basis of zero-point energies alone.[91] Here again the effect of steric hindrance may be important. Earlier work on abstraction from liquid n-heptane by methyl radicals[92] showed rather smaller isotope effects, but gave the high values $E_A^T - E_A^H = 3.4$ kcal mol$^{-1}$, $A_A^T/A_A^H = 5.0$.

Results of particular interest have been obtained by Bromberg and his

[84] R. Stewart and D. G. Lee, *Canad. J. Chem.*, **42**, 439 (1964).
[85] R. Stewart and M. M. Mocek, *Canad. J. Chem.*, **41**, 1161 (1963).
[86] E. S. Lewis, J. M. Perry, and R. H. Grinstein, *J. Am. Chem. Soc.*, **92**, 899 (1970).
[87] P. Krumbiegel, *Z. Chem.*, **8**, 328 (1968).
[88] S. Rummel and H. Huebner, *Z. Chem.*, **9**, 150 (1969).
[89] C. Lifshitz and G. Stein, *J. Chem. Soc.*, 3706 (1962).
[90] M. Simonyi and F. Tüdos, *Adv. Phys. Org. Chem.*, **9**, 127 (1970).
[91] L. N. Shishkina and I. V. Berezin, *Zh. Fiz. Khim.*, **39**, 2547 (1965); *Russ. J. Phys. Chem.*, **39**, 1357 (1965).
[92] V. L. Antonovskii and I. V. Berezin, *Zh. Fiz. Khim.*, **34**, 1286 (1960).

collaborators[93-97] on the autoxidation of 4a,4b-dihydrophenanthrene in octane solution. This is a chain reaction, but by making measurements with and without the addition of inhibitors it is possible to measure the separate isotope effects for the two reactions

$$\text{Initiation,} \qquad PH_2 + O_2 \rightarrow PH + HO_2$$

$$\text{Propagation,} \qquad PH_2 + HO_2 \rightarrow PH + H_2O_2$$

Very large deuterium isotope effects were observed for the initiation step. The first measurements[93,94] gave $k^H/k^D$ between 64 and 95 for temperatures between $-10°C$ and $-31°C$, and later[97] a value of about 250 at $-52°C$ was reported; when the isotope effect is as large as this it becomes difficult to make accurate measurements for both isotopes. Moreover, measurements on the undeuteriated compound down to $-82°C$ showed deviations of up to 100% from the Arrhenius equation at the three lowest temperatures.

These findings certainly indicate large tunnelling corrections, and this was substantiated by a detailed theoretical computation of the energy surfaces for both the initiation and the propagation reactions[95,96] including a calculation of tunnel corrections. Although a theoretical treatment of such complex systems involves considerable uncertainty, it was possible to choose parameters which gave a good representation of the experimental results for both isotopes over the whole temperature range. The calculated tunnel corrections are large, ranging from 120 to 11,000 for the undeuteriated compound at $-10°C$ and $-82°C$; the corresponding figures for the deuteriated material are 15 and 250. The large isotope effects are attributable to a combination of the semiclassical and tunnelling factors, which at $-31°C$ are calculated as 7.6 and 11.8 respectively. It is significant that application of the same theoretical treatment to the propagation step predicts only a small tunnelling contribution, and hence a 'normal' isotope effect; this is in agreement with the observed value of $k^H/k^D = 7.2$ at $-10°C$, and supports the general basis of the calculations. The physical reason for the difference in tunnel corrections between the two reactions is that the initiation reaction is almost thermoneutral, while the propagation reaction is strongly exo-

[93] A. Bromberg, K. A. Muszkat, and E. Fischer, *Chem. Comm.*, 1352 (1968).
[94] A. Bromberg and K. A. Muszkat, *J. Am. Chem. Soc.*, **91**, 2860 (1969).
[95] A. Warshel and A. Bromberg, *J. Chem. Phys.*, **52**, 1262 (1970).
[96] A. Bromberg, K. A. Muszkat, and A. Warshel, *J. Chem. Phys.*, **52**, 5952 (1970).
[97] A. Bromberg, K. A. Muszkat, E. Fischer, and F. S. Klein, *J. Chem. Soc., Perk. Trans.* II, 588 (1972).

thermic ($\Delta H \sim -40 \text{ kcal mol}^{-1}$). As we have already seen (cf. Figure 24) tunnelling is favoured in the former case.*

The most striking example hitherto reported of tunnelling by hydrogen atoms relates to a reaction in the solid phase. If solid acetonitrile is irradiated with $\gamma$-rays, electrons (associated with molecules of the matrix) are produced, and on further irradiation with visible light these are converted into methyl radicals, which then react at a measurable rate according to the process $\cdot CH_3 + CH_3 CN \rightarrow CH_4 + \cdot CH_2 CN$. This reaction can be followed by electron spin resonance observation of either the disappearing methyl radicals or the appearing $\cdot CH_2 CN$ radicals, and it can be studied either after the visible light has been cut off, or during continuous illumination. All these procedures yielded essentially the same rate constants, which makes it virtually certain that the observed rates do refer to the above reaction. Results were first reported[99] for 77 K and 87 K, and subsequently[100] for 69 K, 100 K, and 112 K. They gave a strongly curved Arrhenius plot with an apparent activation energy varying from 1.2 to 2.8 kcal mol$^{-1}$, while the activation energy for the same process in the gas phase[101] at 373–573 K is $10.0 \pm 0.5$ kcal mol$^{-1}$. The authors[100] have succeeded in explaining these facts quantitatively by a one-dimensional tunnelling treatment, using the computational methods already referred to.[66] They assume that the true barrier height in the solid is equal to the activation energy of the high-temperature gas reaction, and that the classical frequency factor for the solid phase reaction is equal to the C—H stretching vibration frequency in $CH_3 CN$; they then seek the form and dimensions of the energy barrier which will best fit the experimental results. Calculations were made with both parabolic and Eckart barriers (cf. Equation 177), but the best fit was obtained with a Gaussian barrier, $V(x) = E \exp(-x^2/a^2)$, $a$ having the physically reasonable value

---

* It should be noted that the experimental results for the initiation reaction give a value of $A_A^D/A_A^H = 1.6$, i.e., not much greater than unity, in spite of the large tunnel correction. Although part of this discrepancy may be due to the sensitivity of the pre-exponential factors to experimental error, it should be emphasized that the criterion $A_A^D/A_A^H \gg 1$ [and also $E_A^D - E_A^H > (E_A)_s^D - (E_A^H)_s$] applies only to tunnel corrections of moderate magnitude. Detailed calculation shows[96,98] that when the tunnel correction becomes very large (i.e., at low temperatures or for narrow barriers) the value of $A_A^D/A_A^H$ will decrease again and will eventually become less than unity. This is physically obvious in the extreme case of a tunnelling process which takes place without any thermal activation, since in such a process the pre-exponential factor will be just the tunnelling rate from the lowest energy level, and this will always be greater for the lighter isotope.

[98] Unpublished calculations by R. L. Tranter, and by M. J. Stern.
[99] E. D. Sprague and F. Williams, *J. Am. Chem. Soc.*, **93**, 787 (1971).
[100] R. J. Le Roy, E. D. Sprague, and F. Williams, *J. Phys. Chem.*, **76**, 546 (1972).
[101] M. H. J. Wijnen, *J. Chem. Phys.*, **22**, 1074 (1954).

0.636 pm. The tunnel 'corrections' are extremely high at these low temperatures, ranging from about $10^5$ to $10^{15}$, and it is reasonable to describe the reaction as taking place entirely by a tunnelling mechanism. Very large hydrogen isotope effects would also be expected ($k^H/k^D = 10^3$–$10^6$ in the temperature range investigated); experiment cannot confirm this quantitatively, but provides the consistent information that in partially deuterated acetonitrile only the light atoms appear to react.

There are of course many simple gas reactions which involve the transfer of hydrogen atoms, and in principle a study of their kinetics (and in particular of isotope effects) should provide a good test of calculations of tunnelling corrections. In practice this turns out to be somewhat inconclusive partly because it is often difficult to obtain accurate measurements over a large temperature range, and partly because when several light atoms are moving simultaneously the 'reaction co-ordinate' is a complicated function of atomic positions. A recent paper[102] examines the *a priori* calculation of energy surfaces for simple gas reactions, and concludes that, even for very simple reactions such as $H + H_2$ or $Cl + H_2$ (where H and $H_2$ may represent any of the hydrogen isotopes), the most thorough calculations so far carried out are not good enough for computing tunnelling corrections. Nevertheless, many authors have used theoretical or semi-empirical energy profiles for calculating tunnel corrections, especially for the reactions $H + H_2$,[103–108] $Cl + H_2$,[109,110] $CF_3 + CH_4$,[111,112] $CH_3 + H_2$,[113,114] and $CF_3 + H_2$,[115] in each case with various combinations of H, D, or T. The general conclusion is that considerable tunnelling corrections are necessary, even at moderately high temperatures, but that it is not yet possible to decide whether a one-dimensional treatment is adequate, or to obtain much information about the shape of the energy surface.

[102] C. A. Parr and D. G. Truhlar, *J. Phys. Chem.*, **75**, 1844 (1971).

[103] W. R. Schulz and D. J. Le Roy, *Canad. J. Chem.*, **42**, 2480 (1964); *J. Chem. Phys.*, **42**, 3869 (1965).

[104] B. A. Ridley, W. R. Schulz, and D. J. Le Roy, *J. Chem. Phys.*, **44**, 3344 (1966).

[105] D. J. Le Roy, B. A. Ridley, and K. A. Quickert, *Disc. Faraday Soc.*, **44**, 97 (1967).

[106] A. A. Westenburg and N. de Haas, *J. Chem. Phys.*, **47**, 1393 (1967).

[107] I. Shavitt, *J. Chem. Phys.*, **49**, 4048 (1968).

[108] K. A. Quickert and D. J. Le Roy, *J. Chem. Phys.*, **53**, 1325 (1970).

[109] J. Bigeleisen, F. S. Klein, R. E. Weston, and M. Wolfsberg, *J. Chem. Phys.*, **30**, 1340 (1959).

[110] A. Persky and F. S. Klein, *J. Chem. Phys.*, **44**, 3617 (1966).

[111] T. E. Sharp and H. S. Johnston, *J. Chem. Phys.*, **37**, 1541 (1962).

[112] H. S. Johnston and E. Tschuikow-Roux, *J. Chem. Phys.*, **36**, 463 (1962).

[113] H. S. Johnston, *Adv. Chem. Phys.*, **3**, 131 (1961).

[114] J. S. Shapiro and R. E. Weston, *J. Phys. Chem.*, **76**, 1669 (1972).

[115] C. L. Kibby and R. E. Weston, *J. Chem. Phys.*, **49**, 1193 (1968).

We return now to the problem of isotope effects for proton-transfer reactions in solution. The examples considered so far relate to primary effects in a common solvent, but much work has also been done on *solvent isotope effects*, i.e., the effect of changing the solvent from $H_2O$ to $D_2O$ or to some mixture of $H_2O$ and $D_2O$. Such effects often appear as an unavoidable accompaniment to primary effects when one or more of the reactants exchanges hydrogen rapidly with the solvent; this is the case for all O—H and N—H acids, and obviously applies, in particular, to the lyonium and lyate ions derived from the solvent. In other instances it is possible to investigate solvent isotope effects in an essentially 'pure' state. For example, we may carry out the reaction $MeCH(NO_2)_2 + C_6H_5O^- \rightarrow$ $[MeC(NO_2)_2]^- + C_6H_5OH$ in either $H_2O$ or $D_2O$ without making any formal isotopic changes in either reactant.* This kind of solvent isotope effect will of course operate in reactions which do not involve any proton transfers, and it has been investigated very extensively for solvolytic reactions, in which the solvent enters the rate-determining step either by stabilizing an incipient carbonium ion ($S_N1$ reactions), or as a nucleophile ($S_N2$ reactions).[116] In these reactions $k^{H_2O}/k^{D_2O}$ is usually between unity and 1.30 though higher values are sometimes found.

Considerably larger kinetic solvent isotope effects are encountered for reactions involving proton-transfer. We have seen in Chapter 11 that our understanding of solvent isotope effects on acid-base equilibria is at best a semi-quantitative one, largely because of difficulties in interpreting the structure of water and the way in which it is modified by solutes. These difficulties are aggravated in the corresponding kinetic problem because we now have to consider the interaction of the solvent with transition states of largely unknown structure and charge distribution. The magnitude of the solvent isotope effect is often used, in conjunction with other information, for drawing conclusions about transition state structure, but the subject is still controversial, and only a few examples will be given here. We shall confine ourselves to *acid catalysis*, since this has been most fully investigated.

The three simplest mechanisms for catalysis by hydronium ions, with their usual designations, are as follows:

---

*The product $C_6H_5OH$ will, of course, undergo rapid isotopic exchange with the solvent, but this does not affect the reaction as written, and would be significant only if the process observed were appreciably reversible.

[116] For a review, see P. M. Laughton and R. E. Robertson in *Solute-Solvent Interactions* (ed. J. F. Coetzee and C. D. Ritchie), Dekker, New York, 1969.

A-1      $S + H_3O^+ \rightleftharpoons SH^+ + H_2O$ (fast)

$SH^+ \rightarrow (SH^+)^\ddagger \rightarrow$ products (slow)

A-2      $S + H_3O^+ \rightleftharpoons SH^+ + H_2O$ (fast)

$SH^+ + H_2O \rightarrow (H_2O \cdots SH^+)^\ddagger \rightarrow$ products (slow)

A-$S_E$2      $S + H_3O^+ \rightarrow (S \cdots H \cdots OH_2^+)^\ddagger \rightarrow$ products (slow)

In the A-1 mechanism the second (slow) step does not involve participation of the solvent or movement of a proton, and the main effect of changing from $H_2O$ to $D_2O$ lies in its influence on the first equilibrium. Since we have seen (p. 234) that most acids investigated are stronger in $H_2O$ than in $D_2O$ by a factor of 2–4, the concentration of $SH^+$ in $H_2O$ will be smaller than that of $SD^+$ in $D_2O$ under the same conditions, and we should expect $k^{H_2O}/k^{D_2O}$ for this class of reaction to lie in the range 0.25–0.5. This is in fact the case for the hydrolysis of a number of acetals and ketals, which are believed on other grounds to follow A-1 mechanisms such as

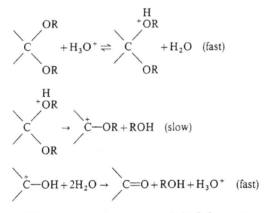

Examples are[117] 1,1-dimethoxyethane, $k^{H_2O}/k^{D_2O} = 0.37$, 2-methyl-1,3-dioxolane, $k^{H_2O}/k^{D_2O} = 0.36$. Solvent isotope effects of similar magnitude are found in the acid-catalysed hydrolysis of epoxides. For example, ethylene oxide[118] has $k^{H_2O}/k^{D_2O} = 0.45$. There is some dispute as to whether these reactions should be classified as A-1 or A-2, but in any case the involvement of a water molecule as a nucleophile in the transition

[117] For references, and a review of other mechanistic evidence, see E. H. Cordes, *Progr. Phys. Org. Chem.*, **4**, 1 (1967).
[118] J. G. Pritchard and F. A. Long, *J. Am. Chem. Soc.*, **78**, 6008 (1956).

state would not be expected to change the isotope effect appreciably;
certainly the question cannot be resolved on the basis of the magnitude
of $k^{H_2O}/k^{D_2O}$ alone.

We have already seen (pp. 164–171) that in the hydrolysis of many
aliphatic diazo-compounds, notably ethyl diazoacetate, there is good
evidence for a pre-equilibrium of the type

$$\bar{N}{=}\overset{+}{N}{=}CRR' + H_3O^+ \rightleftharpoons N{\equiv}\overset{+}{N}CHRR' + H_2O$$

followed by further slow reaction of the cation produced. This is confirmed
by the solvent isotope effect, which for ethyl diazoacetate[119] is
$k^{H_2O}/k^{D_2O} = 0.35$. Once more this figure alone gives no information as to
whether the rate-determining step in the further reaction of the cation
involves water (or some other nucleophile), but attempts have been made
to settle this question by studying the way in which the rate varies with
isotopic composition in $H_2O$–$D_2O$ mixtures.[120] The theory is essentially
the same as that already given (pp. 237–249) for acid-base equilibria in
these mixtures, and equations such as (151) can be derived by considering
the equilibrium between the reactants and the transition state. For an $A$-1
reaction this gives

$$\frac{k^H}{k^x} = \frac{(1-x+xl)^3}{1-x+x\phi^{\ddagger}} = \frac{(1-x+xl)^3}{1-x+xl^3 k^D/k^H} \tag{185}$$

In this equation $\phi^{\ddagger}$ is the fractionation factor for the single proton in the
transition state ($SH^+$); it cannot of course be measured independently
even in principle, but it can be expressed in terms of the observed value of
$k^H/k^D$, since when $x = 1$, $k^H/k^D = l^3/\phi^{\ddagger}$.* If, on the other hand, the reaction
follows an $A$-2 mechanism with a transition state ($H_2O{\cdots}SH^+$), the
corresponding equation is

$$\frac{k^H}{k^x} = \frac{(1-x+xl)^3}{(1-x+x\phi_1)(1-x+x\phi_2)^2} \tag{186}$$

where $\phi_1$ and $\phi_2$ are the fractionation factors for the two kinds of proton
in the transition state, one of which (but not both) can be eliminated by

[119] P. Gross, H. Steiner, and F. Krauss, *Trans. Faraday Soc.*, **34**, 351 (1938).
[120] W. J. Albery and M. H. Davies, *Trans. Faraday Soc.*, **65**, 1066 (1969).
*In this and succeeding equations $k^H$ and $k^D$ have been used in place of $k^{H_2O}$ and
$k^{D_2O}$ for the sake of brevity. It should be noted that the factor $(1-x+xl)^3$ now occurs
in the numerator, instead of in the denominator as in the corresponding equations
for dissociation constants (Equations 149–153); this is because the hydronium ion
is now being considered as a reactant rather than as a product.

using the observed value of $k^H/k^D$. If $\phi_2$ is very close to unity (186) becomes indistinguishable from (185), but if the water molecule is intimately involved in the transition state it may be possible to distinguish experimentally between the two possibilities. Such a distinction demands considerable faith both in the accuracy of the measurements and in the simplifying assumptions which underly the theory, namely, the neglect of transfer effects and the validity of the rule of the geometric mean (Equation 142). However, in the work on ethyl diazoacetate[120] it appears that this faith is justified, since the conclusion that the reaction is *A*-2 rather than *A*-1 is supported by other evidence.

The $A$-$S_E2$ mechanism differs from those already discussed in that a proton is partly transferred from the hydronium ion in the transition state. This implies a primary isotope effect and hence $k^{H_2O} > k^{D_2O}$, in contrast to the $A$-1 and $A$-2 cases. In terms of fractionation theory the solvent isotope effect is given by $k^{H_2O}/k^{D_2O} = l^3/\phi_1\phi_2^2$, where $\phi_1$ and $\phi_2$ refer to the numbered protons in the transition state

$$(H_2\overset{+}{O}\cdots H\cdots S)^{\ddagger} \qquad (187)$$
$$\quad (2) \qquad (1)$$

Since $\phi_1$ refers to a proton 'in flight', which is loosely bound, it will be considerably less than unity, leading to $k^{H_2O}/k^{D_2O} > 1$, although the factor $l^3$ (with $l = 0.69$) will produce a solvent isotope effect which is smaller than that commonly found for pure primary effects. This is in fact what is found for a number of reactions in which the rate-determining step is believed to involve proton-transfer from hydronium ion to carbon, typical values being 1.7 for the reaction of the enolate ion of 2-acetylcyclohexanone with hydronium ion,[121] 1.7–3.0 for the hydrolysis of vinyl ethers by strong aqueous acids,[122] 1.7–3.2 for the acid-catalysed cleavage of alkyl-mercuric iodides,[123], and 1.7–2.5 for the hydrolysis of a number of secondary diazo-ketones.[124]

Since the $A$-$S_E2$ group of reactions involve rate-determining proton transfer they are subject to general acid catalysis, and it is possible to study the solvent isotope effect for acid catalysts other than the hydronium ion. If proton transfer takes place directly (i.e., without the intervention

[121] T. Riley and F. A. Long, *J. Am. Chem. Soc.*, **84**, 522 (1962).
[122] P. Salomaa, A. Kankaanperä, and M. Lajunen, *Acta Chem. Scand.*, **20**, 1790 (1966); A. J. Kresge and Y. Chiang, *J. Chem. Soc.*, B, 58 (1967); M. M. Kreevoy and R. Eliason, *J. Phys. Chem.*, **72**, 1313 (1968).
[123] For a summary, see J. M. Williams and M. M. Kreevoy, *Adv. Phys. Org. Chem.*, **6**, 63 (1968).
[124] H. Dahn and M. Ballenegger, *Helv. Chim. Acta*, **52**, 2417 (1952).

of solvent molecules) the transition state for a monoprotic catalyst HA becomes simply $(A \cdots H \cdots S)$, and the isotope effect becomes essentially a primary one, for which fairly high values are anticipated. This has been observed in a number of reactions; for example, in the hydrolysis of ethyl vinyl ether[122] $k^{H_2O}/k^{D_2O}$ is 6.8 for catalysis by undissociated formic acid compared with 2.5 for hydronium ion catalysis, while the corresponding values for the hydrolysis of cyanoketen dimethylacetal[125] are 5.4 and 3.0.

Much interesting information can be obtained, at least in principle, by the study of rate-determining proton-transfers in $H_2O$–$D_2O$ mixtures. Considering first transfer from hydronium ions, the transition state (187) leads to an expression formally identical with (186), obtained for the $A$-2 mechanism, since in each case the transition state contains one proton of one kind and two of a different nature. However, the implications are different for the two mechanisms, and further progress can be made in the $A$-$S_E2$ case. The proton (1) ('in flight') becomes incorporated in the product, and provided that it does not undergo subsequent isotopic exchange with the solvent, $\phi_1$ can be determined by comparing the isotopic composition of the product with that of the solvent. In fact

$$(D/H) \text{ product}/(D/H) \text{ solvent} = r = \phi_1 \qquad (188)$$

where $r$ (or its reciprocal) is commonly termed the *product isotope effect*[126] or *discrimination isotope effect*, and is not the same as $k^{H_2O}/k^{D_2O}$. Moreover, $\phi_1$ and $\phi_2$ are related by $k^H/k^D = l^3/\phi_1\phi_2^2$, so that when $r$ can be measured Equation (186) can be rewritten as

$$\frac{k^H}{k^x} = \frac{(1-x+xl)^3}{(1-x+rx)\{1-x+x(l^3k^D/k^Hr)^{\frac{1}{3}}\}^2} \qquad (189)$$

This expression has been found to fit the experimental facts for a number of reactions, for example, the addition of water to isobutene.[127]

There is another way in which observations of $k^H/k^x$ for slow proton transfer from hydronium ions can be related to other parameters of the reaction even when the product isotope effect cannot be measured. The fractionation factor $\phi_2$ refers to the two protons of the hydronium ion which are not being transferred, and whose properties should be somewhere between those of the protons of water and those of the protons in

[125] V. Gold and D. C. A. Waterman, *J. Chem. Soc.*, B, 839 (1968).
[126] M. M. Kreevoy and R. A. Kretchmer, *J. Am. Chem. Soc.*, **86**, 2435 (1964); V. Gold and M. A. Kessick, *Pure Appl. Chem.*, **8**, 273 (1964); *Proc. Chem. Soc.*, 295 (1964).
[127] V. Gold and M. A. Kessick, *Disc. Faraday Soc.*, **39**, 84 (1965); *J. Chem. Soc.*, 6718 (1965).

the hydronium ion. $\phi_2$ should therefore be between unity and $l$, and Kresge[128] has suggested the quantitative relation

$$\phi_2 = l^{1-\alpha} \tag{190}$$

where $\alpha$ is a parameter less than unity which characterizes the degree of proton transfer in the transition state. The relation $k^H/k^D = l^3/\phi_1\phi_2^2$ can then be used to eliminate $\phi_1$, and (186) becomes

$$\frac{k^H}{k^x} = \frac{(1-x+xl)^3}{(1-x+xl^{1-\alpha})^2(1-x+xl^{1+2\alpha}k^D/k^H)} \tag{191}$$

In favourable cases (191) can be used to derive the value of $\alpha$ for a given reaction from the experimental results, and it is interesting to compare it with the exponent of the Brönsted relation between catalytic power and acid strength, which may also be regarded as a measure of the extent of proton transfer in the transition state. For a number of reactions these two values of $\alpha$ turn out to be indistinguishable, for example, in the hydrolysis of cyanoketen dimethylacetal[125] and of 2-dichloromethylene-1,3-dioxolan,[129] the addition of water to $p$-methoxy-$\alpha$-methylstyrene,[130] and the hydrolysis of ethyl vinyl ether.[131] However, the values of $\alpha$ are not known with any accuracy, and it seems unlikely that they will always coincide; the concept of 'degree of proton transfer' is a vague one, and in any case might differ considerably when the catalyst is changed from hydronium ion to the weak acids which are normally used to determine the value of the Brönsted exponent. There are in fact some reactions for which a definite discrepancy between the two values of $\alpha$ has been reported, for example, the acid-catalysed decomposition of $p$-nitrophenyldiazomethane.[132]

The variation of rate with isotopic composition follows simpler laws when the proton is transferred from a weak monoprotic acid AH, since the theoretical expression now becomes

$$\frac{k^H}{k^x} = \frac{1-x+x\phi_{AH}}{1-x+x\phi^\ddagger} \tag{192}$$

[128] A. J. Kresge, *Pure Appl. Chem.*, **8**, 243 (1964).
[129] V. Gold and D. C. A. Waterman, *J. Chem. Soc.*, B, 839 (1968).
[130] J. C. Simandoux, B. Torck, M. Hellin, and F. Coussemant, *Tetrahedron Letters*, No. 31, 2971 (1967).
[131] M. M. Kreevoy and R. Eliason, *J. Phys. Chem.*, **72**, 1313 (1968).
[132] H. Dahn and G. Diderich, *Helv. Chim. Acta*, **54**, 1950 (1971); G. Diderich and H. Dahn, *Helv. Chim. Acta*, **55**, 1 (1972).

where $\phi^{\ddagger}$ is now the fractionation factor for the single proton in the transition state $(A \cdots H \cdots S)$. When AH is an oxyacid it is likely that $\phi_{AH}$ will be close to unity (for example, it is $0.96 \pm 0.02$ for acetic acid[133]). In that event Equation (192) predicts that $k^x$ should be a linear function of $x$, and this has been found to be the case in a number of reactions, for example, in the reaction of weak acids with the anion of nitroethane,[134] and the hydrolysis of cyanoketen dimethylacetal catalysed by undissociated acetic acid.[129] The contrast between this linear behaviour and the strongly curved relation for hydronium ion catalysis, predicted by (186)–(191) and confirmed by experiment, constitutes one of the clearest demonstrations of the essential correctness of the interpretation of rates in $H_2O$–$D_2O$ mixtures in terms of fractionation factors.

[133] V. Gold and B. M. Lowe, *J. Chem. Soc.*, A, 1923 (1968).
[134] D. M. Goodall and F. A. Long, *J. Am. Chem. Soc.*, **90**, 238 (1968).

# Author Index

Aalto, T., 232
Aalto, V., 239
Abel, E., 233
Ackermann, T., 21, 22
Ahrens, M. L., 131, 184, 195
Albert, A., 27
Albery, W. J., 112, 131, 165, 168, 171, 245, 266, 292
Alexander, R., 69
Allen, I., 168
Allred, A. L., 239, 244
Alquier, R., 188
Anbar, M., 23, 25, 186
Anderson, F. H., 176, 189
Andon, R. J. L., 32
Andrew, E. R., 14
Antonovskii, V. L., 286
Aprahamian, N. S., 261
Arenberg, C. A., 50, 57
Armstrong, R., 168
Arnal, N., 109
Arnett, E. M., 246
Arrhenius, S., 6
Aston, J. G., 13
Aue, D. H., 219
Avorio, A., 160

Backer, H. J., 101
Bader, R. F. W., 21, 234, 269
Bagster, L. S., 20
Baliga, B. T., 261
Ballenegger, M., 171, 293
Ballinger, P., 31
Banks, B. E., 150
Barclay, I. M., 81
Barker, G. C., 129
Barnes, D. J., 101, 131, 176, 262
Barrow, G. M., 66
Bartlett, P., 52, 175
Barton, G. W., 245
Bascombe, K. N., 23, 29
Bates, R. G., 88, 232, 233
Bates, S. J., 88
Batson, F. M., 66
Baughan, E. C., 24, 152, 162, 188, 208

Bawn, C. E. H., 273
Bayles, J. W., 219
Beauchamp, J. L., 18
Beckey, H. D., 17
Beetlestone, J. G., 223
Bell, R. P., 9, 23, 29, 31, 66, 81, 88, 93, 100, 101, 105, 116, 131, 134, 148, 150, 152, 157, 162, 163, 168, 170, 172, 173, 175, 176, 178, 181, 182, 183, 184, 185, 187, 188, 202, 203, 204, 212, 213, 216, 217, 219, 220, 222, 223, 224, 230, 234, 235, 236, 254, 260, 262, 264, 265, 266, 269, 270, 273, 274, 282, 285
Bender, M. L., 159, 174, 178, 183
Benkovic, S. J., 178
Benson, S. W., 198
Beretta, V., 162
Berezin, I. V., 286
Berg, D., 39
Berglund-Larsson, U., 191
Bergmann, K., 116
Berkowitz, B. J., 68
Bernatek, E., 222
Bernstein, H. J., 120
Bertran, J., 109
Berzelius, J. J., 5
Bethell, D. E., 15
Bhowmik, S., 65
Bigeleisen, J., 87, 229, 238, 256, 257, 266, 276, 289
Bishop, D. M., 18, 198
Bjerrum, N., 63, 96
Black, E. D., 195
Blackall, E. L., 136
Blair, L. K., 85, 89, 219
Bockris, J. O'M., 21
Bolinger, E. D., 169
Bonhoeffer, K. F., 182
Booth, V. H., 222
Borcic, S., 250
Bordwell, F. G., 216
Bosch, N. F., 171
Bourgin, D. G., 273
Bourns, A. N., 261
Bourre-Maladière, P., 15
Bowden, K., 36, 65, 169

Bowers, M. T., 219
Boyd, R. H., 32
Boyle, W. J., 216
Bratu, E., 233
Braumann, J. I., 85, 89, 103, 219
Brdicka, R., 117, 185
Bredig, G., 19, 165
Breyer, F., 20
Brière, G., 50, 129
Briggs, A. G., 35
Broche, A., 182
Brönsted, J. N., 4, 7, 8, 135, 137, 148, 160, 161, 162, 163, 166, 189, 195, 198, 217
Bromberg, A., 287
Brown, D. A., 269
Brown, H. C., 86
Brown, H. T., 184
Brown, J. F., 148, 155
Bruckenstein, S., 52, 61, 64, 66
Bruice, T. C., 178, 217
Buckley, A., 169
Buddenbaum, W. E., 258
Bunnett, J. F., 35
Bunton, C. A., 234
Burnett, R. le G., 88
Burske, N. W., 103
Busing, W. R., 21
Butler, J. A. V., 81, 220, 238, 241
Butterill, S. E., 18

Cadogan, J. I. G., 240
Caldin, E. F., 111, 114, 120, 148, 157, 163, 164, 177, 273, 280, 283
Callander, D. D., 164
Campbell, D. E., 23
Campbell-Crawford, A. N., 171
Canady, W. J., 73
Cardinaud, R., 230
Carnie, W. W., 163
Carpenter, G. B., 15
Carter, J. S., 172
Challis, B. C., 193
Chalvet, O., 109
Chantooni, M. K., 64, 65
Chapman, N. B., 169
Chen, A., 104
Chen, D. T. Y., 73
Chen, H. J., 201
Chen, H. L., 220
Chetwyn, A., 219
Chiang, Y., 177, 193, 220, 224, 293
Christen, M., 261
Christensen, J. J., 73, 74
Christov, S. G., 273
Ciuffarin, E., 103
Clark, H. M., 23
Clarke, J. H. R., 29

Claus, K. G., 136
Claussen, W. F., 78
Clifford, I. L., 37
Clunie, J. C., 184
Clusius, K., 161
Clutter, D. R., 121
Coetzee, J. F., 32, 64
Cohen, A. O., 214, 216
Colapietro, J., 193
Cole, A. G., 87
Coller, B. A. W., 101
Comisarow, M. B., 89
Conley, H. L., 136
Conner, T. M., 121
Conway, B. E., 14, 21
Cooling, G., 20
Copenhafer, D. T., 66
Copp, J. L., 38
Cordes, E. H., 291
Coté, G. L., 253
Coulson, C. A., 78
Coulter, L. V., 87
Coussemant, F., 295
Covington, A. K., 28, 29, 78, 232, 233
Covitz, F., 189, 220
Coward, J. K., 178
Cox, B. G., 120, 162, 178, 180, 212, 222, 264
Cox, J. D., 32
Cram, D. J., 106, 154
Critchlow, J. E., 157, 187
Crochon, B., 50
Crooks, J. E., 66, 114, 176, 230, 236, 262
Cszimadia, I. G., 19
Czerlinski, G., 111
Czmidia, I. G., 106

Dack, M. R. J., 169
Dahlberg, D. B., 246
Dahn, H., 171, 293, 295
Danckwerts, P. V., 148, 187
Darwent, B. de B., 184, 188
Darwish, D., 136
Daudel, R., 109
Davies, C. W., 135
Davies, M., 160
Davies, M. H., 165, 245, 260, 292
Davis, G. G., 180
Davis, J. C., 9
Davis, M. M., 66, 67
Davy, H., 5, 6
Dawson, H. M., 137, 150, 172, 173
Dean, N. C., 173
Debye, P., 125
De Haas, N., 289
De la Mare, P. B. D., 88
Delbanco, A., 163, 217
De Maeyer, L., 12, 21, 112, 114, 116, 129

De Maria, P., 176
De Paz, M., 24
De Salas, E., 154
Dessy, R. E., 104
De Tar, D. F., 66
Deumie, M., 109
Deyrup, A. J., 46
Diamond, R. M., 23
Dickerson, D. G., 184
Diderich, G., 171, 295
Dietz, N., 46
Dillon, R. L., 201
Dogliotti, L., 195
Dogonadze, R. R., 215
Donzel, A., 171
Dubois, J. E., 177
Duerst, R. W., 28
Duncan, P. M., 163
Duus, H. C., 162, 166, 217

Eastham, A. M., 136, 155
Ebert, L., 88
Eckart, C., 276
Edsall, J. T., 187
Edward, J. T., 25
Edwards, J. O., 9, 93, 223
Ehrenson, S., 204
Eigen, M., 12, 21, 22, 112, 114, 116, 123, 129, 130, 131, 186, 194, 195, 223
Eisenberg, D., 78
Eley, D. D., 78
Eliason, R., 177, 293, 295
Elliot, J. H., 50, 57, 58
Elliot, L. D., 37
Emmons, W. D., 260
Engberts, J. B. F. N., 171
Engel, P., 180
Epprecht, A., 233
Erlenmeyer, H., 233
Erley, D. S., 222
Ernst, R., 192
Ertl, G., 129
Evans, A. G., 218
Evans, M. G., 78, 81, 206, 208
Evans, P. G., 184, 185
Everett, D. H., 38, 78, 83, 176
Eyler, J. R., 89
Eyring, H., 273

Fainberg, A. H., 136
Fajans, K., 21
Falk, M., 20, 233, 247
Faulkner, I. J., 155
Faurholt, C., 39
Feates, F. S., 31, 73
Feather, J. A., 174, 176, 220
Felder, F., 41

Felici, N., 50
Fendley, J. A., 282
Ferriso, C. C., 15
Fettis, G. C., 163, 164
Finch, N. D., 14
Fisher, E., 287
Flanagan, P. W. K., 216
Fluendy, M. A. D., 178
Förster, T., 107
Fong, D. W., 71
Fowler, R. H., 272
Fraenkel, W., 165
Frainier, L., 216
Frank, H. S., 78
Frankevich, E. L., 18
Franklin, J. L., 18
Frazer, M. J., 8
Freeman, J. G., 29
French, D. M., 39
Friedman, L., 24, 245
Fujiwara, S., 38
Fukuyama, M., 216
Funderburk, L. H., 262
Fuoss, R. M., 63, 66
Futtrell, J. H., 16

Gamow, G., 272
Gane, R., 98
Gaspar, R., 19
Gaspard, F., 129
Gay-Lussac, J. L., 5
Gelles, E., 136, 176, 180, 220
Gerischer, H., 129
German, E. D., 215
Giardini, A. G., 24
Giauque, W. F., 37
Gibbons, B. H., 187
Gibert, R., 182
Giguère, P. A., 14, 20, 23, 233, 247
Gilbert, A. S., 23
Gilbert, J. M., 65, 103
Gilkerson, W. R., 116
Gillard, R. D., 23
Gillespie, R. J., 46, 92
Glueckauf, E., 22, 23
Gold, H., 171
Gold, H. J., 157
Gold, V., 174, 176, 198, 199, 220, 232, 233, 238, 239, 240, 241, 244, 245, 294, 295, 296
Goldblatt, M., 233
Goldschmidt, H., 19, 218
Goldsmith, H. L., 176
Goldstern, A., 167
Goodall, D. M., 131, 260, 265, 296
Goodhue, L. D., 50
Gordy, W., 235
Green, A. J., 181

Greenspan, J., 163
Greenzaid, P., 184
Gregory, M. J., 217
Gresser, M. J., 106
Grimm, H. G., 16
Grinstein, R. H., 286
Grist, S., 244
Gross, P., 19, 167, 237, 292
Grovenstein, E., 261
Gruen, L. C., 182, 184
Grunwald, E., 29, 68, 70, 71, 112, 186, 204
Guggenheim, E. A., 4, 137, 189
Gurney, R. W., 78
Guthrie, R. D., 154
Gutowsky, H. S., 38

Häfliger, O., 86
Hague, J. N., 111
Hakala, R., 232
Hakka, L. E., 201
Halevi, E. A., 236
Haley, J. F., 29
Hall, G. V., 173
Hall, L., 21
Hallada, C. J., 48
Haller, G. L., 198
Halliwell, H. F., 25
Hamann, S. D., 218
Hamer, W. J., 21
Hammett, L. P., 46, 64, 146, 177, 204
Hammons, J. H., 103
Haney, M. A., 18
Hansen, L. D., 73
Hansson, J., 176
Hantzsch, A., 10, 11, 13, 88, 131, 160
Harbron, E., 280
Harker, D., 92
Harmony, M. D., 273
Harned, H. S., 77
Harper, E. T., 178
Harrison, A. G., 18
Harrison, M. C., 19
Hartley, H. B., 19, 167
Hartter, D. R., 35
Hautala, J. A., 216
Hawthorne, M. F., 260
Hay, R. W., 136
Heicklen, J., 276
Heinziger, K., 28, 238, 244
Heiszwolf, G. J., 105
Hellin, M., 295
Hénaff, P. L., 185
Henshall, J. B., 162, 178, 222
Hepler, L. G., 74, 82
Herington, E. F. G., 32
Herschbach, D. R., 236
Hetzer, H. B., 67

Heubel, J., 160
Hibbert, F., 212
Higginson, W. C. E., 184, 216
Hildenbrand, D. L., 37
Hill, D. G., 189
Hill, D. L., 274
Hillier, G. R., 101, 202
Hine, J., 52, 103, 184
Hine, M., 52, 103
Hirst, J. P. H., 188
Hixon, R. M., 48
Ho, C., 187
Hobbs, K. S., 171
Hochberg, S., 161
Högfeldt, E., 29
Holbrook, N. K., 19
Hood, G. C., 28
Hopkinson, A. C., 19
Horiuti, J., 204
Horne, D. G., 188
Hornel, J. C., 238
Hornig, D. F., 15, 21, 37
Hoskins, C. R., 173
Houston, J. G., 184
Hsü, S. K., 154
Huang, H. H., 190
Huang, T. T. S., 258
Huebner, H., 286
Hughes, E. D., 154
Hulett, J. R., 279
Hund, F., 273
Hunter, E., 37
Hunter, E. C. E., 160
Hutchins, J. E. C., 165
Hyde, R. M., 165
Hyman, H. H., 48

Ilgenfritz, G., 223
Ingold, C. K., 86, 98, 101, 154, 172
Isbell, H. S., 188
Ives, D. J. G., 31, 73, 82
Ivin, K. J., 114
Izatt, R. M., 73
Izmailov, N. A., 25

Jackson, G., 109
Jamock, A., 19
Janiak, P. S., 106
Jencks, W. P., 157, 159, 178, 179, 183, 232, 234
Jensen, M. B., 187
Jermini, C., 192
Johnson, C. D., 35
Johnson, P., 165, 189
Johnson, R. H., 165
Johnson, S., 183
Johnston, H. L., 289
Johnston, H. S., 215, 273, 276

Jolly, W. J., 48, 87
Jonathan, N., 160
Jones, J. G., 48
Jones, J. R., 175, 282, 283, 285
Jones, L. H., 233
Jones, P., 78, 150
Jones, R. B., 9, 93, 223
Jones, W. M., 233
Joos, G., 21
Jorgenson, M. J., 35
Josephs, J. J., 171
Jumper, C. F., 186

Kagan, J., 222
Kakihana, H., 232
Kakiuchi, Y., 14
Kalb, M., 13
Kankaaperä, A., 177, 293
Kasparian, M., 280
Kass, W. J., 258
Katritzky, A. R., 35
Katz, J. J., 48
Kaufman, M., 236
Kauzmann, W., 78
Kavanau, J. L., 78
Kawasaki, A., 183
Kebarle, P., 24
Kemble, E. C., 274
Kemp, T. J., 181
Kendrew, J. C., 189
Kernohan, J. C., 187
Kerr, J. A., 163
Kessick, M. A., 239, 294
Ketelaar, J. A., 37
Key, A., 173
Khakham, I. B., 37
Kharkats, Yu. I., 215
Kibby, C. L., 289
Kilde, G., 189
Kilpatrick, M., 48, 50, 57, 58
King, C. V., 161, 166, 169, 171
King, E. L., 152
King, H. F., 269
Kirkwood, J. G., 97, 102
Kirsanova, J., 183
Kirschman, H. D., 88
Kitt, G. P., 23
Klein, F. S., 287, 289
Klein, H. S., 235
Klemperer, W., 236
Kloosterziel, H., 105
Kneipp, K. G., 268
Knewstubb, P. F., 17
Koeppl, G. W., 21, 234
Koffer, H., 114
Koller, S., 192
Kolthoff, I. M., 52, 61, 65

Konasewich, D. E., 170, 199, 216
Kondratiev, V., 16
Kossiakoff, A., 92
Kouba, J., 268
Kouba, J. E., 201
Kraus, C. A., 63, 64, 66
Krauss, F., 167, 292
Kreevoy, M. M., 119, 170, 177, 184, 199, 216, 293, 294, 295
Kresge, A. J., 21, 177, 193, 201, 220, 224, 234, 239, 240, 243, 244, 293, 295
Kretchmer, R. A., 294
Krishnan, C. V., 246
Kroll, H., 136
Kruglyak, V., 19
Krumbiegel, P., 250, 286
Kruse, W., 112, 130, 131, 195, 223
Kuhn, A. T., 234
Kuhn, R., 104
Kuhn, S. J., 190
Kulka, P., 171
Kuppermann, A., 277
Kuznetsov, A. M., 215

Lachmann, A., 160
Lachs, H., 165
Lagowski, J. J., 48
Laidler, K. J., 73, 198
Lajunen, M., 177, 293
La Mer, V. K., 161, 162, 163, 234, 246
Lampe, F. W., 16
Landesman, H., 9
Landsman, D. A., 83
Langer, R. M., 273
Langford, P. B., 103
Langmuir, M. E., 195
Lapworth, A., 172
Larsen, J. W., 74
Larson, D. W., 136
Latimer, W. M., 78, 87
Latremouille, G. A., 136
Laughton, P. M., 231, 290
Laurence, A. H., 23
Laurie, V. W., 236, 237
Lavoisier, A. L., 5
Lee, D. G., 286
Lee, F. S., 15
Lee, W. H., 46
Leffler, J. E., 81, 204
Lenoir, J., 171
Leong, K. N., 136
Le Roy, D. J., 277, 289
Le Roy, R. J., 277, 288
Leventhal, J. J., 24
Levich, V. G., 215
Lewis, E. S., 262, 286
Lewis, G. N., 4, 6, 7

Lewis, W., 5
Li, R.-R., 175
Lidwell, O. M., 148, 173, 176, 177, 203
Liebig, J., 6
Lienhard, G. E., 176, 189
Lifshitz, C., 286
Lilley, T. H., 29
Lindsey, R. V., 102
Linton, H., 21
Liotta, S., 163
Lister, M. W., 136
Littler, J. S., 181
Lo, H. H., 121
Loewenstein, A., 121, 186
Long, F. A., 31, 32, 131, 176, 190, 193, 212, 241, 248, 291, 293, 296
Long, J., 18
Longuet-Higgins, H. C., 175, 176, 269
Los, J. M., 189
Lowe, B. M., 233, 241, 296
Lowry, T. M., 4, 7, 137, 155, 189
Lundgren, J. O., 23
Luz, Z., 184
Luzzati, V., 15

Maas, G., 112, 131, 195
McBain, J. W., 39
McCauley, C. E., 171
McClure, A., 163
McCollum, J. D., 52
McCormack, W. E., 156
McCoubrey, J. C., 90, 162, 224
McDaniel, D. H., 86
McElhill, E. A., 168
McEwen, W. K., 103
McGarvey, J. J., 114
McKelvey, D. R., 246
McKenzie, A., 188
McMahon, R. F., 168
McTigue, P. T., 170, 182, 183, 184
Malherbe, R., 171
Mansfield, J. W., 101, 202
Maranville, L. F., 28
Marcus, R. A., 214, 216, 273
Marks, R. E., 283
Marks, S. B., 121
Marlies, C. A., 162
Maron, S. H., 42
Marsden, P. D., 31, 73, 82
Marshall, D. R., 181
Martin, B., 282
Martin, D. L., 154
Martin, R. B., 136
Marum, E., 19
Maryott, A. A., 66
Mathiesen, E., 218
Matsu, H., 14

Mayer, M. G., 229
Mazur, R. H., 168
Mead, C. A., 119
Meadows, G. W., 188
Meany, J. E., 136, 187
Mebane, A., 171
Meiboom, S., 186
Meier, J., 39
Melander, L., 191, 250
Merbach, A., 171
Meriwether, L., 136
Middleton, W. J., 102
Millen, D. J., 15
Miller, G. R., 184
Miller, H. C., 9
Miller, S. A., 216
Miller, S. I., 169, 175
Miller, W. B. T., 235
Miller, W. S., 19
Millington, J. F., 187
Mills, G. A., 39
Mitchell, A. G., 188
Mocek, M. M., 102, 286
Moczygemba, G. A., 48
Möller, E., 176, 220
Moelwyn-Hughes, E. A., 165, 189
Monoszon, A. M., 87
Monse, E. V., 258
Moore, T. S., 37
Moreno, E. C., 42, 87, 223
More, O'Ferrall, R., 21, 165, 169, 234, 268
Morris, P. J., 136
Morrison, G. C., 9, 93
Mortensen, E. M., 277
Moruzzi, J. L., 25
Moseley, P. G. W., 73
Moskowitz, J. W., 19
Muenter, J. S., 236, 237
Muetterties, E. L., 9
Mulder, H. D., 87
Mullhaupt, J. T., 15
Munson, B., 18, 85, 219
Murdoch, J. R., 215
Murrell, J. N., 198
Murrill, E., 220
Muszkat, K. A., 287
Mylonakis, S. G., 201

Nancollas, G. N., 135
Nelson, W. E., 238
Newbury, R. S., 245
Newman, M. S., 188
Nicholson, A. L., 163, 217
Nonhebel, G., 19
Noonan, E., 246
Nordheim, L., 272
Nordman, C. E., 15

Northcott, D., 235
Novak, R. W., 66
Noyes, R. M., 125
Nürnberg, H. W., 118, 127, 131
Nuttall, R. H., 231
Nyburg, S. C., 25

Oakes, B. D., 103
O'Donnell, J. P., 23
Ogata, Y., 183
Ogden, G., 273
Ogg, R. A., 120, 160, 206
O'Hara, F., 82
Okuzumi, Y., 104
Olah, G. A., 190
Olovsson, I., 15, 23
Olsen, F. P., 35
Olson, A. R., 160, 162, 222
Onak, T. P., 9
Onsager, L., 125
Onwood, D. P., 31
O'Reilly, D. E., 14
Orr, W. J. C., 238, 241
Oscarson, J. L., 73
Ostwald, W., 6
Owen, B. B., 77

Paabo, M., 67, 232
Page, M. I., 179
Papee, H. M., 73
Parker, A. J., 63, 69
Parr, C. A., 289
Partington, J. R., 160
Patat, F., 19
Patel, J. M., 42, 87, 223
Patel, P. R., 42, 87, 223
Patterson, A., 39
Paul, M. A., 32
Pauling, L., 92
Pavia, A. C., 23
Payne, M. A., 220
Peacock, J., 164
Pearson, R. G., 8, 116, 201, 213, 217, 219
Peck, R. C., 103
Pedersen, K. J., 11, 136, 137, 138, 151, 160, 161, 175, 195, 208
Pentz, L., 233, 242
Perez Ossorio, R., 154
Perkin, W. H., 184
Perlmutter-Hayman, B., 162
Perry, J. M., 286
Persky, A., 289
Petersen, S. W., 23
Peterson, E. M., 14, 23
Pethybridge, A. D., 135
Pfluger, H. L., 217
Phelps, A. V., 25

Phillips, W. D., 9
Pickering, P. S. U., 184
Pigman, W., 188
Pink, J. M., 187
Pinsent, B. R. W., 83
Pitzer, K. S., 277
Plane, R. A., 136
Platt, J. R., 269
Pleskov, V. A., 87
Pocker, Y., 155, 184, 187
Polanyi, M., 81, 204, 206, 208
Polglase, M. F., 78
Popev, A. I., 45
Pople, J. A., 120
Porter, G., 109
Porter, L. J., 136
Powell, R. E., 78
Powis, F., 137, 172, 173
Prasad, D., 73
Price, E., 70
Pritchard, J. G., 291
Prue, J. E., 135, 136, 175
Pryor, J. H., 31
Pryor, W. A., 268
Pshenichnov, E. I., 16
Pudjaatmaka, A. H., 103
Purlee, E. L., 238
Pyper, J. W., 245

Quickert, K. A., 277, 289

Raisin, C. G., 154
Rand, M. H., 184, 220
Rao, B. P., 37
Rapp, D., 198, 276
Rauk, A., 106
Rawlinson, D. A., 101
Rawlinson, D. J., 180, 181
Ray, J. D., 160
Redlich, O., 28, 229, 233
Regan, C. M., 168
Reilly, C. A., 28
Reiman, C. K., 173
Reitz, O., 177
Renaud, M., 87
Reuwer, J. F., 257
Rewicki, D., 104
Ricci, J. E., 92
Rice, F. O., 173
Richards, E. M., 155
Richards, R. E., 14
Riddell, F. G., 120, 212
Ridgewell, H. F. F., 176
Ridley, B. A., 289
Riley, T., 176, 293
Ripley, R. F., 165
Ritchie, C. D., 32, 65, 81, 103, 204, 269

Ritchie, P. D., 188
Riveros, J. M., 85, 219
Roberts, J. D., 167, 168
Roberts, T. R., 181
Robertson, P. W., 88
Robertson, R. E., 231, 235, 290
Robinson, E. A., 46
Robinson, J. K., 262
Robinson, R. A., 32, 88, 232, 233
Robinson, R. R., 116, 176, 190
Rochester, C. H., 32, 36
Roginsky, S., 273
Rony, P. R., 155, 156
Roper, G. C., 87
Rosenfeld, J. L. J., 18
Rosenkewitsch, L., 273
Ross, V. F., 9, 93
Rossotti, F. J. C., 154
Rothrock, D. A., 66
Roughton, F. J. W., 39, 222
Rudolph, J., 23, 234
Rule, C. K., 234
Rummel, S., 286
Rumpf, P., 100
Rund, J. V., 136
Rybicka, S. M., 148, 157

Sachs, W. H., 265, 270
Sagatys, D. S., 220
Sager, W. F., 81
Saito, A., 66
Salama, A., 136
Salem, L., 269
Salomaa, P., 177, 232, 233, 239, 241, 243, 246, 247, 293
Salvesen, K., 232, 234
Sammon, D. C., 129
Samuel, D., 184
Sanders, W. H., 250
Satchell, D. P. N., 8, 240
Satchell, R. S., 8
Sato, Y., 201
Sayre, W. G., 121
Schaad, L. J., 257
Schachtschneider, J. H., 257
Schaefgen, J. R., 188
Schaleger, L. I., 241
Schechter, H., 216
Scheider, M., 20
Scheider, W. G., 120
Scheie, C. E., 23
Schlag, E. W., 198
Schmidt, F. C., 87
Schmidt, H., 117
Schmitz, K. S., 125
Schneider, M. E., 258
Schoen, J., 12

Schultz, J. W., 9, 93
Schulz, W. R., 289
Schulze, J., 193
Schurr, J. M., 125
Schwarzenbach, G., 39, 41, 86, 233
Serjeant, E. P., 27
Shannon, K., 163
Shapiro, I., 9, 35
Shapiro, J. S., 289
Sharma, A., 187
Sharp, D. W. A., 231
Sharp, T. E., 276, 289
Shavitt, I., 289
Sheppard, N., 15, 23
Sherman, J., 16
Sherred, J. A., 148, 157
Shin, H., 277
Shiner, V. J., 234, 245, 250, 282
Shishkina, L. N., 286
Shono, H., 14
Shorter, J., 169
Sidgwick, N. V., 37
Simandoux, J. C., 295
Simmons, E. L., 114
Simonyi, M., 286
Simpson, L. B., 189
Sinclair, J. R., 87
Skinner, B. G., 148
Skrabal, A., 138
Slade, M. D., 73
Slae, S., 193
Slater, J. S., 163
Smith, D. E., 73
Smith, G. F., 137, 173, 189, 208
Smith, H. M., 28
Smith, J. A. S., 14
Smith, J. E., 173
Smith, M., 208
Smith, M. C., 189
Smith, M. J., 182
Smith, M. L., 282
Smith, R. D., 176
Smith, S., 136
Smoluchowski, A., 124
Smyth, K. C., 89
Snethlage, H. C. S., 166
Snyder, R. G., 257
Sörensen, P. E., 185, 187
Sokolov, N. D., 16
Spencer, T., 176
Spindel, W., 258
Spiro, M., 180
Spitalsky, E., 165
Spivey, E., 150
Sprague, E. D., 288
Stamm, O. A., 192
Stanford, S. C., 235

Starkey, J. D., 65
Steele, B. D., 20
Steele, C., 163
Stein, G., 286
Steinberger, R., 136
Steiner, E. C., 65, 103
Steiner, H., 167, 237, 292
Stern, M. J., 258, 288
Stewart, R., 23, 36, 102, 285, 286
Stivers, E. C., 257
Strating, J., 171
Street, D. G., 101, 202
Strehlow, H., 117, 120, 184
Streitwieser, A., 103, 235, 260
Strube, S., 39
Sturtevant, J. M., 187
Subba Rao, S. C., 283
Suess, H., 237
Suhrmann, R., 20
Sunko, D. E., 250
Swain, C. G., 21, 151, 155, 234, 257
Swift, T. J., 121

Taessler, I., 15
Tait, M. J., 28
Tal'Rose, V. L., 18
Tamassy-Lentei, I., 19
Tamm, K., 111
Tanford, C., 102
Tankey, H., 183
Tantram, A. D. S., 148, 157
Taylor, D. W., 193
Taylor, R. C., 15
Teeter, C. E., 166
Tellier-Pollon, S., 160
Thiel, A., 39
Thiele, J., 160
Thomas, L., 19
Thomas, R. J., 193
Thompson, H. W., 253
Thompson, R. J., 9
Thornton, E. R., 21, 233, 237, 242
Thumm, B. A., 189
Thyagarajan, B. S., 106
Tickle, P., 35
Tickner, A. W., 17
Timimi, B. A., 162, 178, 222
Tomalin, G., 280, 283
Tomlinson, C., 232
Tong, L. K. J., 161, 162
Torck, B., 295
Toullec, J., 177
Tranter, R. L., 265, 270, 288
Trepka, R. D., 106
Trotman-Dickenson, A. F., 163, 164, 217, 218
Truhlar, D. G., 277, 289
Tsang, J., 73

Tschuikow-Roux, E., 289
Tuck, D. G., 23
Tüdos, F., 286
Tung-Chia Wang, 176
Turnbull, D., 42
Tyler, J. K., 160

Udby, O., 19
Unterecker, D. F., 66
Urey, H. C., 39, 173, 229
Uschold, R. E., 65, 103

Vaal, E. G., 15
Vance, J. E., 163, 217
Van Dam, W., 101
Vander Donckt, E., 107
Van der Linde, W., 235
Van der Linden, R., 285
Van der Raalte, D., 18
Van Duuren, B. L., 107
Van Looy, H., 64
Van Sickle, D. E., 260
Van Velden, P. F., 37
Vasenko, E. N., 55
Vast, P., 161
Vaughan-Jackson, M. W., 148
Verhoek, F. H., 54, 188
Vesala, A., 232, 247
Vesala, S., 232, 243, 247
Vetchinkin, S. I., 16
Viallet, P., 109
Vidale, G. L., 15
Vincent, J. R., 175
Vitullo, V. P., 201
Vogel, P. C., 258
Vogelsong, D. C., 180, 219
Voipio, A., 163
Volmer, A., 14
Volqvartz, K., 162, 166
Von Stackelberg, M., 117
Vorländer, D., 39

Waddington, T. C., 42, 231
Waind, G. M., 162
Walden, P., 4
Waldron, R. D., 37
Walrafen, G. E., 233
Walters, E. A., 212, 248
Walters, W. D., 182
Wang, I. C., 25
Warshel, A., 287
Watanabe, W., 167, 168
Waterman, D. C. A., 199, 294, 295
Waters, W. A., 181
Webb, H. M., 219
Weidemann, E. G., 21
Weiss, J., 273

Weiss, J. J., 285
Weller, A., 107
Wells, P. R., 204
Werner, A., 7, 13
Westenburg, A. A., 289
Westheimer, F. H., 97, 102, 136, 189, 198, 220, 266
Weston, R. E., 28, 87, 198, 236, 239, 244, 289
Wettermark, G., 195
Wheeler, D. D., 222
Wheeler, J. A., 274
Wheland, G. W., 86
White, M. J., 89
Wiberg, E., 43
Wiberley, S. E., 23
Wicke, E., 22
Wiesner, K., 189
Wigner, E., 273
Wijnen, M. H. J., 288
Wilhelmy, L. F., 133
Wilkinson, G., 23
Willi, A. V., 266, 268
Williams, A., 174
Williams, D. A. R., 120, 212
Williams, F., 288
Williams, F. T., 216
Williams, F. V., 217
Williams, J. M., 14, 23, 293
Williams, R. E., 9
Wilson, C. L., 154
Wilson, G. L., 163, 217
Wilson, J. M., 35
Winmill, T. F., 37
Winstein, S., 136
Wischin, A., 237

Wissbrunn, K. F., 39
Witschonke, C. R., 64
Wo Kong Kwok, 169
Wolfe, S., 106
Wolfsberg, M., 226, 245, 258, 266, 289
Woodward, L. A., 29, 176
Woolcock, J. W., 167
Wright, G. A., 100, 101, 176
Wright, J., 148, 157, 177
Wulff, M. A., 162
Wunderly, S. W., 156
Wyman, J., 187
Wynne-Jones, K. M. A., 184, 220
Wynne-Jones, W. F. K. (Lord Wynne-Jones), 28, 57, 88, 189, 233

Yagil, G., 23, 25, 36
Yankwich, P. E., 258
Yates, K., 19, 180
Yee, K. C., 216
Yerger, E. A., 66
Yokoi, K., 183
Yoon, Y. K., 15
Youle, P. V., 222
Young, D. C., 222
Young, H. S., 66
Young, T. F., 28

Zawidzki, T. W., 73
Zimmermann, H., 23, 234
Zollinger, H., 190, 192, 261
Zucker, L., 177
Zuidema, G., 171
Zundel, G., 21
Zwanenburg, B., 171
Zwolinski, B. J., 273

# Subject Index

Acetaldehyde:
  addition to methanol, 188
  aldol condensation of, 181
  hydration of, 184, 220
Acetals, hydrolysis of, 291
Acetic acid:
  in $H_2O–D_2O$ mixtures, 241, 247
  as solvent, 45
Acetone:
  aldol condensation of, 182
  halogenation of, 137, 150, 157, 172, 180
Acetonitrile,
  reactions in solid, 288
  as solvent, 64
Acetophenone, 105
Acetylacetone, 105
o-Acetylbenzoic acid, 222
2-Acetylcyclohexanone, 293
Acidity functions, 32
Acids:
  carbon, 103, 131, 195
  hard and soft, 8
  Lewis, 7, 8
  normal, 130, 194
  pseudo, 9, 132, 216
Acid strength,
  definition of, 14
  in different solvents, 36
  measurement of, 24, 49
  of carbon acids, 66
  of chloro-acids, 62
  of oxy-acids, 57
  of simple hydrides, 54
Aci-nitro isomers, acid strengths of, 41
Acridine, fluorescence spectrum of, 108
Activity coefficients:
  degenerate, 68, 240, 246
  of individual ions, 69
Alcohols:
  acid strengths of, 85
  as solvents, 44
Aldol condensation, 181
Alkylmercuric iodides, cleavage of, 293
Alpha particle emission, 272
Amidines, 95

Amines:
  as catalysts, 217
  proton affinities of, 85
  rates of proton transfer, 121
Amino-acids, 98
Aminoketones, 178
Ammonia:
  acid strength of, 86
  dissociation of in solution, 36
  magnetic resonance in, 120
  as a solvent, 48
Ammonium hydroxide, 36
Aprotic solvents, 49, 65, 147
  dipolar, 63
Arrhenius equation, deviations from, 280
Arsenic acid, 92, 243
Association of ions, 61, 64, 135
Azobenzene, 118

Bases:
  anhydro, 13
  aquo, 13
  carbinol, 13
  hard and soft, 8
  normal, 130, 194
  pseudo, 9, 12
Benzene, acid-base equilibria in, 66
Benzoic acid, 241
Benzoylphenyldiazomethane, 171
Bicarbonate ion, as catalyst, 222
Bifluoride ion, 42, 253
  as catalyst, 224
Bifunctional catalysis, 155
Borate ion, structure of, 9, 93
Boric acid, 9, 93, 223, 245
Brönsted relation, 195
  deviations from, 217

Carbon acids:
  isotope effects for, 261
  rate of reaction of, 131, 195, 201
  strengths of, 103
Carbon dioxide, hydration and dissociation
  of, 38, 93, 187, 222

Carbonic acid:
  acid strength of, 38, 93, 222
  in $H_2O-D_2O$ mixtures, 247
Carbonyl compounds:
  addition reactions of, 183–190, 214
  halogenation of, 179, 220
  ionization and enolization of, 171–183, 202
o-Carboxyacetophenone, 178
o-Carboxyisobutyrophenone, 178
Carboxylic acids, 95
  halogen-substituted, 102
  sulpho-, 98
Catalysis:
  acid-base, 133
  bifunctional, 155
  concerted, 149
  electrolyte, 137
  general acid-base, 137
  intramolecular, 178
  nucleophilic, 159
  tautomeric, 157
Chloral, 188
Chloranil, 286
p-Chlorodiazobenzene, 192, 261
Concerted mechanisms, 149
Conjugate acid-base pairs, 7
Crystal violet, 13
Cyano-compounds:
  acidity of, 106
  reaction with bases, 211
Cyanoketen dimethylacetal, 294, 295
Cyclohexanone, 181

De Broglie relation, 272
Decarboxylation, 135
Deuterium oxide:
  ionic product of, 233
  mixtures with $H_2O$, 237
Diazoacetate ion, 169, 199
Diazo-compounds, 164–171
Diazo coupling reactions, 191, 261
Diazoketones, 171, 293
2-Diazophenol-4-sulphonate, 192
Diazosulphones, 171
Dicarboxylic acids, 96
1,3-Dichloroacetone, 186
2-Dichloromethylene-1,3-dioxolan, 295
Diffusion-controlled reactions, 124
4a,4b-Dihydrophenanthrene, 287
Dihydroxyacetone, depolymerisation of, 152, 188
β-Diketones, 105, 176
Dimedone, 41
1,1-Dimethoxyethane, 291
Dimethylformamide, as solvent, 63
Dimethyl sulphoxide, as solvent, 65, 264

Dimethylamine, 122
Dimethylanthranilic acid, 131
2,6-Dimethylpyridine, 264, 281
2,4-Dinitrophenol, 230, 242
Diphenyldiazomethane, 168
Dispersion forces, 70
Dissociation field effect, 39
Disulphones, 211

Electron emission, 272
Electron-deficient compounds, 2
Electrophilic aromatic substitution, 190
Encounters, 123, 186
Entropy changes:
  in carbonyl hydration reactions, 187
  in intramolecular processes, 179
  in protolytic reactions, 77
Epoxides, hydrolysis of, 291
Enzymes, 178, 187
Ethanol, as solvent, 45, 51–54
2-Ethoxycarbonylcyclohexanone, 180
2-Ethoxycarbonylcyclopentanone, 282
Ethyl acetoacetate, 41
Ethyl diazoacetate, 19, 165, 292
Ethyl malonate, 176, 180
Ethyl nitroacetate, 176, 262
Ethyl α-methylacetoacetate, 263
Ethyl vinyl ether, 294, 295
Ethylene oxide, 291
Ethylenedinitramine, 213
Excited states, acid-base properties of, 107

Fast reactions, 111
Fluorene, 103
Fluorescence spectra, 107
Fluoro-alcohols, 102
Formaldehyde:
  aldol condensation of, 183
  hydration of, 117
Formamide, as solvent, 54
Formic acid:
  dissociation of $HCO_2H$ and $DCO_2H$, 236
  in $H_2O-D_2O$ mixtures, 241
  as solvent, 41
o-Formylbenzoic acid, 222
Fractionation factors (isotopic), 239

Geometric mean, rule of, 238, 260
Glucose, mutarotation of, 137, 152, 188, 214, 220
Glycolaldehyde:
  hydration of, 185
  depolymerisation of, 188
Guanidine, 95

Hard and soft acids and bases, 8

Heat capacity changes in protolytic reactions, 77
Homoconjugation, 64
Hydrates:
  of carbonyl compounds, 183
  of strong acids, 14
Hydrazoic acid, 241
Hydrides, acid strengths of, 86
Hydrocarbons, acidic properties of, 103
Hydrogen bonding, 2, 37, 131, 231
Hydrogen bromide, 19, 20
Hydrogen chloride, 20
Hydrogen fluoride:
  as catalyst, 224
  dissociation of, 42, 87
  as solvent, 48
Hydrogen halides, acid strengths of, 87–91
Hydrophobic bonding, 220
Hydronium ion, 13–25
  further hydration of, 22
  heat of solution of, 25
  life-time in solution, 21
  structure of, 14, 18
Hydroxide ion, 25, 248
Hydroxonium ion, *see* hydronium ion
2-Hydroxypyridine, 155
Hypophosphorous acid, 93

Indene, 103
Intramolecular catalysis, 178
Isotope exchange, 145, 192, 201, 262

Keto-compounds, *see* Carbonyl compounds
Keto-enol isomers:
  acid strengths of, 40
  interconversion, 142, 171
Keto-carboxylic acids, 178
α-Keto-esters, 188
β-Keto-esters, 105, 176

Levelling effect, 49
Lewis acids, 7, 8
Life-time methods, 119
Line broadening, 119
2,6-Lutidine, *see* 2,6-Dimethylpyridine
Lyate ion, 13
Lyonium ion, 13

Menthone, inversion of, 220, 264
Mesityl oxide oxalic ester, 157
Methane, acid strength of, 86
Methanesulphonic acid, 29
Methanol:
  addition to aldehydes, 188
  as solvent, 45, 51–54
*p*-Methoxyacetophenone, 283
*p*-Methoxybenzenesulphonate ion, 261

*p*-Methoxy-α-methylstyrene, 295
Methyl methanetricarboxylate, *see*
  Tricarbomethoxymethane
*o*-Methylacetophenone, 283
Methylamine, 121
2-Methyl-1,3-dioxolane, 291
Methylene glycol, 185
Methyleneazomethines, 154
Monochloroacetic acid, 102, 241
Mutarotation:
  of glucose, 137, 152, 188
  of nitrocamphor, 157
  of tetra-acetyl- and tetramethyl-glucose,
    136, 155, 188, 214, 220

1-Naphthol sulphonates, 192, 261
Nitramide, 12
  decomposition of, 137, 160–164, 213, 222
Nitric acid:
  acid strength of, 28
  hydrate of, 14
Nitroacetone, 176
Nitrobenzene, as solvent, 63
4-Nitrobenzyl cyanide, 280, 283
Nitrocamphor, 157
Nitroethane, 41, 264, 296
Nitromethane, 10, 106
  as solvent, 63
2-Nitrophenol, 242
3-Nitrophenol, 261
*p*-Nitrophenyldiazomethane, 295
2-Nitropropane, 260, 262, 265, 281, 283
Nitrourethane, 222
Nucleophilic additions, 183
Nucleophilic catalysis, 159

Oxyacids, strength of, 92
Oxonium ion, *see* hydronium ion

Perchloric acid:
  acid strength of, 28
  hydrate of, 14
Periodic acid, 93, 248
  in $H_2O–D_2O$ mixtures, 243
Phenyldiazomethanes, 171
1-Phenyl-2,2,2-trifluoroethanol, 285
Phosphine, acid strength of, 87
Phosphoric acid, 20, 243
Phosphorous acid, 93
Picric acid:
  in ethanol, 19, 166
  in $H_2O–D_2O$ mixtures, 247
  in various solvents, 70
Polarography, 117
Potential energy curves, 204
Pressure jump technique, 114
Product rules, 228

Propan-2-one-1-sulphonate ion, 263
Propylene carbonate, as solvent, 63
Proton magnetic resonance, 3
  for measuring acid strengths, 27
  in $H_2O$–$D_2O$ mixtures, 239
  for studying rate processes, 120, 184
Prototropic isomerisations, 141
Pseudo-acids, 9, 132, 216
Pseudo-bases, 9, 12
Push-pull mechanisms, *see* Concerted
  mechanisms
Pyruvic acid, 118

Racemization, 145, 177
Raman spectra:
  for measuring acid strengths, 27
  for studying rate processes, 119
  and water structure, 233
Redox systems, 43
Relaxation methods, 113

Salicylic acid, 131
Salt effects, 134
Secondary isotope effects, 235, 259
Singlet states, acid-base properties of, 108
Solvent isotope effects, 235, 259
Statistical corrections:
  in equilibria, 97, 229, 264
  in rates, 197
Steric hindrance, 131, 219, 282, 286
Stopped flow technique, 113
Sulpholane, as solvent, 63
Sulphonate group, 98
Sulphones:
  acidity of, 106
  reaction with bases, 211
Sulphur dioxide:
  hydration in solution, 93, 247
  as solvent, 19

Sulphuric acid:
  hydrate of, 14
  as solvent, 46
Sulphurous acid, 93
  in $H_2O$–$D_2O$ mixtures, 247
Synchronous mechanisms, *see* Concerted
  mechanisms

Tautomeric catalysis, 157
Telluric acid, 93
Temperature jump technique, 113
Ternary mechanisms, *see* Concerted
  mechanisms
Tetra-acetylglucose, 136
Tetramethylglucose, 136
Toluene, 85, 260
Transfer effects in isotopic equilibria, 246
2,4,6-Tri-t-butylphenol, 286
Tricarbomethoxymethane, 176, 180, 265
Trifluoroacetic acid, 29, 119
Trimethylamine, 122, 235
2,4,6-Trimethylpyridine, 262, 281, 283
Trinitrobenzyl anion, 280
Triphenylmethane, 103
Triplet states, acid-base properties of, 108
Tritium oxide, 233
Tunnel effect, 270–289

Ultrasonic absorption, 116
Uncertainty principle, 119, 273

Vinyl ethers, hydrolysis of, 177, 293

Water:
  ionic product of, 233
  proton affinity of, 15
  structure of, 78

Zero-point energy, 227, 251